AF522654

Electricity and Magnetism

Electricity and Magnetism

Prof. J.D. Dubey

Raj K. Tiwari

KNOWLEDGE BOOK DISTRIBUTORS
NEW DELHI (INDIA)

ISBN : 978–93–80350–20–2

Published By :
KNOWLEDGE BOOK DISTRIBUTORS
B-72/1 IInd Floor, Old Jankipuri,
Uttam Nagar, New Delhi–110059
E-mail : knoledgebookdistributors@yahoo.in

Edition : **2010**

Printed in India

Preface

Electricity and Magnetism are conventional specialized subjects. Both have a significant role to play in the field of science & technology. Electro-magnetism is the support base for many recently developed subjects such as electronics and telecommunications, computer science etc.

The book starts with the vector analysis which is going to be helpful to understand mathematical applications in electricity and magnetism. Various other chapters also elaborately discuss the important portions of electricity and magnetism. Electric charges and fields have been given due importance in the second chapter of the book. Electric potential as well as electric fields in material space have been placed in the middle portion of the book. Boundary value problems pertaining to electro-magnetism is covered widely in the last very portion of the book.

The book contain two parts : One is Electricity and the other one is Magnetism and meets the year long demand of students of various engineering colleges and universities. Mathematical tools have been applied properly to solve the numericals of the book. The book contains both solved and unsolved sections of numericals.

The book has been written as per the syllabus prescribed by 'UGC' for various universities and institutions. The book is very beneficial for the students of B.Sc., M.Sc., B.Tech., M.Tech. etc. The book will also provide adequate cognizance to the students preparing for civil service examination. It will also be helpful for those students who are preparing for IES & IFS too. The book will certainly be fruitful for the academicians to explore the new methods of research. It will also be useful for the special courses in Electricity and Magnetism.

Contents include : Vectors; Electric Charges and Fields; Electric Potential; Electric Fields in Material Space; Boundary Conditions : Boundary Value Problems and Moving Charges.

The authors are grateful to a number of friends who have inspired them for preparing this book.

Generally, the book writing-whether original or compilation is a complicated job requiring much academic labour. The present work is a research compilation of required material from authoritative publications/readings/excerpts/researches/ examples, written and compiled by various authors in this field. Assistance from various Websites through internet has also been taken.

The authors, therefore humbly, acknowledge the contributions of all those eminent scholars alongwith their respective publishers and Websites from whose learned writings/displays/references/literature we either have quoted or consulted freely while preparing this book.

Prof. J.D. Dubey
Raj K. Tiwari

Contents

1

Vectors

Introduction

The technique of vector Algebra is applied for better understanding of concepts of electricity and magnetism in Physics. Mathematical convenience of expressing equations/formulae in a compact form is provided to us by vector analysis. That's why we are first of all going to develop some mathematical tools required to deal with vectors.

Scalars and Vectors

Scalar quantities are those which are completely defined by magnitude alone Examples: mass, time, length, temperature, charge, electrostatic potential etc. These quantities obey the ordinary laws of algebra.

Vector quantities on the other hand, are those which are completely defined only when both the magnitude and the direction in which they act are known. Examples displacement, velocity, acceleration, force, electric field, magnetic field etc. These quantities cannot be added using simple algebraic rules, except when they are acting along the same straight line.

Symbolic Representation of a Vector

A bold face letter or a light face letter with an arrow head is used for the representation of vector quantities. For example, velocity, a vector quantity, may be represented by V or $\vec{V}$. The magnitude of the velocity vector is represented by V or by $|V|$ or $|\vec{V}|$ called modulus of the vector.

Graphical Representation of a Vector

A line with an arrowhead at one end is used for graphical representation graphical representation of Vector. The magnitude of the vector quantity is given by the length of the line chosen on a suitable scale, while the direction in space is indicated by arrow head mark on the line.

Let's assume that 1 cm represents a velocity of 1 km/s for the representation of a velocity $\vec{v}$ of 6 kms. Now a line OP, 6 cm long drawn along north east with arrow head at p represents completely the given velocity. The end P where the arrow head is marked is called head or terminus of the vecotr, the other end is called tail or initial point. Thus the vector $|\vec{V}|$ is represented by $\vec{OP}$ as shown in Fig. 1.1.

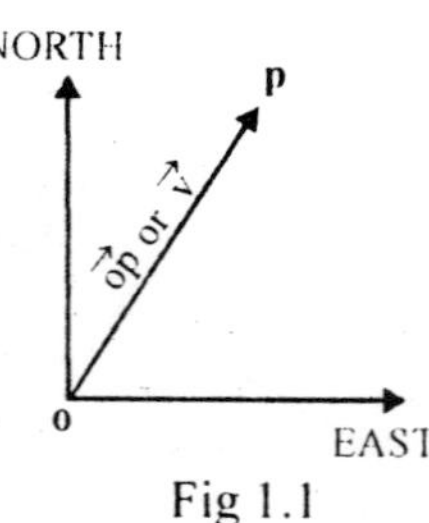

Fig 1.1

Equality of Vectors

If two vector have the same length and are drawn parallel to each other having the same sense then they are said to be equal. In Fig. 1.2, the straight lines AB, EF and CD are parallel and equal. They represent the same vector in terms of definition.

Hence, it can be shown $\vec{AB} = \vec{CD} = \vec{EF}$. It should be kept in mind that equality of vectors is independent of the position to their initial points. thus it a vector is displaced parallel to itself, it will remain unchanged.

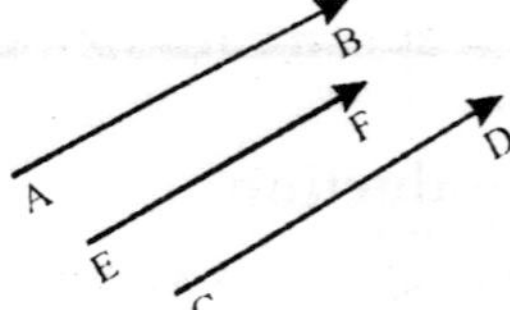

Fig. 1.2

Negative Vector

Another vector having the smae length but opposite in direction is called the negative of a vector $\vec{A}$. It is denoted by — A as shown in Fig. 1·3.

Angle between two Vectors

As shown in Fig 1·4, there are two angles θ_1 and θ_2 made by two vectors $\vec{A}$ and $\vec{B}$ at their common point of origin. If $\theta_1 \neq \theta_2$ then one of the angles i.e., $\angle\theta_1$ is less then 180°. Then the angle θ_1 which is less then 180° is taken as the angle between the two vectors $\vec{A}$ and $\vec{B}$ If $\theta_1 = \theta_2 = 180^o$, and the other angle θ_2 is greater than 180° then any of the two is the angle between the two vectors.

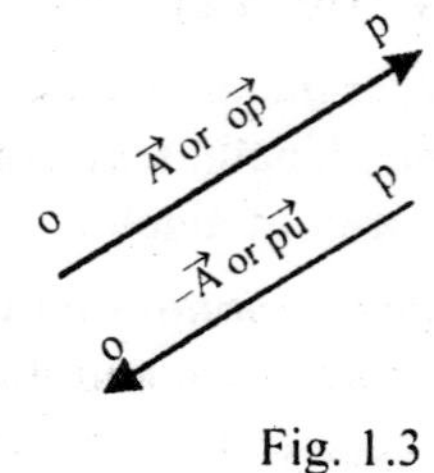

Fig. 1.3

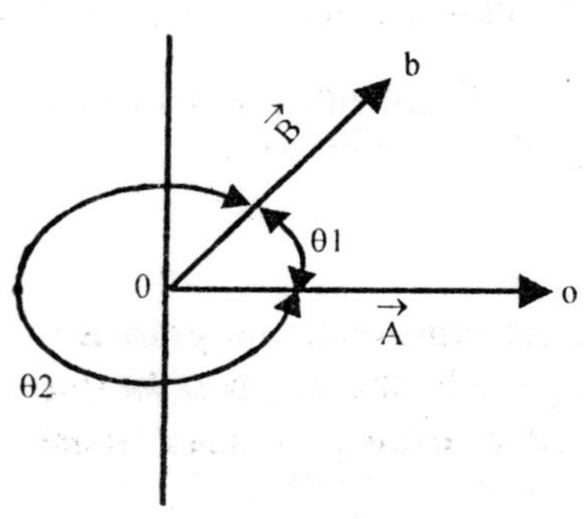

Fig 1.4.

(a) Vectors Addition :

(i) Assume $\vec{A}$ and $\vec{B}$, two vertors Such that terminus of $\vec{A}$ is the initial point of $\vec{B}$ as shown in fig. 1·5, then the vector $\vec{C}$ which is obtained by joining the initial point of $\vec{A}$ to the terminal point of $\vec{B}$ is called the resultant or sum of the two vectors $\vec{A}$ and $\vec{B}$. $\vec{A} + \vec{B}$ is the summation representation of two vectors.

$$\vec{C} = \vec{A} + \vec{B}$$

(ii) If the point of origin of two vectors $\vec{A}$ and $\vec{B}$ is same [as shown in Fig. 1·6 (a)] then their sum is obtained by setting

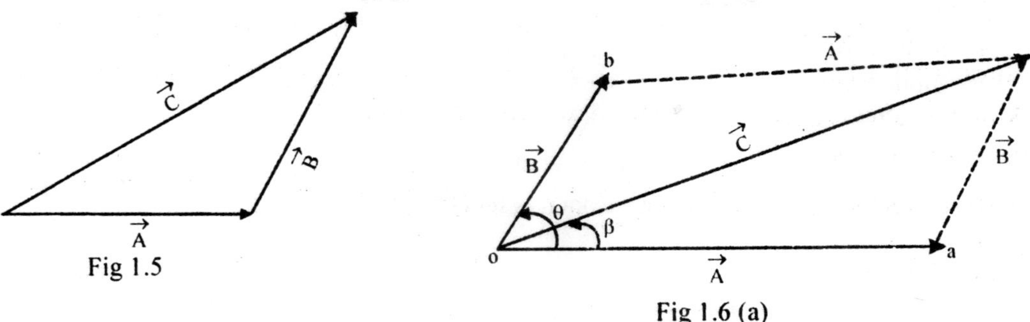

Fig 1.5

Fig 1.6 (a)

off the vector $\vec{B}$ at the end of vector $\vec{A}$ and drawing the vector $\vec{C}$ by joining 'O' the beginning of $\vec{A}$ to 'C' the end of $\vec{B}$,

Hence,

$$\vec{C} = \vec{A} + \vec{B} \qquad ...(1\cdot 1)$$

A similar result is obtained by setting off vector $\vec{A}$ at the end of vector $\vec{B}$, so that

$$\vec{C} = \vec{B} + \vec{A} \qquad ...(1\cdot 2)$$

Comparing (1·1) and (1·2), we have

$$\vec{A} + \vec{B} = \vec{B} + \vec{A} \qquad ...(1\cdot 3)$$

therefore, we can say that the diagonal of the parallelogram with the two vectors as its sides represents the sum of two vectors.

Equation (1·3) shows that sum of the vectors remains same in whatever order they are added. Therefore, we must say that vector addition is commutative.

The magnitude of the resultant of two vectors $\vec{A}$ and $\vec{B}$ inclined to each other at an angle θ is given as follows

$$C = \sqrt{A^2 + B^2 + 2AB\cos\theta}$$

If β is the angle which the resultant $\vec{C}$ makes with the direction of $\vec{A}$, then

$$\tan\beta = \frac{B\sin\theta}{A + B\cos\theta}$$

(b) Difference of two Vectors

$(\vec{A} - \vec{B})$ represents the difference between two vectros $\vec{A}$ and $\vec{B}$ shown in Fig 1·6 (b) by $\vec{D}$ which when added to $\vec{B}$ given vector $\vec{A}$. Equivalently, $\vec{A} - \vec{B}$ may also be defined as the sum $\vec{A}+(-\vec{B})$. Thus subtracting a vector $\vec{B}$ from vector $\vec{A}$ is the same as adding its negative. To illustrate it, take two vectors $\vec{oa}=\vec{A}$ and $\vec{ab}=\vec{B}$ as shown in Fig. 1·6 (b). Produce ba backward, so that ab = ab, then vector drawn from a to d i.e. $\vec{ad}$ represents vector $-\vec{B}$.

Hence vector $\vec{od}$ which is drawn from *o* is sum of vectors $\vec{oa}$ and $\vec{ad}$.

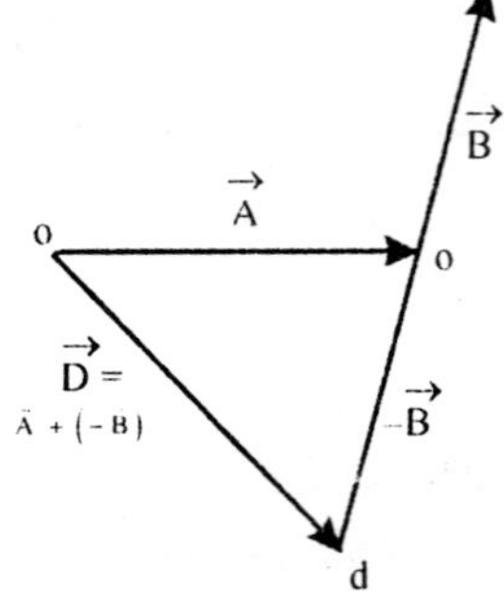

Fig.1.6.(b)

$$\vec{od} = \vec{oa} + \vec{od}$$

$$= \vec{A} + (-\vec{B}) = \vec{A} - \vec{B}$$

Difference of two vectors is not commutative i.e.

$$\vec{A} - \vec{B} \neq \vec{B} - \vec{A}$$

Zero or Null Vector

If head (Terminal) point and Tale (initial) point of a vector are coincident then the vector is called null or zero vector. It is represented by the symbol $\vec{0}$. It has zero megnitude and no specific direction.

Multiplication of a Vector by a Scalar

Assume $\vec{B}$, whose magnitude is m times the magnitude of $\vec{A}$ represents the product of a vector $\vec{A}$ by a scalar m and having direction the same as or opposite to that of $\vec{A}$ as 'm' is positive or negative. It is demonstrated as follows:

$$\boxed{\vec{B} = m\vec{A}}$$

Let's go through some of the example of multiplication of vectors by a scalar. When we multiply mass, a scalar, with velocity, a vector, we obtain a new physical quantity known as momentum. Since scalar has no direction, the multiplication of a scalar with a vector will give a new vector which will have the same direction as that of the original vector. Momentum will have the direction of velocity.

Unit Vector

If a vector has unit magnitude that is called unit vector. If $\vec{A}$ is a vector with magnitude $A \neq 0$., the unit vector is defined as the ratio of vector by its scalar i.e. $\frac{\vec{A}}{A}$ and has the same direction as that of $\vec{A}$. Symbol for unit vector in the directin of $\vec{A}$ is $\hat{A}$ and is read as 'A carat' or 'A cap'. Therefore, we can demonstrate it in following ways:

$$\frac{\vec{A}}{A} = \hat{A}$$

or
$$\vec{A} = A\,\hat{A}$$

As shown in the Fig 1·7, $\hat{i}, \hat{j}$ *and* $\hat{k}$ denote unit vectors which have the directions of the positive x, y, and z axes of a three dimensional rectangular co-ordinate system respectively,

A vector $\vec{A}$ taken along any one of the three axes may be written as scalar multiple of magnitude of vector $\vec{A}$ and the unit vector along that axis. Apart from this $A\hat{i}$ can be the demonstrative representation of vector $\vec{A}$. For example, a vector $\vec{A}$ taken along x-axis, can be written as $\vec{A}\hat{i}$.

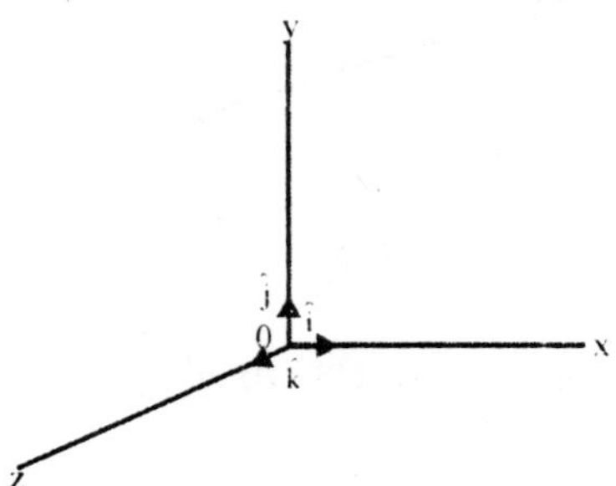

Fig 1.7.

Components of a Vector

Components of a vector $\vec{A}$ are any vectors whose sum is vector $\vec{A}$. The components most frequently used are those parallel to x, y and z axes mutually perpendicular to each other in three dimensional space. These are called rectangular components of a vector.

Assume a vector $\vec{A}$ with its initial point at the origin O of a rectangular co-ordinate system. Now draw a rectangular parallelopiped with the three edges, which met at O, lying along the three rectangular co-ordinate axes and such that the vector $\vec{A}$ constitutes one of the diagonals acreoss the parallelopiped. Now from Fig. 1·8.

Here the vector components of $\vec{A}$ along ox, oy and oz axis are $\vec{A}x, \vec{A}y$ *and* $\vec{A}z$. While Ax, Ay and Az are the magnitudes of vectors $\vec{A}x, \vec{A}y\,\vec{A}z$ if $\hat{i}, \hat{i}$ and $\hat{k}$ are the unit Vectors along ox, oy and oz axis. Now, we can demonstrate it as follows.

$$\vec{A} = \overline{O}\vec{P} = \overline{O}\vec{B} + \overline{B}\vec{P}$$

$$\text{[illegible]} + \overline{H}\vec{B} + \overline{B}\vec{P}$$

$$= \vec{OH} + \vec{OC} + \vec{OE}$$

$$= \vec{OC} + \vec{OE} + \vec{OH}$$

$$= \vec{A}x + \vec{A}y + \vec{A}z \qquad ...(1\cdot4)$$

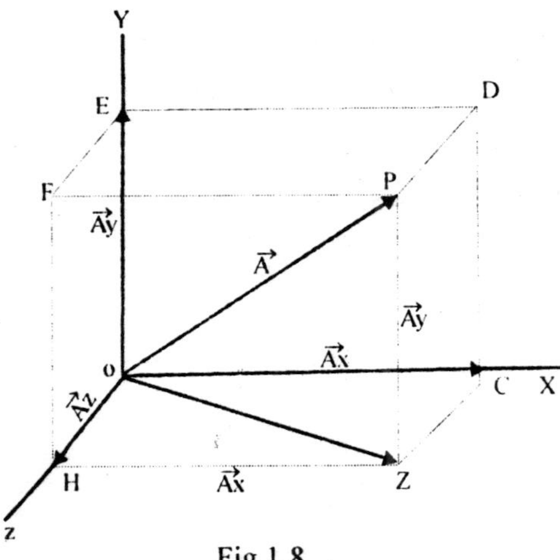

Fig 1.8

Note: At the place of vectors $\hat{i}, \hat{j}$ and $\hat{k}$, ax, ay abd az has also been used in this book at some places.

$$\vec{A} = Ax\hat{i} + Ay\hat{j} + Az\hat{k} \qquad ...(1\cdot5)$$

according to Pathagorean Theorem in three dimensions, we have

$$Ax^2 + Ay^2 + Az^2 = HB^2 + BP^2 + OH^2$$

$$= OB^2 + BP^2 = OP^2 = A^2$$

Therefore,

$$\text{magnitude of } \vec{A} = A = \sqrt{(Ax)^2 + (Ay)^2 + (Az)^2} \qquad ...(1\cdot6)$$

If the angles between vector $\vec{A}$ and x, y and z axes are (A, x), (A, y) and (A, z) respectively, then from Fig, 1·8, we have

$$\left.\begin{aligned} Ax &= A\cos(A, x) \\ Ay &= A\cos(A, y) \\ Az &= A\cos(A, z) \end{aligned}\right| \qquad ...(1\cdot7)$$

Now add the square of the equation P on 1·7.

$$A^2_x + A^2_y + A^2_z = A^2[\cos^2(A,x) + \cos^2(A,y) + \cos^2(A,z)] \qquad ...(1\cdot8)$$

convertes as follows after putting $(A^2 = Ax^2 + Ay^2 + Az^2)$ from eq. (1·6).

$$\cos^2(A,x) + \cos^2(A,y) + \cos^2(A,z) = 1$$

This is called vector directional cosine law, cos (A, x), cos (A, y) and cos (A, z) are called directional cosines of vector $\vec{A}$ in x, y and z directions respectively. Besides it can be demonstrated as follows if they are demonstrated by i, m, n.

$$A = A_x l + A_y m + A_z n$$

and unit vector $\hat{A}$ along the direction $\vec{A}$ is given by

$$\hat{A} = \frac{\vec{A}}{A} = \frac{A_x\hat{i} + A_y\hat{j} + A_z\hat{k}}{A} = \frac{A_x}{A}\hat{i} + \frac{A_y}{A}\hat{j} + \frac{A_z}{A}\hat{k},$$

$$= \hat{i}l + \hat{j}m + \hat{k}n$$

(a) *Position Vector of a Point and its Components*

Lets assume that a set of axes ox, oy, oz define a cartesion system of reference and P (x, y, z) be a point in space. *The vector drawn from the origin to this point i.e.* $\overline{OP}$ *is called position vector for this point. Position vector is often denoted by* $\vec{r}$ *As per the fig. 1·9 we can demonstrate* $\vec{r}$ *in following way*

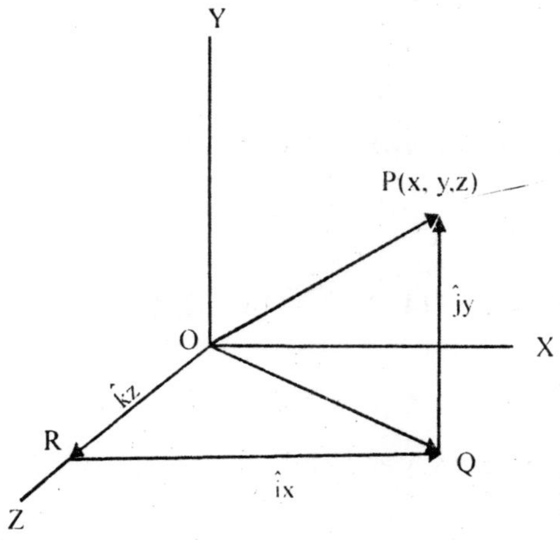

Fig 1.9

$$\vec{r} = x\hat{i} + y\hat{j} + z\hat{k} \qquad ...(1·9)$$

and its magnitude as

$$r = \sqrt{x^2 + y^2 + z^2}$$

(b) *Displacement Vector*

Assume two points $P(x_1, y_1, z_1)$ and $Q(x_2, y_2, z_2)$ in space and $\vec{r}_1$ and $\vec{r}_2$ be the position vectors of these points w.r.t. arbitrary chosen origin O [Fig. 1·10].Then $\vec{r}_2 - \vec{r}_1$ represents the position vector of Q w.r.t. P. This difference is called the displacement vector for P to Q.

Hence, the displacement Vector is the difference between two position vectors.

Example:1. If $\vec{r}_1 = 10\hat{i} - 4\hat{j} + 6\hat{k}$ and $\vec{r}_2 = 2\hat{i} + \hat{j}$

then find,

(a) The component of $\vec{r}_1$ along $\hat{j}$.

(b) The magnitude of $3\vec{r}_1 - \vec{r}_2$

(c) Unit Vector along $\vec{r}_1 + 2\vec{r}_2$

Solution :

(a) The component of $\vec{r}_1$ along $\hat{j}$ is - 4.

(b)
$$(3\vec{r}_1 - \vec{r}_2) = 3(10, -4, 6) - (2, 1, 0)$$
$$= (30, -12, 18) - (2, 1, 0)$$
$$= (28, -13, 18)$$

Hence,

$$|3\vec{r}_1 - \vec{r}_2| = \sqrt{28^2 + (-13)^2 + (18)^2} = \sqrt{1277}$$
$$= 35.75$$

(c) Let $\vec{R} = \vec{r}_1 + 2\vec{r}_2 = (10, -4, 6) + (4, 2, 0) = (14, -2, 6)$.

A unit vector along $\vec{R}$ is

$$= \frac{\vec{R}}{R} = \frac{(14, -2, 6)}{\sqrt{14^2 + (-2)^2 + 6^2}}$$

or

$$= 0.9113\,\hat{i} - 0.1302\,\hat{j} + 0.3906\,\hat{k}$$

Exercise 1

Given vectors $\vec{r}_2 = \hat{i} + 3\hat{k}$ and $\vec{r}_2 = 5\hat{i} + 2\hat{j} - 6\hat{k}$, determine

(a) $|\vec{r}_1 + \vec{r}_2|$

(b) $5\vec{r}_1 - \vec{r}_2$

(c) The component of $\vec{r}_1$ along $\hat{j}$

(d) A unit vector parallel to $3\vec{r}_1 + \vec{r}_2$

Answer: (a) 7, (b) (0,—2, 21), (c), 0, (d) ± (0.2279, 0.3419).

Example 2. Points P and Q are located at (0, 2, 4) and (—3, 1, 5). Calculate

(a) The position vector P

(b) The distance vector from P to Q

(c) The distance between P and Q

(d) A vector parallel to PQ with magntude of 10

Solution:

(a) $r_p = o\hat{i} + 2j + 4\hat{k} = 2\hat{j} + 4\hat{k}$

(b) $r_{PQ} = r_Q - r_P = (-3,\ 1,\ 5) - (0,\ 2,\ 4) = (-3,\ -1,\ 1)$

or $r_{PQ} = -3\hat{i} - 3\hat{j} + \hat{k}$

(c) Since r_{PQ} is the distance vector from P to Q, the distance between P and Q is the magnitude of this vector; that is,

$$d = |r_{PQ}| = \sqrt{9+1+1} = 3.317$$

Alternatively:

$$d = \sqrt{(x_Q - x_P)^2 + (y_Q - y_P)^2 (z_Q - z_P)^2}$$

$$= \sqrt{9+1+1} = 3.317$$

(d) Let the required vector be $\vec{R}$, then

$$\vec{R} = \vec{r}_1\ \hat{r}_2$$

where $\vec{r}_1$ = 10 is the magnitude of $\vec{R}$ Since $\vec{R}$ is parallel to PQ, it must hve the same unit vector as r_{PQ} or r_{QP} Hence,

$$\vec{r}_1\,\hat{r}_1 = \pm\frac{r_{PQ}}{r_{PQ}} = \pm\frac{(-3,-1,\ 1)}{3.317}$$

and

$$\vec{R} = \pm\frac{10(-3,-1+1)}{3.317} = \pm(-9.045a_x - 3.015a_y + 3.015a_z)$$

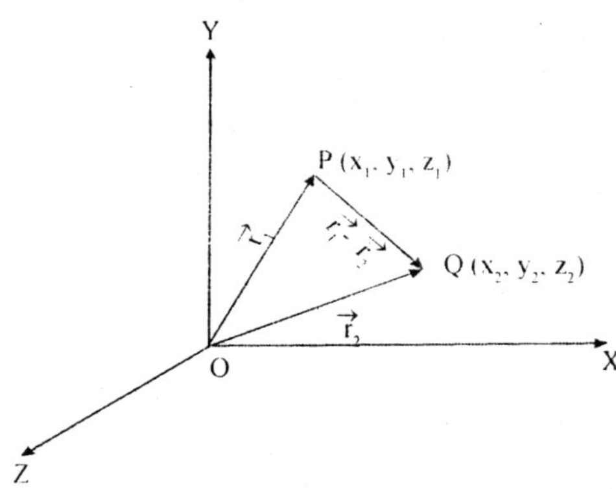

Fig 1.10

Exercise 2. Given points P(1, —3, 5), Q(2, 4, 6), and R(0, 3, 8), find: (a) the position vectors of P and R, (b) the distance vector r_{QR}, (c) the distance between Q and R.

Answer: (a) $\hat{i} - 3\hat{j} + 5\hat{k}$, $3\hat{i} + 3\hat{j}$

(b) $-2\hat{i} - \hat{j} + 2\hat{k}$

Example 3. Given that $\vec{r}_2 = 2\hat{i} + 4\hat{j} - 5\hat{k}$

$$\vec{r}_2 = \hat{i} + 2\hat{j} + 3\hat{k}$$

find (a) lenght of $\vec{r}_2$ and $\vec{r}_1$

(b) unit vectors along $\vec{r}_1$ and $\vec{r}_2$

(c) unit vector parallel to resultant of $\vec{r}_1$ and $\vec{r}_2$

(d) difference of $\vec{r}_1$ and $\vec{r}_2$

Solution:

(a) length of $\vec{r}_1 = r_1 = \sqrt{2^2 + 4^2 + (-5)^2} = \sqrt{4+16+25}$

$= \sqrt{45}$

(b) length of $\vec{r}_2 = r_2 = \sqrt{1^2 + 2^2 + 3^2} = \sqrt{1 + 4 + 9}$

$$= \sqrt{14}$$

(c) unit vector along r_1 i.e.

$$\hat{r}_1 = \frac{\hat{r}_1}{r_1} = \frac{2\hat{i} + 4\hat{j} - 5\hat{k}}{\sqrt{45}}$$

$$= \frac{2}{\sqrt{45}}\hat{i} + \frac{4}{\sqrt{45}}\hat{j} - \frac{5}{\sqrt{45}}\hat{k}$$

Unit vector along $\vec{r}_2$ i.e.

$$\hat{r}_2 = \frac{\vec{r}_2}{r_2} = \frac{\hat{i} + 2\hat{j} + 3\hat{k}}{\sqrt{14}} = \frac{1}{\sqrt{14}}\hat{i} + \frac{2}{\sqrt{14}}\hat{j} + \frac{3}{\sqrt{14}}\hat{k}$$

(c) Resultant of $\vec{r}_1$ and $\vec{r}_2$

$$= \vec{R} = \vec{r}_1 + \vec{r}_2$$

$$= 2\hat{i} + 4\hat{j} - 5\hat{k} + \hat{i} + 2\hat{j} + 3\hat{k}$$

$$= 3\hat{i} + 6\hat{j} - 2\hat{k}$$

Magnitude of $\vec{R}$ =

$$R = \sqrt{3^2 + 6^2 + (-2)^2} = \sqrt{9 + 36 + 4} = \sqrt{49} = 7$$

Unit vector parallel to $\vec{R}$ =

$$= \frac{\vec{R}}{R} = \frac{3\hat{i} + 6\hat{j} - 2\hat{k}}{7} = \frac{3}{7}\hat{i} + \frac{6}{7}\hat{j} - \frac{2}{7}\hat{k}$$

(d) Difference of $\vec{r}_1$ and $\vec{r}_2$

$$= \vec{D} = \vec{r}_1 - \vec{r}_2$$

$$= \vec{r}_1 + (-\vec{r}_2)$$

$$= 2\hat{i} + 4\hat{j} - 5\hat{k} + \left[-\hat{i} - 2\hat{j} - 3\hat{k}\right]$$

$$= \hat{i} + 2\hat{j} - 8\hat{k}$$

Multiplication of Vectors

There are two type of vector multiplication, one is called *scalar or dot product* and the other is called *vector or cross product.*

(a) *Scalar or dot product of two vectors*

The dot product of two vectors $\vec{A}$ and $\vec{B}$, written as $\vec{A}.\vec{B}$, is defined geometrically as the product of the magnitudes of $\vec{A}$ and $\vec{B}$ and the cosine of the angles between them. Hence, the scalar product of two vectors $\vec{A}$ and $\vec{B}$ inclined at an $\angle\theta$ as shown in Fig. 1·11, is given by the equation

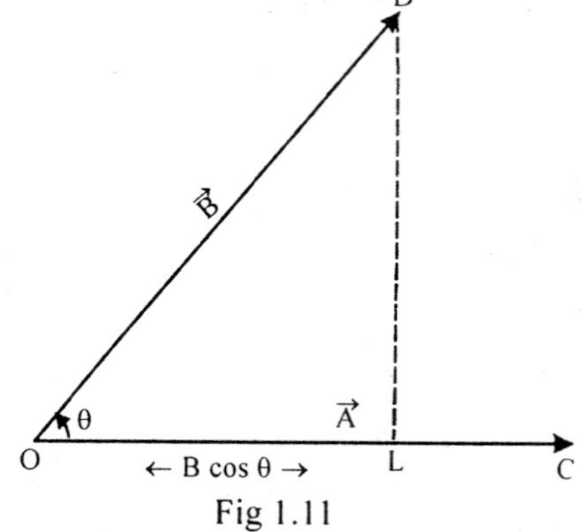

Fig 1.11

$$\vec{A}.\vec{B} = |\vec{A}||\vec{B}|\cos\theta \quad \text{...(1·9a)}$$

$$z = OC \times OD \ cos\ \theta$$

(as OC is magnitude of vector $\vec{A}$ and OD is magnitude of $\vec{B}$).

$$= OC.OL$$

As OD cos $cos\theta = |\vec{B}|cos\theta$ is the magnitude of the component of $\vec{B}$ along the direction of $\vec{A}$ and is equal to OL.

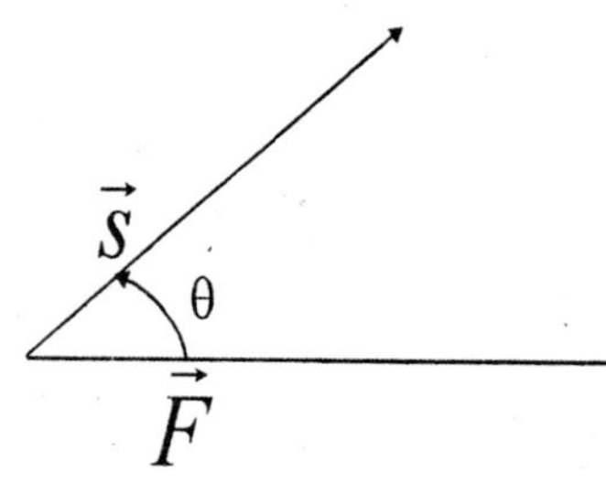

Fig. 1.12

Therefore; we can say that *the product of the size of one vector with magnitude of the component of the other in the direction of first is the scalar product of two vectors.*

The work done by a force in displacing a body through a certain distance is one of the most important example of dot product of two vectors, Let a particle acted on by a force vector $\vec{F}$ by displaced throught a certain distance $\vec{S}$ (Fig. 1·12). The work done W by the force $\vec{F}$ in this process is given by

$$W = \vec{F}.\vec{S} = FS\cos\theta$$

where θ is the angle between $\vec{F}$ and $\vec{S}$. Although the two quantities i.e. force and displacement are vectors. ye ± work done is scalar.

Important Features of Scalar Product

(i) Commutative Law

$$\text{Also } \vec{B}.\vec{A} = |\vec{B}||\vec{A}| \ \cos\left(\vec{B},\vec{A}\right)$$

$$= \left|\vec{B}\right|\left|\vec{A}\right| \cos(-\theta) = \left|\vec{B}\right|\left|\vec{A}\right| \cos\theta \qquad ...(1\cdot10)$$

If, we compare this equation with equation 1·9, we find :

$$\vec{A}.\vec{B} = \vec{B}.\vec{A}$$

Therefore, the scalar product of two vectors is commutative

(ii) Scalar product of two parallel/Collinear vectors:

We know that the angle between two collinear or vector is zero so that,

i.e. $\angle\theta = 0$.

$$\vec{A}.\vec{B} = \left|\vec{A}\right|\left|\vec{B}\right| \cos 0^\circ = \left|\vec{A}\right|\left|\vec{B}\right| \qquad (\because \cos 0^\circ = 1)$$

Thus the dot product of two vectors which are parallel, is equal to product of their magnitudes.

Assume $\vec{B} = \vec{A}$, the scalar product of a vector with itself, is as follows:

$$\vec{A}.\vec{A} = |\vec{A}|\ |\vec{A}| \cos 0^\circ = |\vec{A}|\ |\vec{A}| = A^2$$

$$\text{magnitude of vector } \vec{A} = A = \sqrt{\vec{A}.\vec{A}} \qquad ....(1\cdot11)$$

(iii) Scalor product of two perpendicular vectors

We know that the angle between two perpendicular vector is 90°

Hence

$$\vec{A}.\vec{B} = |\vec{A}|\ |\vec{B}| \cos 90^\circ = 0 \qquad (\because \cos 90^\circ = 0)$$

Thus, two vectors are said to be perpendicular to each other if their dot product is equal to zero.

(iv) Dot products of unit orthogonal vectors $\hat{i}, \hat{j}, \hat{k}$

We know that $\hat{i}, \hat{j}$ and $\hat{k}$ are unit vectors along the three axes of rectangular co-ordinate system. These vectors form 90° angle with each other. Now, apply the (ii) & (iii) features of scalar product to these vectors, we find:

$$\hat{i}\,.\,\hat{i} = \hat{j}\,.\,\hat{j} = \hat{k}\,.\,\hat{k} = (1)\,(1) \cos 0^\circ = 1$$

or

$$i^2 = j^2 = k^2 = 1 \qquad ...(1\cdot12)$$

and

$$\hat{i}\,.\,\hat{j} = \hat{j}\,.\,\hat{k} = \hat{k}\,.\,\hat{i} = (1)\,(1) \cos 90^\circ = 0 \qquad ...(1\cdot13)$$

(*b*) *Vector or Cross Product of two Vectors*

The cross product of two vectors, $\vec{A}$ and $\vec{B}$, written as $\vec{A} \times \vec{B}$, is a vector quantity whose magnitude is the are of the parallelopiped formed by $\vec{A}$ and $\vec{B}$ and is in the direction of advance of a right handed screw as $\vec{A}$ is turned into $\vec{B}$.

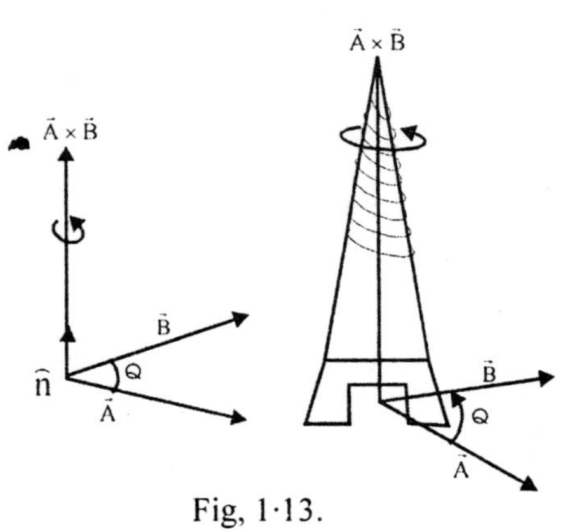

Fig, 1·13.

We must say that the magnitude is equal to the product of the magnitudes of the vectors and sine of the smaller angle θ between their direction (Fig. 1·13) and whose direction is perpendicular to the plane containing vectors $\vec{A}$ and $\vec{B}$. The screw should be placed with its axis perpendicular to the plane containing the two vectors.

In Fig. (1·13), Vectors $\vec{A}$ and $\vec{B}$ are drawn in the plane of the paper. If vector $\vec{A}$ is rotated towards vector $\vec{B}$ $(\theta \to 0)$ by means of a right handed screw, the screw will move normal to the plane of the paper in the outward direction (i.e. towards the reader). If $\hat{n}$ is a unit vector in this direction then according to defintion

$$\vec{A} \times \vec{B} = |\vec{A}|\,|\vec{B}|\,\sin\theta\,\hat{n} \qquad ...(1\cdot14)$$

(*v*) *distributive Law:*

The Scalar product of a vector $\vec{A}$ with sum of the two vector $\vec{B}$ and $\vec{C}$ is distributive which can be shown as follows:

i.e.

$$\vec{A}\,.(\vec{B}+\vec{C}) = \vec{A}\,.\,\vec{B}+\vec{A}\,.\,\vec{C} \qquad (1\cdot16)$$

or

$$\vec{A}\,.\,\vec{A} = |\vec{A}|^2 = \vec{A}^2$$

(*vi*) *Scalar product of two vectors in terms of rectangular components.*

Vector $\vec{A}$ and $\vec{B}$ in terms of their rectangular components can be written as follows:

$$\vec{A} = A_x\hat{i} + A_y\hat{j} + A_Z\hat{k}$$

and

$$\vec{B} = B_x\hat{i} + B_y\hat{j} + B_z\hat{k}$$

Now, $\hat{A}.\hat{B} = (A_x\hat{i} + A_y\hat{j} + A_z\hat{k}).(B_x\hat{i} + B_y\hat{j} + B_Z\hat{k})$

If, we apply distributive Law then:

$$\vec{A}.\vec{B} = A_xB_x\hat{i}\,.\,\hat{i} + A_xB_y\hat{i}\,.\,\hat{j} + A_xB_z\hat{i}\,.\,\hat{k}$$

$$+A_yB_x\hat{j}\,.\,\hat{i} + A_yB_y\hat{i}\,.\,\hat{j} + A_yB_z\hat{i}\,.\,\hat{k}$$

$$+A_zB_x\hat{k}\,.\,\hat{i} + A_zB_y\hat{k}\,.\,\hat{j} + A_zB_z\hat{k}\,.\,\hat{k}$$

Now, putting the value of product of $\hat{i}$, $\hat{j}$ and $\hat{k}$ from equation (1·12) and (1·13), we find:

$$\vec{A}\,.\,\vec{B} = A_xB_x + A_yB_y + A_zB_z \qquad ...(1\cdot16)$$

Therefore, the scalar product of two vector is the sum of the products of their components along each of the rectangular co-ordinate axis.

Also $$\vec{A}\,.\,\vec{A} = A^2 = A^2{}_x + A^2{}_v + A^2{}_z$$

Now, the magnitude of a vector in terms of its rectangular components is shown as follows:

$$A = \sqrt{A^2{}_x + A^2{}_v + A^2{}_z}$$

The vector multiplication of eq. (1·16) is called cross product due to the cross sign; it is also called vector product because the result is a vector. If $A = (A_x, A_y, A_z)$ and $B = (B_x, B_y, B_z)$ then

$$A \times B = \begin{vmatrix} \hat{i} & \hat{j} & \hat{k} \\ A_x & A_y & A_z \\ B_x & B_y & B_z \end{vmatrix}$$

$$= (A_yB_z - A_zB_y)\,\hat{i} + (A_zB_x - A_xB_z)\,\hat{j} + (A_xB_y - A_yB_x)\,\hat{k}$$

which is obtained by "crossing" terms in cyclic permutation, hence the name cross product.

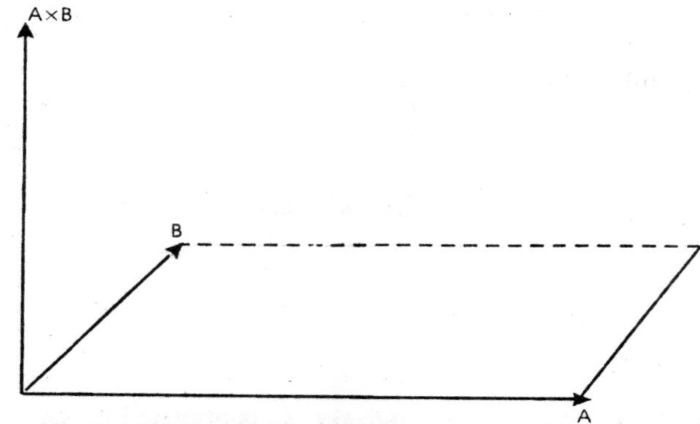

Figure 1·14. The cross product of A and B is a vector with magnitude equal to the area of the parallelogram and direction as indicated.

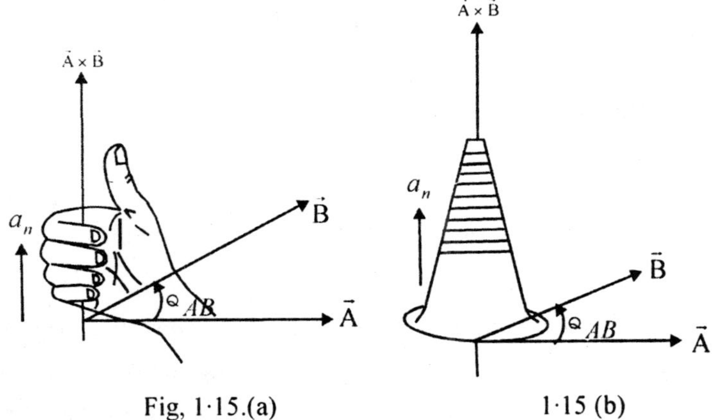

Fig, 1·15.(a) 1·15 (b)

Important Features of the Vector Pro-duct:

(i) It is not commutative

$$A \times B \neq B \times A$$

But, It is but anticommutative

$$A \times B = -B \times A \qquad ...(1\cdot17)$$

Let's prove it;

Assume that the magnitude of $\vec{B} \times \vec{A}$ is $|\vec{B}|\,|\vec{A}|\,\sin\theta$, which is same as the magnitude of $\vec{A} \times \vec{B}$. Now, we have to turn vector. $\vec{B}$ towards vector $\vec{A}$ to obtain its direction. This motion is clockwise while the rotation of $\vec{A}$ towards $\vec{B}$ is anti-clockwise because the screw has to be rotated into right-hand direction.

Hence, the screw on the paper seems to move away from reader or we can say that screw will move in direction opposite to that of $\hat{n}$.

Therefore, the unit vector along $\vec{B} \times \vec{A}$ is $\hat{n}$.

so that,

$$\vec{B} \times \vec{A} = -|\vec{B}|\,|\vec{A}|\,\sin\theta\,\hat{n} \qquad ...(1\cdot18)$$

Comparing (1·17) and (1·18), we have

$$\vec{A} \times \vec{B} = -\vec{B} \times \vec{A}$$

Therefore, the direction of $\vec{A} \times \vec{B}$ and $\vec{B} \times \vec{A}$ is oppsite but magnitude is same thus the vector muitiplication is anit-commutative rather than commutative.

(ii) Area of parallelogram and vector area=Base × Vertical

area of parallelogram = Base × Vertical height

$$= |\vec{A}|\, h$$

$$= |\vec{A}|\left(|\vec{B}|\ sin\theta\right) \qquad \left(\because h = |\vec{B}| sin\,\theta\right)$$

$$= |\vec{A} \times \vec{B}|.$$

Where $\vec{A}$ and $\vec{B}$ are the two adjacent sides of a para-llelogram. Thus, the magnitude of the vector product is equal to the area of the parallelogram formed by the two vectors as sides.

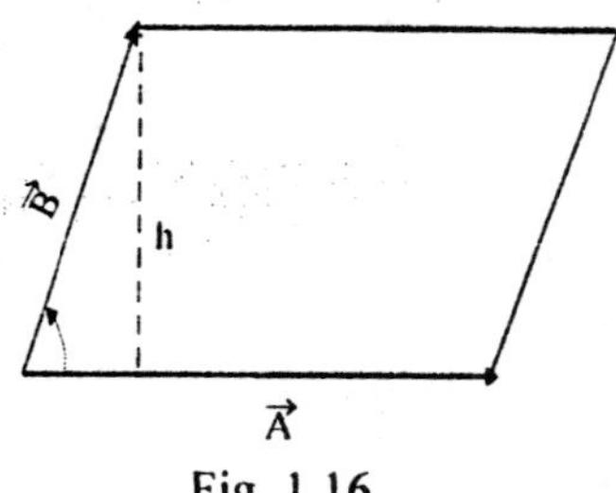

Fig. 1.16

With a plane area, a direction is associated perpendicular to the plane of the area. Since a perpendicular to a plane may have two direction, one out of the plane and other into the plane, so these two direction can be associated with a plane. Now, the area vector can be mentioned as cross product in both the cases discussed above.

Hence, the vector area of parallelogram is represented by $\vec{A} \times \vec{B}$ where $\vec{A}$ and $\vec{B}$ are the adjacent sides of parallelogram.

(iii) Cross product of two parallel/collinear vectors

We know that the angle between two parallel and anti-parallel vector is 0° and 180° respectively and in both the cases

$$sin\,\theta = 0$$

$$\therefore \quad \vec{A} \times \vec{B} = \hat{n}\, |\vec{A}|\, |\vec{B}|\, \sin 0° = 0 \qquad (\because\ sin\ 0° = 0)$$

Thus two vector are said to be parallel to each other if their cross product vanishes.

Obviouly cross product of a vector by itself is zero i.e.

$$\vec{A} \times \vec{A} = 0$$

(iv) Cross product of two perpendicular vectors

We know that the angle between two perpendicular vectors is 90°

Therefore;

$$\vec{A} \times \vec{B} = |\vec{A}|\ |\vec{B}|\ sin\ 90° = \hat{n}$$

$$= |\vec{A}|\ |\vec{B}| = \hat{n}$$

Thus, the cross product of two perpendicular vectors is equal to the product of the magnitudes of the two vectors and having the direction of unit vector $\hat{n}$ where $\hat{n}$ is perpendicular to the plane containing vectors $\vec{A}$ and $\vec{B}$ and whose sense determined by right handed screw.

(v) Vector products of unit normal vectors $\hat{i}, \hat{j}, \hat{k}$.

We know that unit vectors $\hat{i}, \hat{j}$ and $\hat{k}$ form 90° with each other and fixed in a orthogonal frame. If we apply above features of vector product to these vector, we get:

$$\hat{i} \times \hat{i} = \hat{j} \times \hat{j} = \hat{k} \times \hat{k} = 0 \qquad ...(1\cdot19)$$

$$\hat{i} \times \hat{j} = \hat{k},\ \hat{j} \times \hat{k} = \hat{i}\ \hat{k} \times \hat{i} = \hat{j}$$

$$\hat{j} \times \hat{i} = -\hat{k},\ \hat{k} \times \hat{j} = -\hat{i},\ \hat{i} \times \hat{k} = -\hat{j} \qquad ...(1\cdot20)$$

There is one significant point to be noticed that in these products, the cyclic order $\hat{i}\,\hat{j}\,\hat{k}\,\hat{i}\,\hat{j}$ ....(Fig. 1·17) should be properly observed. Moving clock wise leads to positive results (for instance $\hat{i} \times \hat{j} = \hat{k}$). While moving counter-clock wise leads to negative results (as an example $\hat{k} \times \hat{j} = -\hat{i}$).

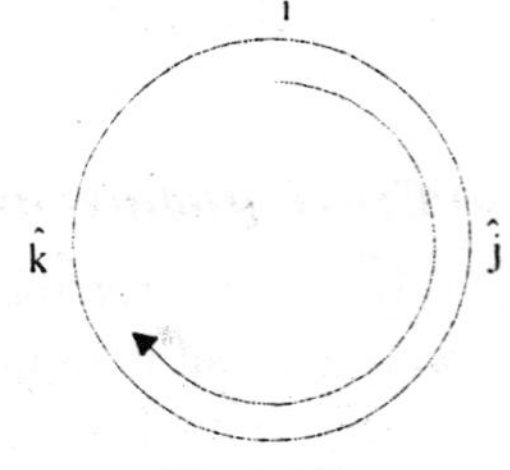

Fig 1.17

(vi) *distributive Law:*

Vector product of a vector $\vec{A}$ with sum of two vectors $\vec{B}$ and $\vec{C}$ is distributive and it can be Fig. 1·17. demonstrated in following way

i.e.
$$\vec{A} \times \left(\vec{B} + \vec{C}\right) = \vec{A} \times \vec{B} + \vec{A} \times \vec{C} \qquad ...(1\cdot21)$$

Note that the order/sequence of factors in cross product is important. The usual laws of algebra will apply only when proper order is maintained . The usual laws of algebra will apply only when proper order is maintained.

(vii) Vector Product in some Physical Quantities

Cross product of vectors may be used to Present some physical quantities such as momentum, torque etc. Now, assume a particle of mass m situated at the point p (Fig. 1·18), at a particular instant of time and this particle has a velocity vector $\vec{V}$ and is acted upon by a foce vector $\vec{F}$ at that instant. Then its linear momentum vector $\vec{p}$ is given

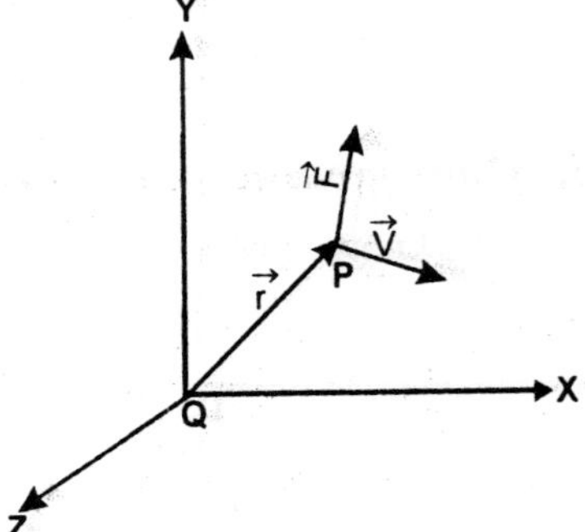

Fig 1.18

by the product $m\vec{V}$. Further angular momentum $\vec{L}$ of the particle is defined as the vector product of position vector $\vec{r}$ of the particle and its linear momentum $\vec{p}$

i.e. $$\vec{L} = \vec{r} \times \vec{p}$$

The moment of a force $\vec{F}$ acting at a position vector $\vec{r}$ relative to the origin i.e. Torqueon a particle is expressed as follows

$$\vec{\tau} = \vec{r} \times \vec{F}.$$

Example 4. Two vectors are given below find the angle between two.

$$\vec{A} = 5\hat{i} + 6\hat{j} - 4\hat{k} \; and \; \vec{B} = 2\hat{i} + 3\hat{j}.$$

Solution. Here $A_x = 5, A_y = 6, A_z = -4$

and $$B_x = 2, B_y = 3, B_z = 0$$

$\therefore$ $$\vec{A}.\vec{B} = A_x B_x + A_y B_y + A_z B_z$$

$$= 5 \times 2 + 6 \times 3 + (-4) \times 0$$

$$= 10 + 18 + 0 = 28$$

Magnitude of

$$\vec{A} = |\vec{A}| = \sqrt{A^2{}_x + A^2{}_y + A^2{}_z} = \sqrt{5^2 + 6^2 + (-4)^2}$$

$$= \sqrt{25 + 36 + 16} = \sqrt{77}$$

Magnitude of

$$\vec{B} = |\vec{B}| = \sqrt{B_x^2 + B_y^2 + B_z^2} = \sqrt{2^2 + 3^2 + 0^2}$$

$$\sqrt{4 + 9} = \sqrt{13}$$

Now $\vec{A}\,.\,\vec{B} = |\vec{A}|\;|\vec{B}|\;cos\theta$ where θ is the angle between the vectors $\vec{A}$ and $\vec{B}$.

$$28 = \sqrt{77}\,\sqrt{13}\;cos\theta = \sqrt{1001}\,cos\theta$$

$$\cos\theta = \frac{28}{\sqrt{1001}}, \text{ or } \theta = \cos^{-1}\left(\frac{28}{\sqrt{1001}}\right).$$

Example.5

If; $|\vec{A}+\vec{B}|=|\vec{A}-\vec{B}|$ then prove that vector $\vec{A}$ is perpendicular to $\vec{B}$.

Solution. $|\vec{A}+\vec{B}|=|\vec{A}-\vec{B}|$

$$|\vec{A}+\vec{B}|^2 = |\vec{A}-\vec{B}|^2$$

or $$(\vec{A}+\vec{B}).(\vec{A}+\vec{B}) = (\vec{A}-\vec{B}).(\vec{A}-\vec{B})$$

or $$\vec{A}\,.\,\vec{A}+\vec{A}\,.\,\vec{B}+\vec{B}\,.\,\vec{A}+\vec{B}\,.\vec{B}$$

or $$=\vec{A}\,.\,\vec{A}-\vec{A}.\,\vec{B}-\vec{B}.\,\vec{A}+\vec{B}.\,\vec{B}$$

or $$A^2+2\vec{A}\,.\,\vec{B}+B^2 = A^2-2\vec{A}\,.\,\vec{B}+B^2 \quad (\because \vec{A}.\vec{B}=\vec{B}.\vec{A})$$

or $$4\vec{A}\,.\,\vec{B}=0$$

or $\vec{A}\,.\,\vec{B}=0$ Hence; $\vec{A}\perp\vec{B}$

Example 6. If $\vec{A}=2\hat{i}-3\hat{j}-\hat{k}$ *and* $\vec{B}=\hat{i}+4\hat{j}-2\hat{k}$, *find*

(*i*) $\vec{A}\times\vec{B}$ *and* (*ii*) $\vec{B}\times\vec{A}$.

Solution. (i) $\vec{A}\times\vec{B} = (2\hat{i}-3\hat{j}-\hat{k})\times(\hat{i}+4\hat{j}-2\hat{k})$

$$=\begin{vmatrix}\hat{i} & \hat{j} & \hat{k}\\ 2 & -3 & -1\\ 1 & 4 & -2\end{vmatrix} = \hat{i}\begin{vmatrix}-3 & -1\\ 4 & -2\end{vmatrix} - \hat{j}\begin{vmatrix}2 & -1\\ 1 & -2\end{vmatrix} + \hat{k}\begin{vmatrix}2 & -3\\ 1 & 4\end{vmatrix}$$

$$=\hat{i}(6+4)-\hat{j}(-4+1)+\hat{k}(8+3)$$

$$=10\hat{i}+3\hat{j}+11\hat{k}$$

(ii) $\vec{B}\times\vec{A} = (\hat{i}+4\hat{j}-2\hat{k})\times(2\hat{i}-3\hat{j}-\hat{k})$

$$=\begin{vmatrix}\hat{i} & \hat{j} & \hat{k}\\ 1 & 4 & -2\\ 2 & -3 & -1\end{vmatrix} = \hat{i}\begin{vmatrix}4 & -2\\ -3 & -1\end{vmatrix} - \hat{j}\begin{vmatrix}1 & -2\\ 2 & -2\end{vmatrix} + \hat{k}\begin{vmatrix}1 & 4\\ 2 & -3\end{vmatrix}$$

$$= \hat{i}(-4-6) - \hat{j}(-1+4)\,\hat{k}(-3-8)$$

$$= -10\hat{i} - 3\hat{j} - 11\hat{k}.$$

Example7. Show that vector $\vec{A} = 4\hat{i} - 6\hat{j} - -2\hat{k}$ and

$B = -12\hat{i} + 18\hat{j} - 6\hat{k}$ are parallel.

Solution. Vector $\vec{A}$ and $\vec{B}$ are parallel if $\vec{A} \times \vec{B} = 0$, so

$$\vec{A} \times \vec{B} = \begin{vmatrix} \hat{i} & \hat{j} & \hat{k} \\ 4 & -6 & -2 \\ -12 & 18 & 6 \end{vmatrix} = \hat{i}\begin{vmatrix} -6 & -2 \\ 18 & 6 \end{vmatrix} - \hat{j}\begin{vmatrix} 4 & -2 \\ -12 & 6 \end{vmatrix} + \hat{k}\begin{vmatrix} 4 & -6 \\ -12 & 18 \end{vmatrix}$$

$$= \hat{i}\ (-36+36) - \hat{j}\ (24-24) + \hat{k}\ (72-72)$$

$$= 0\hat{i} + 0\hat{j} + 0\hat{k} = 0$$

Now, it is proved, that vector $\vec{A} \parallel \vec{B}$.

Product of three Vectors

The vector product of two vectors $\vec{B}$ and $\vec{C}$ is a vector. So it can be multiplied either scalarly or vectorally with a third vector $\vec{A}$. Thus there are two types of triple products: one $\vec{A} \cdot (\vec{B} \times \vec{C})$ is known as the scalar triple product, and the other $\vec{A} \times (\vec{B} \times \vec{C})$ is called the vector triple product .

(a) *Scalar Triple Product*

Given three vectors **A, B,** and **C,** we define the scalar triple product as.

$$\boxed{A \,.\, (B \times C) = B \,.\, (C \times A) = C \,.\, (A \times B)} \qquad (1{\cdot}22)$$

Obtained in cyclic permutation. If $A = (A_x, A_y, A_z)$, $B = (B_z, B_y, B_z)$ and $C = (C_x, C_y, C_z)$, then A . (B ×C) is the volume of a parallelepiped having **A, B,** and **C** as edges and is easily obtained by finding the determinant of the 3×3 matrix formed by **A, B** and **C**; that is,

$$A \,.\, (B \times C) = \begin{vmatrix} A_x & A_y & A_z \\ B_x & B_y & B_z \\ C_x & C_y & C_z \end{vmatrix} \qquad ...(1.23)$$

Since the result of this vector multiplication is scalar, eq. (1·28) or (1·29) is called the *Scalar triple product.*

(b) *Vector Triple Product*

For vectors **A, B,** and **C,** we define the vector triple product as

$$\boxed{A \times (B \times C) = B\,(A \,.\, C) - C\,(A \,.\, B)} \qquad ...(1.24)$$

obtained using the "bac-cab" rule. It should be noted that

$$(A \,.\, B)C \neq A(B \,.\, C) \qquad ...(1.25)$$

but

$$(A \,.\, B)C = C(A \,.\, B). \qquad ...(1.26)$$

If (A_x, A_y, A_z), (B_x, B_y, B_z) and (C_x, C_y, C_z) are the magnitudes of the rectangular components of vectors $\vec{A}$, $\vec{B}$ and $\vec{C}$ respectively, then

$$\vec{B} \times \vec{C} = \begin{vmatrix} \hat{i} & \hat{j} & \hat{k} \\ B_x & B_y & B_z \\ C_x & C_y & C_z \end{vmatrix} = (B_yC_z - B_zC_y)\hat{i}$$

$$+(B_zC_x - B_xC_z)\hat{j} + (B_xC_y - B_yC_x)\hat{k} \qquad ...(1.27)$$

Applying the same relation to

$$\vec{A} \times (\vec{B} \times \vec{C}) = \begin{vmatrix} \hat{i} & \hat{j} & \hat{k} \\ A_x & A_y & A_z \\ (\vec{B} \times \vec{C})_x & (\vec{B} \times \vec{C})_y & (\vec{B} \times \vec{C})_z \end{vmatrix}$$

Putting the component of $(\vec{B} \times \vec{C})$ from (1·24), we get

$$\vec{A} \times (\vec{B} \times \vec{C}) = \begin{vmatrix} \hat{i} & \hat{j} & \hat{k} \\ A_x & A_y & A_z \\ B_yC_z - B_zC_y & B_zC_x - B_yC_z & B_xC_y - B_yC_x \end{vmatrix}$$

Considering the first term of this determinant, we have

$$\hat{i}\left[A_y(B_xC_y - B_yC_x) - A_z(B_zC_x - B_xC_z)\right]$$

$$= \hat{i}\left[A_yB_xC_y - A_yB_yC_x - A_zB_zC_x + A_zB_xC_z\right]$$

Adding and Subtracting $A_xB_xC_x$

$$= \hat{i}\left[A_y B_x C_y - A_y B_y C_x - A_z B_z C_x + A_z B_x C_z + A_x B_x C_x - A_x B_x C_x\right]$$

Separating positive terms all of which contain B_x and separating the negative terms all of which contain C_x, we get

First term, $= \hat{i}\ [A_y B_x C_y + A_z B_x C_z + A_x B_x C_x$

$- A_y B_y C_x - A_z B_z C_x - A_x B_x C_x]$

$= \hat{i}\ [B_x (A_x C_x + A_y C_y + A_z C_z) -$

$C_x(A_x B_x + A_y B_y + A_z B_z)]$

$$= B_x \hat{i}\left(\vec{A} \cdot \vec{C}\right) - C_x \hat{i}\left(\vec{A} \cdot \vec{B}\right)$$

Similarly, the second and the third terms of the above determinant can be written as

Second term $= B_y \hat{j}\left(\vec{A} \cdot \vec{C}\right) - C_y \hat{j}\left(\vec{A} \cdot \vec{B}\right)$

Third term $= B_z \hat{K}\left(\vec{A} \cdot \vec{C}\right) - C_z \hat{k}\left(\vec{A} \cdot \vec{B}\right)$

Adding all the terms, we get.

$$\vec{A} \times \left(\vec{B} \times \vec{C}\right) = \left(B_x \hat{i} + B_y \hat{j} + B_z \hat{k}\right)\left(\vec{A} \cdot \vec{C}\right) - \left(C_x \hat{i} + C_y \hat{j} + C_z \hat{K}\right)\left(\vec{A} \cdot \vec{B}\right)$$

$$= \vec{B}\left(\vec{A} \cdot \vec{C}\right) - \vec{C}\left(\vec{A} \cdot \vec{B}\right) \qquad ...(1{\cdot}28)$$

Calling $\vec{A}$ as external vector, $\vec{B}$ as the middle and $\vec{C}$ as the extreme vector, the right hand side of equation (1·24) may be *remembered* as.

[*external dot extreme*] *middle* -[*external dot middle*] extreme

Now consider the expression $\left(\vec{A} \times \vec{B}\right) \times \vec{C}$. Using the above result this can be written as.

$$\left(\vec{A} \times \vec{B}\right) \times \vec{C} = \left(\vec{C} \cdot \vec{A}\right) \vec{B} - \left(\vec{C} \cdot \vec{B}\right) \vec{A}$$

Thus $\vec{A} \times \left(\vec{B} \times \vec{C}\right) \neq \left(\vec{A} \times \vec{B}\right) \times \vec{C}$

Hence vector triple product is neither commutative nor associative i.e. the value of the vector triple product is quite different if we interchange the order of factors.

Example 8. Prove that

$$A \times (B \times C) + B \times (C \times A) \times C \times (A \times B) = 0 \qquad \text{(K.U. 1992 A)}$$

Solution. From triple product of three vector, we get.

$$\vec{A}\times(\vec{B}\times\vec{C})=(\vec{A}\,.\,\vec{C})\,\vec{B}-(\vec{A}\,.\,\vec{B})\,\vec{C}$$

$$\vec{B}\times(\vec{C}\times\vec{A})=(\vec{B}\,.\,\vec{A})\,\vec{C}-(\vec{B}\,.\,\vec{C})\,\vec{A}$$

$$\vec{C}\times(\vec{A}\times\vec{B})=(\vec{C}\,.\,\vec{B})\,\vec{A}-(\vec{C}\,.\,\vec{A})\,\vec{B}$$

Adding the above three equations.

$$\vec{A}\times(\vec{B}\times\vec{C})+\vec{B}\times(\vec{C}\times\vec{A})+\vec{C}\times(\vec{A}\times\vec{B})$$

$$=(\vec{A}\,.\,\vec{C})\,\vec{B}-(\vec{A}\,.\,\vec{B})\,\vec{C}+(\vec{B}\,.\,\vec{A})\,\vec{C}$$

$$-(\vec{B}\,.\,\vec{C})\,\vec{A}+(\vec{C}\,.\,\vec{B})\vec{A}-(\vec{C}\,.\,\vec{A})\,\vec{B}$$

$$=(\vec{A}\,.\,\vec{C})\,\vec{B}-(\vec{A}\,.\,\vec{B})\,\vec{C}+(\vec{A}\,.\,\vec{B})\,\vec{C}$$

$$-(\vec{B}\,.\,\vec{C})\,\vec{A}+(\vec{B}\,.\,\vec{C})\vec{A}-(\vec{A}\,.\,\vec{C})\,\vec{B}$$

$$=0$$ ($\because$ Scalar product is commutative)

Polar and Axial Vectors

Vectors which have come across so far are classified into two catagories:

(i) polar vector and (ii) axial vectors.

A polar vector is a vector which corresponds to a linear effect in its direction. The vectors $\vec{r}$ (displacement), $\vec{v}$ (linear velocity), $\vec{a}$ (acceleration), $\vec{F}$ (Force) and $\vec{p}$ (linear momentum) are the examples of polar vectors. This vector has a direction independent of the reference frame e.g. if we consider a polar vector $\vec{r}$ and suppose a boy goes from his room to the laboratory, his displacement vector $\vec{r}$ point from the room to laboratory. this displacement is certain which ever be the reference frame.

On the other hand if the effect of a physical quantity is rotation about a certain axis, it is called axial vector. the vector $\vec{\omega}$ (angular velocity), $\vec{\alpha}$ (Angular acceleration), $\vec{\tau}$ (torque), $\vec{L}$ (angular momentun) and area are the examples of exial vectors.

An axial vector is represented by a straight line indicating the axis with a curved ring like arrow around it indicating the direction of rotation or by a straight arrow such that looking along it the rotation seems to be clockwise i.e sense of direction is given by right handed screw rule. It means that sense of direction of this vector depends on the handedness of the reference frame.

All vectors defined as the vector products of two polar vectors $\left[\text{i.e. } \vec{\tau} = \vec{r} \times \vec{F} \textit{ and } \vec{L} = \vec{r} \times \vec{p} \textit{ etc.}\right]$ are axial vectors, because sense of their direction is given by right handed screw rule. These vector products change sign if we go from a right handed to left handed frame.

Component form of $\vec{A} \cdot \left(\vec{B} \times \vec{C}\right)$

From the formula of scalar products (eqn. 1·15), we know that

$$\vec{A} \cdot \vec{B} = A_x B_x + A_y B_y + A_z B_z$$

Thus,

$$\vec{A} \cdot \left(\vec{B} \times \vec{C}\right) = A_x\left(\vec{B} \times \vec{C}\right)_x + A_y\left(\vec{B} \times \vec{C}\right)_y + A_z\left(\vec{B} \times \vec{C}\right)_z \qquad (1\cdot 29)$$

From the formula of vector product (eqn. 1·21).

$$\vec{B} \times \vec{C} = \begin{vmatrix} \hat{i} & \hat{j} & \hat{k} \\ B_x & B_y & B_z \\ C_x & C_y & C_z \end{vmatrix}$$

$\left(\vec{B} \times \vec{C}\right)_x = B_y C_z - B_z C_y, \left(\vec{B} \times \vec{C}\right)_y = B_z C_x - B_z C_z$ and

$\left(\vec{B} \times \vec{C}\right)_z = \mathrm{B_x} C_y - B_y C_x$

Substituting these results in (1·28), we get.

$$\vec{\mathrm{A}} \cdot \left(\vec{B} \times \vec{C}\right) = \mathrm{A_x}\left(\mathrm{B_y C_z - B_z C_y}\right) + \mathrm{A_y}\left(\mathrm{B_z C_x - B_x C_z}\right)$$

$$+\mathrm{A_z}\left(\mathrm{B_x C_y - B_y C_x}\right)$$

Which can obviously be written in the determinant form as.

$$\vec{A} \cdot \left(\vec{\mathrm{B}} \times \vec{\mathrm{C}}\right) = \begin{vmatrix} A_x & A_y & A_z \\ B_x & B_y & B_z \\ C_x & C_y & C_z \end{vmatrix}$$

Example 9. Prove that $\vec{A} \cdot \left(\vec{B} \times \vec{C}\right) = \vec{B} \cdot \left(\vec{C} \times \vec{A}\right) = \vec{C} \cdot \left(\vec{A} \times \vec{B}\right)$.

Solution
$$\vec{A}\cdot\left(\vec{B}\times\vec{C}\right)=\begin{vmatrix}A_x & A_y & A_z\\ B_x & B_y & B_z\\ C_x & C_y & C_z\end{vmatrix}$$

Using the properties of determinant, once interchanging first and second row with negative sign and then second and theird row with another negative sign, we have.

$$=-\begin{vmatrix}B_x & B_y & B_z\\ A_x & A_y & A_z\\ C_x & C_y & C_z\end{vmatrix}=\begin{vmatrix}B_x & B_y & B_z\\ C_x & C_y & C_z\\ A_x & A_y & A_z\end{vmatrix}=\vec{B}\cdot\left(\vec{C}\times\vec{A}\right)$$

Similarly
$$\vec{A}\cdot\left(\vec{B}\times\vec{C}\right)=\begin{vmatrix}A_x & A_y & A_z\\ B_x & B_y & B_z\\ C_x & C_y & C_z\end{vmatrix}=-\begin{vmatrix}A_x & A_y & A_z\\ C_x & C_y & C_z\\ B_x & B_y & B_z\end{vmatrix}$$

$$=\begin{vmatrix}C_x & C_y & C_z\\ A_x & A_y & A_z\\ B_x & B_y & B_z\end{vmatrix}=\vec{C}\cdot\left(\vec{A}\times\vec{B}\right)$$

Hence $\quad \vec{A}\cdot\left(\vec{B}\times\vec{C}\right)=\vec{B}\cdot\left(\vec{C}\times\vec{A}\right)=\vec{C}\cdot\left(\vec{A}\times\vec{B}\right).$

This shows that the and cross in scalar triple product can be interchanged at will.

Example 10. Prove $\vec{A}\cdot\left(\vec{A}\times\vec{C}\right)=0.$

Solution. $\vec{A}\cdot\left(\vec{A}\times\vec{C}\right)=\left(\vec{A}\times\vec{A}\right)\cdot\vec{C}$ [$\because$ in scalar triple product, dot and cross are interchangeable]

$$=\vec{0}\cdot\vec{C} \qquad \left(\because \vec{A}\times\vec{A}=0\right)$$

$$=0.$$

Example 11. *Find the value of p if the vectors* $\vec{A}=2\hat{i}+p\hat{j}+\hat{k},\ \vec{B}=\hat{i}+2\hat{j}+3\hat{k}$ and $\vec{C}=3\hat{i}+4\hat{j}+5\hat{k}$ are coplanar.

Solution. The three vectors $\vec{A},\vec{B}$ and $\vec{C}$ are coplanar if their scalar triple product, which represents the volume of the parallelopiped formed by the vectors is zero i.e.

$$\vec{A}\cdot\left(\vec{B}\times\vec{C}\right)=\begin{vmatrix}A_x & A_y & A_z\\ B_x & B_y & B_z\\ C_x & C_y & C_z\end{vmatrix}=-\begin{vmatrix}2 & p & 1\\ 1 & 2 & -3\\ 3 & -4 & 5\end{vmatrix}=0$$

or $2(10-2)+p(-9-5)+1(-4-6)=0$

or $-14p-14=0$

or $14p=-14$ or $p=-1$

Example 12. Given vectors A= $3a_x + 4a_y + a_z$ and B=$2a_y$-$5a_z$, find the between A and B.

Solution: The angle θ_{AB} can be found by using either dot product or cross product.

$$A\,.\,B=(3,4,1)\,.\,(0,2,-5)$$

$$=0+8-5=3$$

$$|A|=\sqrt{3^2+4^2+1^2}=\sqrt{26}$$

$$|B|=\sqrt{0^2+2^2+(-5)^2}=\sqrt{29}$$

$$\cos\theta_{AB}=\frac{A\,.\,B}{|A|\,|B|}=\frac{3}{\sqrt{(26)(29)}}=0.1092$$

$$\theta_{AB}=\cos^{-1}\,0.1092\;=\;83.73°$$

Alternatively,

$$A\times B=\begin{vmatrix}a_x & a_y & a_z\\ 3 & 4 & 1\\ 0 & 2 & -5\end{vmatrix}$$

$$=(-20-2)a_x+(0+15)a_y+(6-0)a_z$$

$$=(-22,15,6)$$

$$|A\times B|=\sqrt{(-22)^2+15^2+|6^2}=\sqrt{745}$$

$$\sin\theta_{AB}=\frac{|A\times B|}{|A|\,|B|}=\frac{\sqrt{745}}{\sqrt{(26)(29)}}=0.994$$

$$\theta_{AB}=\cos^{-1}\;0.994=83.73°$$

Exercise 3

If $A = a_x + 3a_z$ and $B = 5a_x + 2a_y - 6a_z$, find θ_{AB}.

Answar: 120.6°

Example 13. There field quantities are given by,

$$P = 2a_x - a_z$$
$$Q = 2a_x - a_y + 2a_z$$
$$R = 2a_x - 3a_y + a_z$$

Determine

(a) $(P + Q) \times (P - Q)$

(b) $Q . R \times P$

(c) $P . Q \times R$

(d) $\sin \theta_{QR}$

(e) $P \times (Q \times R)$

(f) A unit vector perpendicular to both Q and R.

(g) The omponent of P along Q.

Solution:

(a) $(P+Q) \times (P-Q) = P \times (P - Q) + Q \times (P - Q)$

$= P \times P - P \times Q + Q \times P - Q \times Q$

$= 0 + Q \times P + Q \times P - 0$

$= 2Q \times P$

$$= 2\begin{vmatrix} a_x & a_y & a_z \\ 2 & -1 & 2 \\ 2 & 0 & -1 \end{vmatrix}$$

$= 2(1-0)a_x + 2(4 + 2)a_y + 2(0+2)a_z$

$= 2a_x + 12a_y + 4a_z$

(b) The only way $Q. R \times P$ makes sense is

$$Q. (R \times P) = (2, -1, 2). \begin{vmatrix} a_x & a_y & a_z \\ 2 & -3 & 1 \\ 2 & 0 & -1 \end{vmatrix}$$

$= (2, -1, 2). (3, 4, 6)$

$= 6 - 4 + 12 = 14.$

Alternatively,

$$Q\,.\,(R \times P) = \begin{vmatrix} 2 & -1 & 2 \\ 2 & -3 & 1 \\ 2 & 0 & -1 \end{vmatrix}$$

To find the determinant of a 3×3 matrix, we repeat the first two rows and cross multiply; when the cross multiplication is from right to left, the result should be negated as shown below. This technique of finding a determinant applies only to a 3×3 matrix. Hence,

$$Q\,.\,(R \times P) = \begin{array}{cccccc} & 2 & -1 & 2 & \\ & 2 & -3 & 1 & \\ - & 2 & 0 & -1 & + \\ - & 2 & -1 & 2 & + \\ - & 2 & -3 & 1 & + \end{array}$$

$$= +6 + 0 - 2 + 12 - 0 - 2$$

$$= 14$$

as obtained before

(c) From eq. (1.28)

$$P.\,(Q \times R) = Q.\,(R \times P) = 14$$

or $$P.\,(Q \times R) = (2, 0, -1).\,(5, 2, -4)$$

$$= 10 + 0 + 4$$

$$= 14$$

(d) $$\sin\theta_{QR} = \frac{|Q \times R|}{|Q|\,|R|} = \frac{|(5,\ 2, -4)|}{|(2, -1,\ 2)|\ |(2, -3, 1)|}$$

$$= \frac{\sqrt{45}}{\sqrt[3]{14}} = \frac{\sqrt{5}}{\sqrt{14}} = 0.5976$$

(e) $$P \times (Q \times R) = (2, 0, -1) \times (5, 2, -4)$$

$$= (2, 3, 4)$$

Alternatively, using the bac-cab rule,

$$P \times (Q \times R) = Q\,(P\,.\,R) - R(P\,.Q)$$

$$= (2, -1, 2)\,(4 + 0 - 1) - (2, -3, 1)\,(4 + 0 - 2)$$

$$= (2, 3, 4)$$

(f) A unit vector perpendicular to both Q and R is given by

$$a = \frac{\pm Q \times R}{|Q \times R|} = \frac{\pm(5,\ 2, -4)}{\sqrt{45}}$$

$$= \pm\ (0.745,\ 0.298,\ -0.596)$$

Note that $|a| = 1$, a . Q = 0 = a . R. Any of those can be used to check a. (g) The component of P along Q is

$$P_Q = |P| \cos\theta_{PQ} a_Q$$

$$= (P\ .a_Q)a_Q = \frac{(P.Q)Q}{|Q|^2}$$

$$= \frac{(4+0-2)(2,\ -1,\ 2)}{(4+1+4)} = \frac{2}{9}(2,-1,\ 2)$$

$$= 0.4444a_x - 0.2222a_y + 0.4444a_z.$$

Exercise 4

Let $E = 3a_y + 4a_z$ and $F = 4a_x - 10a_y + 5a_z$.

(a) Find the component of E along F.

(b) Determine a unit vector perpendicular to both E and F.

Answer: (a) (–0.2837, 0.7092, – 0.3546), (b) ± (0.9398, 0.2734, –0.205).

Example 14. Derive the cosine formula

$$a^2 = b^2 + c^2 - 2bc \cos A$$

and the sine formula

$$\frac{\sin A}{a} = \frac{\sin B}{b} = \frac{\sin C}{c}$$

using dot product and cross product, respectively.

Solution: Consider a triangle as shown in Figure 1.11. From the figure, we notice that

$$a + b + c = 0$$

that is,

$$b + c = -a$$

Hence,

$$a^2 = a\ .\ a = (b + c)\ .\ (b + c)$$

$$= b \,.\, b + c \,.\, c + 2b \,.\, c$$

$$a^2 = b^2 + c^2 - 2bc \cos A$$

where A is the angle between b and c.

The area of a triangle is half of the product of its height and base. Hence,

$$|\frac{1}{2} a \times b| = |\frac{1}{2} b \times c| = |\frac{1}{2} c \times a|$$

$$ab \sin C = bc \sin A = ca \sin B$$

Dividing through by abc gives

$$\frac{\sin A}{a} = \frac{\sin B}{b} = \frac{\sin C}{c}$$

Figure 1.20 For example 1.4

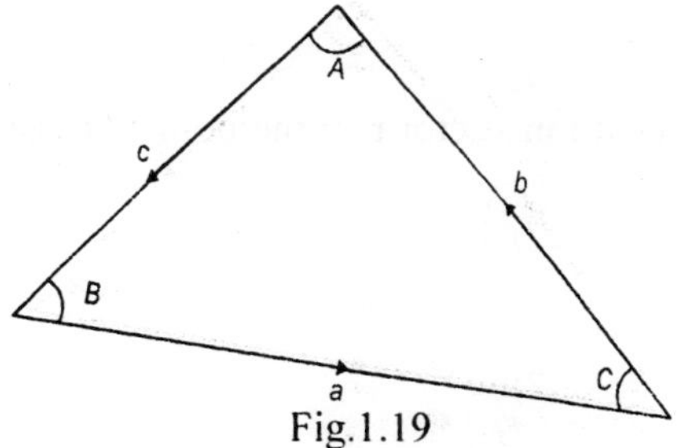

Fig.1.19

Exercise 5

Show that vector a = (4, 0, –1), b = (1, 3, 4), c = (–5, –3, –3) form the sides of a triangle. Is this a right angle triangle? Calculate the area of the triengle.

Answer: yes, 10.5.

Example 15. Show that points $P_1(5, 2, -4)$, $P_2(1, 1, 2)$, and $P_3(-3, 0, 8)$ all lie on a straight line. Determine the shortest distance between the line and point $P_4(3, -1, 0)$.

Solution: The distance vector r_{p1p2} is given by

$$r_{P_1P_2} = r_{P_2} - r_{P_1} = (1,\ 1,\ 2) - (5,\ 2, -4)$$

$$= (-4, -1, 6)$$

Similarly,

$$r_{P_1P_3} = r_{P_3} - r_{P_1} = (-3,\ 0,\ 8) - (-5,\ 2, -4)$$

$$= (-8, -2, 12)$$

$$r_{P_1P_4} = r_{P4} - r_{P_1} = (3,\ -1,\ 0) - (5,\ 2, -4)$$

$$= (-2, -3, 4)$$

$$r_{P_1P_2} \times r_{P_1P_3} = \begin{vmatrix} a_x & a_y & a_z \\ -4 & -1 & 6 \\ -8 & -2 & 12 \end{vmatrix}$$

$$= (0, 0, 0)$$

Showing that the angle between $r_{P_1P_2}$ and $r_{P_1P_3}$ is zero ($\sin\theta = 0$). This implies that P_1, P_2, and P_3 lie on a straight line.

Alternatively, the vector equation of the straight line is easily determined from Figure 1.20(a). for any point P on the line joining P_1 and P_2

$$r_{P_1P_2} = \lambda r_{P_1P_2}$$

where λ is a constant. Hence the position vector $r_p$of the point P must satisfy

$$r_P - r_{P_1} = \lambda(r_{P_2} - r_{P_1})$$

That is,

$$r_P = r_{P_1} + \lambda(r_{P_2} - r_{P_1})$$

$$= (5, 2, -4) - \lambda(4, 1, -6)$$

$$r_P = (5 - 4\lambda, 2 - \lambda, -4 + 6\lambda)$$

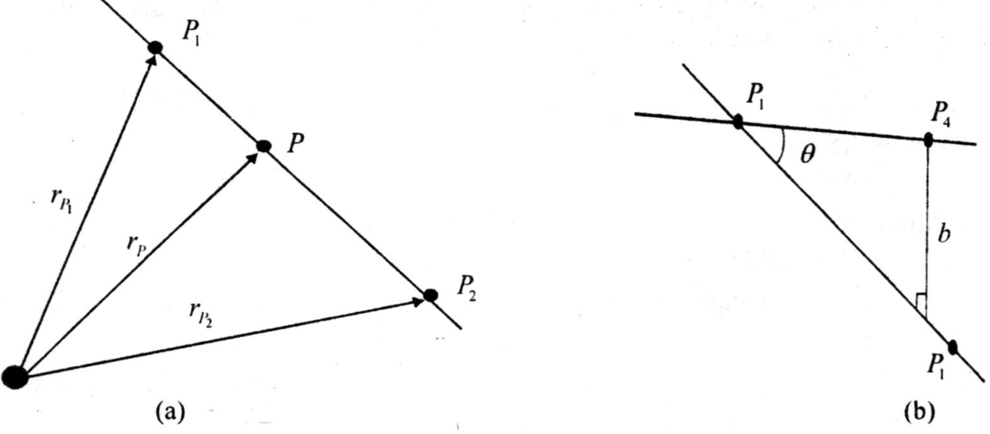

Figure 1.20 For Example 1.21

This is the vector equation of the stratght line joining P_1 and P_2. If P_3 is on this line, the position vector of P_3 must satisfy the equation; r_3 does satisfy the equation when $\lambda = 2$.

The shortest distance between the line and point P4 (3, –1, 0) is the perpendicular distance from the point to the line. From Figure 1.12(b), it is clear that

$$d = r_{P1P4} \sin\theta = |r_{P1P4} \times a_{P1P2}|$$

$$= \frac{|(-2,-3,\ 4) \times (-4,-1,\ 6)|}{|(-4,-1,\ 6)|}$$

$$= \frac{\sqrt{312}}{\sqrt{53}} = 2.426$$

Any point on the line may be used as a reference point. Thus, instead of using P_1 as a reference point, we could use P_3 so that

$$d = |r_{P_3P_4}|\sin\theta = |r_{P_3P_4} \times a_{P_3P_1}|$$

Exercise 6

If P_1 is (1, 2, –3) and P_2 is (–4, 0, 5), find

(a) The distance P_1P_2.

(b) The vector equation of the line P_1P_2

(c) The shortest distance between the line P_1P_2 and point P_3 (7, –1, 2).

Answer: (a) 9.644, (b) $(1-5\lambda)a_x + 2(1-\lambda)a_y + (8\lambda-3)a_z$, (c) 8.2.

Check Yourself

1. Identify which of the following quantities is not a vector: (a) force, (b) momentum, (c) acceleration, (d) work, (e) weight.
2. Which of the following is not a scalar field?
   (a) Displacement of a mosquito in space
   (b) Light intensity in a drawing room
   (c) Temperaturedistribution in your classroom
   (d) Atmospheric pressure in a given region
   (e) Humidity of a city
3. The rectangular coordinate systems shown in Figure 1.13 are right-handed except:
4. Which of these is correct?
   (a) $A \times A = |A|^2$ (b) $A \times B + B \times A = 0$ (c) $A . B . C = B. C. A$
   (d) $a_x . a_y = a_z$ (e) $a_k = a_x - a_y$
   Where a_k is a unit vector.

5. Which of the following identities is not valid?

(a) a (b + c) = ab + bc

(b) a × (b +c) = a × b + a × c

(c) a . b = b . a

(d) c . (a × b) = – b .(a × c)

(e) $a_A \cdot a_B = \cos\theta_{AB}$

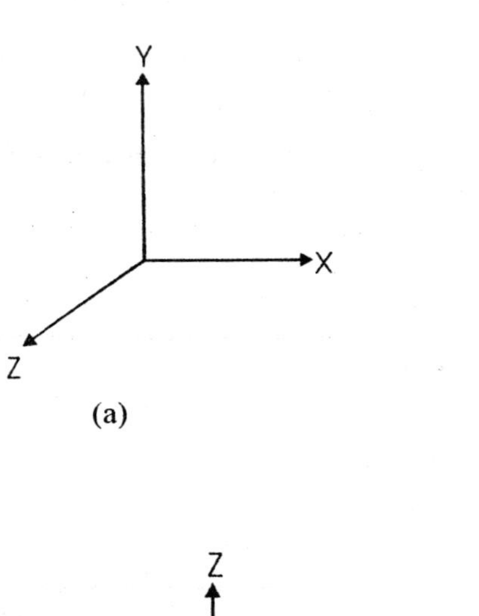

(a)

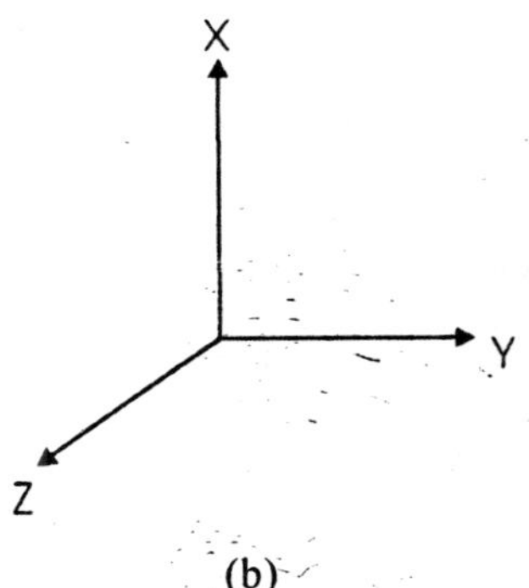

(b)

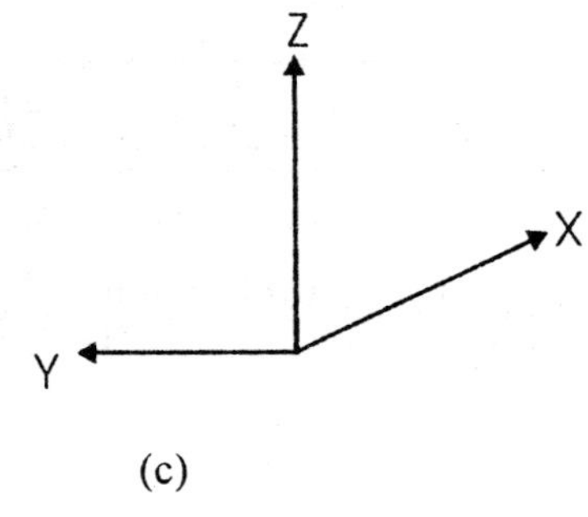

(c)

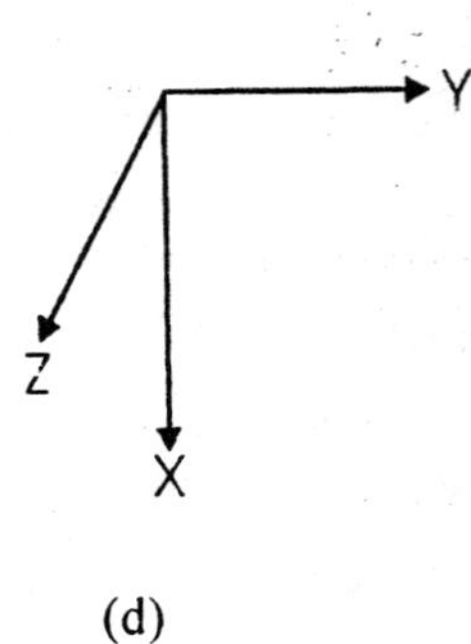

(d)

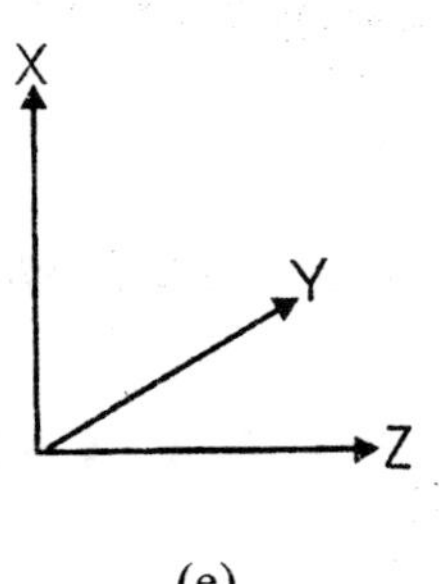

(e)

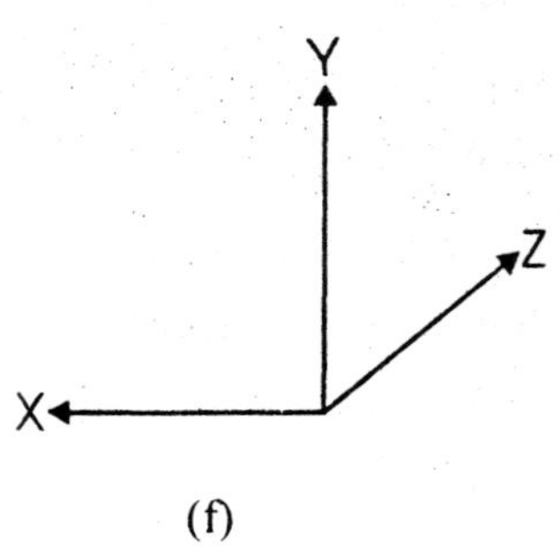

(f)

6. Which of the following statement are meaningless?

   (a) A . B + 2A = 0

   (b) A . B + 5 =2A

   (c) A(A + B) + 2 = 0

   (d) A . A + B . B = 0

7. Let $F = 2a_x - 6a_y + 10a_z$ and $G = a_x + G_y a_y + 5a_z$. If F and G have the same unit vector, G_y is

   (a) 6 (d) 0 (b) –3 (e) 6

8. Given that $A = a_x + \alpha a_y + a_z$ and $B = \alpha a_x + a_y + a_z$, if A and B are normal to each other, α is

   (a) –2 (b) –1/2 (c) 0 (d) 1 (e) 2

9. The component of $6a_x + 2a_y - 3a_z$ along $3a_x - 4a_y$ is

   (a) $-12a_x - 9a_x - 3a_z$

   (b) $30a_x - 40a_y$

   (c) 10/7

   (d) 2

   (e) 10

10. Given $A = -6a_x + 3a_y + 2a_z$, projection of A along ay is

    (a) –2 (b) –4 (c) 3 (d) 7 (e) 12

Answers: 1.1d, 1.2a, 1.3b,e, 1.4b, 1.5a, 1.6b,c, 1.7b, 1.8b, 1.9d, 1.10c.

Problems 11. Find the unit vector along the line joining point (2, 4, 4) to point (–3, 2, 2).

12. Let $A = 2a_x + 5a_y - 3a_z$, $B = 3a_x, -4a_y$, and $C = a_x + a_y + a_z$. (a) Determine A + 2B. (b) Calculate |A –5C|. (c) For what values of *k* is |KB| = 2? (d) Find (A×B)/(A.B).

13. If

    $A = 2a_x + a_y - 3a_z$

    $B = a_y - a_z$

    $C = 3a_x + 5a_y + 7a_z$

Determine:

(a) A – 2B + C

(b) C – 4(A + B)

(c) $\dfrac{2A - 3B}{|C|}$

(d) $A . C - |B|^2$

(e) $\frac{1}{2}B \times \left(\frac{1}{3}A + \frac{1}{4}C\right)$

14. If the position vector of points *T* and *S* are $3a_x - 2a_y + a_z$ and $4a_x + 6a_y + 2a_x$, respectively, find: (a) the coordinates of *T* and *S*, (b) the distance vector from *T* to *S*, (c) the distance between *T* and *S*.

15. If

$A = 5ax + 3ay + 2az$

$B = -ax, + 4ay + 6az$

$C = 8ax + 2ay$

find the values of α and β such that $\alpha A + \beta B + C$ is parallel to the y-axis.

16. Given Vectors

$A = \alpha a_x + a_y + 4a_z$

$B = 3a_x + \beta a_y - 6a_z$

$C = 5a_x - 2a_y + \gamma a_z$

Determine α, β, and such that the vectors are mutually orthogonal.

17. (a) Show

$$(A\,.B)^2 + (A \times B)^2 = (AB)^2$$

(b) Show that

$$a_x = \frac{a_y \times a_z}{a_x.a_y \times a_z},\ a_y = \frac{a_z \times a_x}{a_x.a_y \times a_z},\ a_z = \frac{a_x \times a_y}{a_x.a_y \times a_z}$$

18. Given that

$p = 2ax - ay - 2az$

$Q = 4ax + 3ay + 2az$

$C = -ax + ay + 2az$

Find: (a) $|P + Q - R|$, (b) $P.\ Q \times R$, (c) $Q \times P.\ R$, (d) $(P \times Q) . (Q \times R)$, (e) $(P \times Q) \times (Q \times R)$, (f) $\cos\theta_{PR}$, (g) $\sin\theta_{PQ}$.

19. Given vectors $T = 2ax - 6ay + 3az$ and $S = ax + az$, find: (a) the scalar Projection of T on S, (b) the vector projection of S on T, (c) the smaller angle between T and S.

20. If $A = -a_x + 6a_y + 5a_z$ and $B = a_x + 2a_y + 3a_x$, find : (a) the scalar projections of **A** on **B**, (b) the vector projection of **B** on **A**, (c) the unit vector perpendicular to the plane containing **A** and **B**.

21. Calculate the angles that vector $H = 3a_x + 5a_y - 8a_z$ makes with the x-, y-, and z-axes.

22. Find the triple scalar product of P, Q, and R given that

$P = 2a_x - a_y + a_z$

$Q = a_x + a_y + a_z$ and

$R = 2a_x + 3a_z$

23. Simplify the following expressions:
    (a) A × (A×B)
    (b) A × [A × (A × B)]
24. Show that the dot and cross in the triple scalar product may be interchanged, i.e., A . (B × C) = (A × B) . C.
25. Points P_1 (1, 2, 3), P_2 (–5, 2, 0), and P_3 (2, 7, –3) from a triangle in space. Calculate the area of the triangle.
26. The vertices of a triangle are located at (4, 1, –3), (–2, 5, 4), and (0, 1, 6). Find the three angles of the triangle.
27. Point *P*, *Q* and *R* are located at (–1, 4, 8), (2, –1, 3), and (–1, 2, 3), respectively. Determine: (a) the distance between *P* and *Q*, (b) the distance vector from *P* to *R*, (c) the angle between *QP* and *QR*, (d) the area of triangle *PQR*, (e) the perimeter of triangle *PQR*.
28. If r is the position vector of the point (*x*, *y*, *z*) and A is a constant vector, show that:
    (a) (r –A) . A = 0 is the equation of a constant plane.
    (b) (r – A). r = 0 is the equation of a sphere.

(a) *Differentiation of a Vector w.r.t. Scalar*

The vector quantities are often expressed as function of scalar variables for some of the certain physical problems. For example,

The displacement of a particle can be expressed as a function of single scalar time. The vector may be differentiated w.r.t. this single variable to give a new vector.

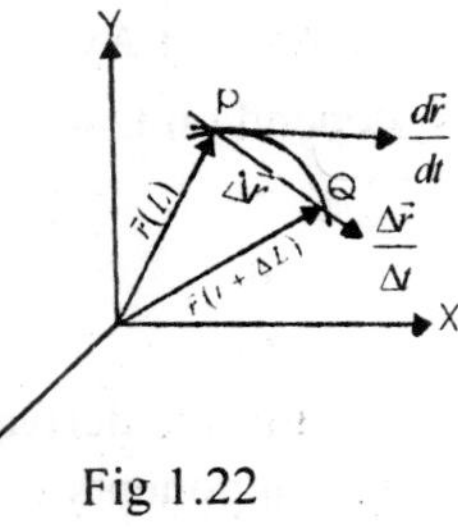

Fig 1.22

For finding the time dericative of position vector w.r.t. time, let us consider a particle moving along a curved path and reaching from P to Q in time Δt (Fig. 1.22). If we denote the position vector of P by $\vec{r}$ (t) and of Q by $\vec{r}(t+\Delta t)$, then the average rate of change of the position vector with respect to time is given by

$$\frac{\vec{r}(t+\Delta t)-\vec{r}(t)}{\Delta t}=\frac{\overrightarrow{OQ}-\overrightarrow{PQ}}{\Delta t}$$

$$=\frac{\overrightarrow{PQ}}{\Delta t}=\frac{\Delta\vec{r}}{\Delta t}$$

and when Δt becomes vanishingly small, the ratio $\frac{\Delta\vec{r}}{\Delta t}$ attains a limiting value which is the rate of increase of $\vec{r}$ i.e the differential of $\vec{r}$ w.r.t. time or $\frac{d\vec{r}}{dt}$ is found by dividing $\vec{\Delta r}$ by Δt and taking the limit as $\Delta t \to 0$. Thus

$$\frac{\vec{dr}}{dt} = Lim_{\Delta t \to 0}\left[\frac{\vec{\Delta r}}{\Delta t}\right]$$

Now we can see that as Δt becomes smaller and smaller. Q approaches closer and closer to P and the vector $\overrightarrow{PQ}$ becomes tangent to the trajectory of the particle at time t.

The differential of $\vec{r}$ *i.e.* $\frac{dr}{dt}$ is a vector because its numerator is a vector and denominator is a scalar.

In mechanics, this represents the instantaneous velocity vector of the particle *i.e.* $\vec{V}\frac{\vec{dr}}{dt}$.

The above argument holds good equally for any arbitrary vector $\vec{A}$ which is a function of scalar t, so we have

$$\frac{\vec{dA}}{dt} = Lt_{\Delta t \to 0}\frac{\vec{A}(t+\Delta t) - \vec{A}(t)}{\Delta t}$$

If we write $\vec{A}$ in the component from in fixed co-ordinate system i.e. $\vec{A} = \hat{i}\,Ax + \hat{j}\,Ay + \hat{k}\,Az$, then the unit vectors $\hat{i}, \hat{j}, \hat{k}$ are constant and the components A_x, A_y, A_z depend on t, so

$$\frac{\vec{dA}}{dt} = \frac{dA_x}{dt}\hat{i} + \frac{dA_y}{dt}\hat{j} + \frac{dA_z}{dt}\hat{k} \quad ....(1.30)$$

Thus the derivative of a vector w.r.t. a scalar is equal to the vector sum of the derivatives of its components w.r.t. the same scalar.

(b) *Derivative of Products*

Limiting process from subtraction of vectors and division by scalar is applied above to obtain the differential of vectors w.r.t. a scalar variable and also the rules of ordinarydifferential calculus can be applied to the differentiation of vectors. So we have the following important results:

$$\frac{d}{dt}\left(a\vec{A}\right) = \frac{da}{dt}\vec{A} + a\frac{d\vec{A}}{dt} \text{ where } a \text{ is a scalar function}$$

$$\frac{d}{dt}\left(\vec{A} \pm \vec{B}\right) = \frac{d\vec{A}}{dt} \pm \frac{d\vec{B}}{dt}$$

$$\frac{d}{dt}\left(\vec{A}\,.\,\vec{B}\right) = \frac{d\vec{A}}{dt}.\vec{B} + \vec{A}.\frac{d\vec{B}}{dt}$$

$$\frac{d}{dt}\left(\vec{A} \times \vec{B}\right) = \frac{d\vec{A}}{dt} \times \vec{B} + \vec{A} \times \frac{d\vec{B}}{dt}$$

Scalar and Vector Field

There have been the situations in which the value of a physical quantity varies from one point to another in a given region of space. Such a physical quantity can be expressed as a continuous function of the position of point in that region of space. Such a function is called point or position function. The region in which the physical quantity takes different values at different point is called the field. Fields are of two types: one is scalar and the other is vector depending on the nature of the varying physical quantity concerned.

A region, where every point is connected with a scalar quantity is called scalar field. It is denoted by a continuous scalar function. Examples of scalar fields are distribution of temperature along a metal rod, distribution of electric potential in a region surrounding a charged body or distribution of any other non-directed (scalar) quantity. A scalar field can be visualized by drawing imaginary surfaces passing through all such point in the region for which the field has the same value. Such surfaces are called equal or level surfaces. Example of such surface, on each point of which scalar quantity electric potential is an equi-potential surface, on each point of which scalar quantity electric potential has the same value. Two level surfaces cannot cross each other, they must be one above the other, because if they cross each other, then the value of scalar will be the same on the two surfaces along their line of intersection. Which is contrary to our definition of level surface.

A *vector field*, on the other hand, is a region in which every point is characterised by a vector quantity. Examples of vector field are distribution of velocity through a stream of water. distribution of electric field intensity round a charged body. Such a field is represented by a continuous vector function. which at any given point is speacified by a vector of definite magnitude and direction, but both of which change continvously from point to point. A vector field can be represented by a set of curves, the tangent at any point of which gives the direction of vector. Such curves are called vector *lines* or lines of *flow or flux lines*. The magnitude of the vectior lines crossing unit area brwan roung that point normal to the direction of lines, No two lines of flow intersect each other, because the direction of the vector at the point of intersection will become indefinite.

A scalar or a vector field may or may not change in time. A field which does not depend on tiem is called stationary or steady state field.

Del Operator and Gradient of a Scalar

Let $\phi\ (\vec{r})$ be a scalar field, where $\vec{r}$ is the position vector of the variable observation point A (Fig. 1·23) in space and is given by $\vec{r} = x\hat{i} + y\hat{j} + z\hat{k}$ where x, y, z are co-ordinates of the observation point. We may therefore consider ϕ to be a function of three variables x, y, z and write it as ϕ (x, y, z). Anychange in the value of this scalar function corresponding to the displacement dr, will depend on the direction of displacement, as the rate of change of scalar

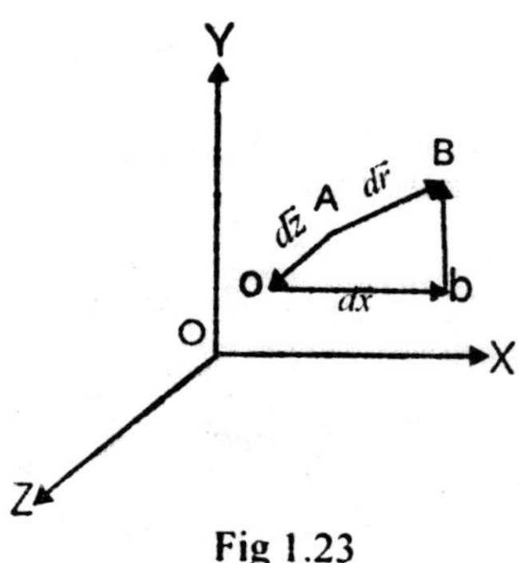

Fig 1.23

function may be different in different directions i.e rate of increase of ϕ may be greater in certain direction than others. It is well known from differential calculus that whenever there is a function of more than one variable, we can define partial derivative of that function w.r.t. one of the independent variables. If y, z remain constant and x changes, the partial derivative $\frac{\partial \phi}{\partial x}$ denotes the rate of change of ϕ along x-direction. Similarly, $\frac{\partial \phi}{\partial y}$ and $\frac{\partial \phi}{\partial z}$ denote the rate of change of ϕ along y and z directions respectively. So we can specify total change in ϕ in any direction, say in moving from A to a neighbouring point B such that $\vec{A}B = \vec{d}r$ and using first order Taylor's approximation as

$$d\phi = \frac{\partial \phi}{\partial x}dx + \frac{\partial \phi}{\partial y}dy + \frac{\partial \phi}{\partial z}dz$$

where dx, dy and dz repreent the change in the co-ordinates x, y and z respectively as shown in (Fig. 1·23), so that

$$d\vec{r} = dx\,\hat{i} + dy\,\hat{j} + dz\,\hat{k}$$

It is evident that $d\phi$ can be written as the scalar product of vectors.

$$\left(\hat{i}\frac{\partial \phi}{\partial x} + \hat{j}\frac{\partial \phi}{\partial y} + \hat{k}\frac{\partial \phi}{\partial z}\right)$$

and

$$\left(dx\,\hat{i} + dy\,\hat{j} + dz\,\hat{k}\right)$$

i.e. $$d\phi = \left(\hat{i}\frac{\partial \phi}{\partial x} + \hat{j}\frac{\partial \phi}{\partial y} + \hat{k}\frac{\partial \phi}{\partial z}\right).\left(\hat{i}\,dx + \hat{j}\,dy + \hat{k}\,dz\right) \qquad ....(1.31)$$

The first vector on the right hand side of equation (1·30) i.e. $\hat{i}\frac{\partial \phi}{\partial x} + \hat{j}\frac{\partial \phi}{\partial y} + \hat{k}\frac{\partial \phi}{\partial z}$ is called the gradient of scalar field ϕ and is written as

$$\text{grad } \phi = \left(\hat{i}\frac{\partial}{\partial x} + \hat{j}\frac{\partial}{\partial y} + \hat{k}\frac{\partial}{\partial z}\right)\phi$$

$$= \vec{\nabla}\phi$$

when $\vec{\nabla} = \hat{i}\frac{\partial}{\partial x} + \hat{j}\frac{\partial}{\partial y} + \hat{k}\frac{\partial}{\partial z}$ is a differential operator and is pronounced as Del or Nabla. So equation (1·30) can be written as

$$d\phi = grad\phi.\vec{d}r \qquad ...(1\cdot32)$$

It may be nated that differential $\vec{\nabla}$ be itself cannot be characterised either by magnitude or by direction. By itself $\vec{\nabla}$ has no meaning just as $\frac{d}{dx}$ alone has no meaning. It assumes a definite value only when it is applied to some field quantity. Is the important to note that $\vec{\nabla}$ abeys the conventions of derivative nation, So whatever is to be differentiated must be placed on the right of $\vec{\nabla}$. Putting an arrow over ∇ to represent its vector nature is not necessary. the symbol ∇ itself is sufficient. *So now onward we shall be using symbol ∇ without arrow to represent del.*

Physical Significance of Grad ϕ

If 'θ' be the angle between the vector $\nabla\ \phi$ and the displacment vector $\vec{d}r$ (Fig 1.24) than the equation (1.25) can be written as

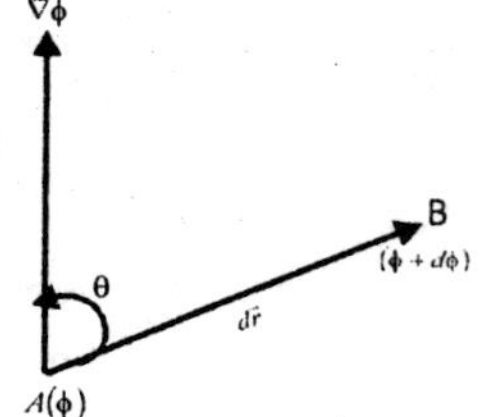

Fig 1.24

$$d\phi = |\nabla\phi|\,|\vec{d}r|\,\cos\theta$$

Where $d\phi$ is the change in ϕ for a displacement $\vec{d}r$. Or the rate of change of scalar function ϕ w.r.t. distance in any direction is given by

$$\frac{d\phi}{dr} = |\nabla\phi|\cos\theta$$

Which depends on the direction of displacement.

Now we may write,

$$\frac{d\theta}{dr} = (\nabla\phi).\,\hat{n} \text{ where } \hat{n} \text{ is a unit vector along } \vec{d}r$$

If $\hat{n}$ is along $\nabla\phi$ i.e., when we move along $\nabla\phi$, then θ=0, cosθ =1 and so $\frac{d\phi}{dr}$ is maximum.

Thus, grad ϕ lies in a direction along which rate of change of ϕ is maximum and

$$\left(\frac{d\theta}{dr}\right)_{max} = |\nabla\phi|$$

Hence, the gradient of a scalar function ϕ is defined as a vector field having a magnitude equal to the maximum space rate of change of ϕ and having a direction identical with the direction of displacement along which the rate of change is maximum.

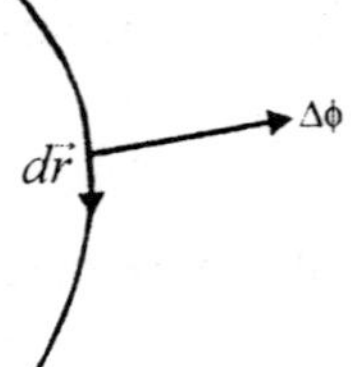

Level Surface
ϕ Constant
Fig 1.25

Further if we move along a direction normal to vector $\nabla\phi$, Fig.1·25, then θ=90°, cosθ=0° and so $d\phi$=0. In that case, there is no change in ϕ or ϕ=constant which defines a level surface. Hence it follows that $\nabla\phi$ is a vector normal to the level surface ϕ = constant.

As an example of a vector being represented by the gradient of a scalar field, is intensity of electric field which can be derived from the scalar potential as its negative gradient.

Integration of a Vector w.r.t. Scalar

The integration of a vector which is function of a single scalar variable is done by the use of usual rules of ordinary integral calculus, as shown below.

Let there be a vector function $\vec{A}$ of some scalar variable t and writing it as $\vec{A}(t) = Ax(t)\hat{i} + Ay(t)\hat{j} + Az(t)\hat{k}$, we define the vector integration as,

$$\int \vec{A}(t)\, dt = \int [Ax(t)\hat{i} + Ay(t)\hat{j} + Az(t)\hat{k}]\, dt + \vec{c}$$

$$= \hat{i}\int Ax(t)\, \mathrm{dt} + \hat{\mathrm{j}}\int \mathrm{Ay(t)}\, dt + \hat{k}\int Az(t)\, dt + \vec{c}$$

where $\vec{c}$ is vector constant of integration.

If we have a definite integral $\int_{t_1}^{t^2} A(t)\, dt$, then it will be evaluated as,

$$\hat{i}\int_{t_2}^{t_2} Ax(t)\, \mathrm{dt} + \hat{\mathrm{j}}\int_{\mathrm{t_2}}^{t_2} \mathrm{Ay(t)}dt + \hat{k}\int_{t_1}^{t_2} Az(t)\, dt.$$

The integrals which are frequently used in vector calculus are (i) line integral (ii) surface integral (iii) volume integral.

(a) Line Integral

Line integral is the integration of a tangential component of vector along a curved path.

Assume: A vector $\vec{A} = \vec{A}$ (x, y, z) defined throughout in some region of spae. Let ab be any curve drawn in this field and $\vec{dr}$ small segment of lenght along it at any point *P* (Fig.1·26). Further let $\vec{A}$ denote the vector

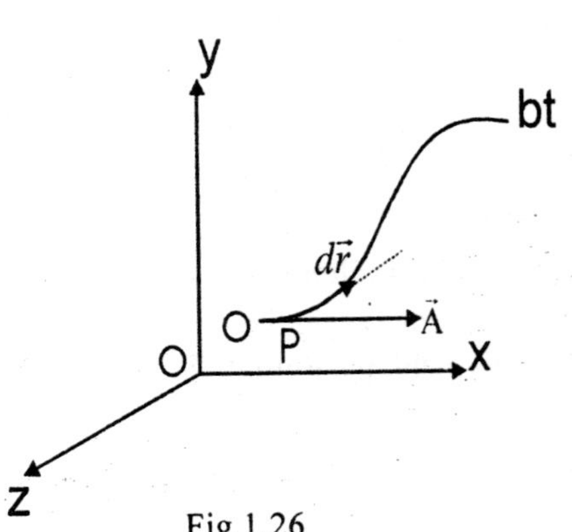

Fig 1.26

field at *P* and its direction makes an angle θ with that of length segment $\vec{dr}$. Then the scalar product given the

$$\vec{A} \cdot \vec{dr} = A\, dr \cos\theta$$

product of the length segment and the component of $\vec{A}$ in its direction i.e tangent to the curve. if the magnitude and direction of $\vec{A}$ varies from point to point along the curve, then the integral is called the line integral of $\vec{A}$ along the curve ab.

$$\int_a^b \vec{A} \cdot \vec{dr} = \int_a^b A\, dr \cos\theta \qquad ...(1{\cdot}32)$$

Concept of line integral is useful in several branches of physics, for example:

(i) If the vector $\vec{A}$ denotes a force, $\vec{F}$, then the line integral of the vector $\vec{F}$ along any path ab gives the amount of work done by the vector in displacing a particle from a to b along the curve.

(ii) If $\vec{A}$ denotes the electric field $\vec{E}$, then the line integral of $\vec{E}$ taken over the path ab, gives potential difference between the points a and b.

(iii) If $\vec{A}$ denoptes the gravitational field $\vec{g}$, then the line interal of $\vec{g}$ taken over the path ab, represents the difference in gravitational potential energies between the two point.

(iv) If $\vec{A}$ represents the vector velocity $\vec{v}$ at any point in a fluid, then the line integral of $\vec{v}$ a taken along a closed path (represented as $\oint(\vec{v}.\vec{dr})$ is callled circulation of the fluid. In general the integral $\oint \vec{v}.\vec{dr}$ is called the circulation of vector field $\vec{A}$ around a loop or closed path.

(b) Surface Integral

Let us consider a vector field $\vec{A} = \vec{A}$(x, y, z) and imagine a surface S (curved or flat) drawn in this vector field. Let us now take a small area element ds upon, this surface and let $\vec{A}$ represent the value of vector field at the middle of $\vec{ds}$ (Fig. 1·27). Further, take $\hat{n}$ to be a unit vector along the positive (outword drawn) normal on the surface element ds. Let θ be the angle between the direction of $\hat{n}$ and $\vec{A}$ as shown in (Fig. 1·23).Then the component of $\vec{A}$ normal to ds is $\vec{A}\hat{n}$ =A cos θ. The scalar product $\vec{A} \cdot \hat{n}ds$, which is equal to the product of area ds

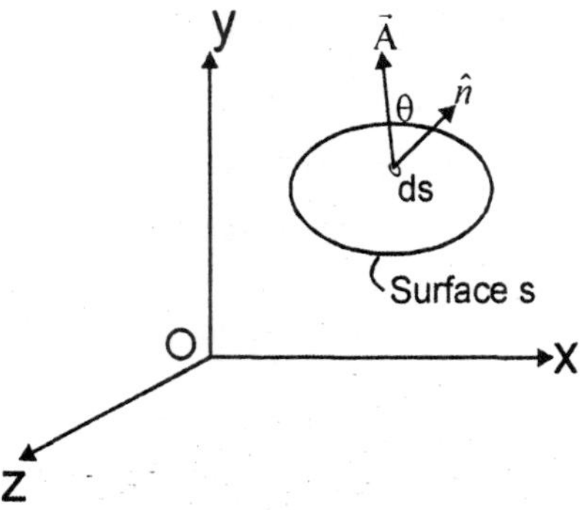

Fig. 1.27

and component of vector $\vec{A}$ normal to $\vec{ds}$, is called flux of vector A through the element of area $\vec{ds}$. An integration of $\vec{A}\,.\,\hat{n}$ ds taken over the entire surface S is defined as the total flux or surface integral of $\vec{A}$ throughwhole surface S, i.e.,

$$\iint_s \vec{A}\,.\,\hat{n}\,ds = \iint_s A\cos\theta\,ds = \iint_s \left(A_x ds_x + A_y ds_y + A_z ds_z\right) \quad ...(1.33)$$

If the integration is over a closed surface S, then it is denoted by $\oint_s \vec{A}\,.\,\hat{n}\,ds.$

Physical Meaning of Surface Integral

Assume: A fixed area $\vec{ds}$ is drawn in a moving fluid either air or Liquid (Fig.1·28). If $\vec{v}$ denotes the vector velocity of fluid at any point (x, y, z), then the volume of the fluid passing normally through an area element ds' in unit time is evidently equal to the Volume of fluid contained in a cylinder of cross-section *ds*' and length *V* i.e. *V ds*'. Some volume of the fluid will also pass through the oblique area element ds in unit time. If $\hat{n}$ be a unit vector drawn normal to ds, and inclined at an angle q with the direction of $\vec{V}$ then the component of ds normal to V i.e. ds'= ds $\cos\theta$

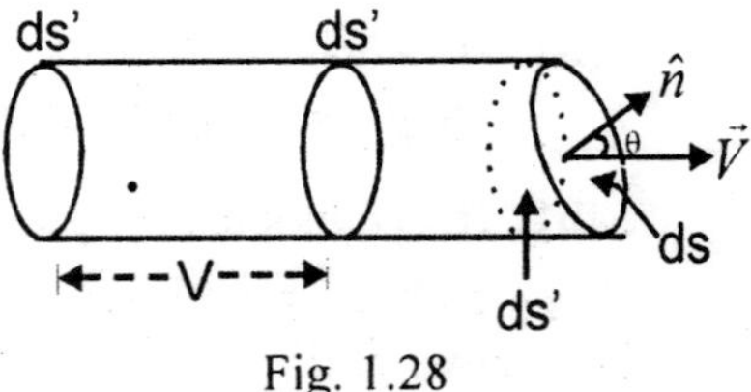

Fig. 1.28

$$= V\cos\theta\,ds = \vec{V}\,.\,\hat{n}\,ds \quad ....(1.34)$$

Integral of this normal contribution i.e $\iint_s V\cos\theta\,ds$ represents the flow of fluid through the whole surface in unit time and the volume of the fluid crossing per unit area per unit time=$\vec{V}\,.\,\hat{n}$ and is called flow flux per unit area.

(c) Volume Integral

Let us consider a closed surface enclosing the volume V in a vector field $\vec{A}$. *If* $dv = dx\,dy\,dz$ represent a small volume element inside this closed surface (Fig. 1·29), then

$\iiint_V \vec{A}\,dv$ represents the *volume integral or space integral* of the vector field field $\vec{A}$ for entire volume V. In cartesian components it is written as,

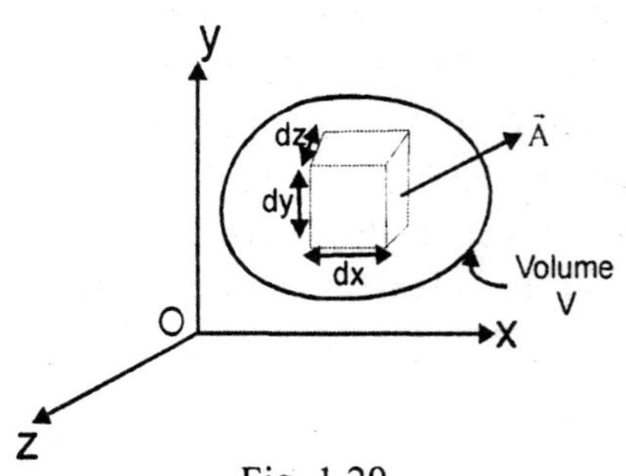

Fig. 1.29

$$= \iiint_V \vec{A}\, dv = \iiint_{xyz} \left(A_x\hat{i} + A_y\hat{j} + A_z\hat{k}\right) d_x\, d_y\, d_z$$

$$= \hat{i} \iiint_{xyz} A_x dxdydz + \hat{i} \iiint_{xyz} A_y dxdydz + \hat{k} \iiint_{xyz} A_x d_x d_y d_z. \qquad ...(1\cdot35)$$

Additional Information About Grand ϕ

Every vector field cannot be expressed as gradient of a scalar field, but if a vector field is represented by the gradient of a scalar, then the following important results will follow:

(i) If vector field $\vec{A}$ is represented as gradient of some scalar function ϕ, then the value of the line integral of $\vec{A}$ between the point a and b i.e. $\int_a^b \vec{A} \cdot \vec{dr}$ is independent of the path joining a and b but depends only on the value of the scalar at the extremities a and b. For proving it we proceed as follows:

From the definition of gradient (eqn. 1·27) $\vec{A} \cdot \vec{dr} = \nabla\phi \cdot \vec{dr} = d\phi$

$$\therefore \qquad \int_a^b \vec{A}.\vec{dr} = \int_a^b d\phi = \phi \int_a^b = \phi(b) - \phi(a)$$

where ϕ (a) and ϕ (b) are values of the scalar function ϕ at a and b respectively. It is evident that ϕ (a)- ϕ (b) is independent of the path joining a and b but depends only on the end points a and b.

(ii) The line integral of $\vec{A}$ = grad ϕ over a closed path is zezo. For proving it we proceed as follow:

Let us consider two points a and b in a vector field $\vec{A}$ and take two paths marked as 1 and 2 between a and b as shown in Fig. 1·30. Now if after moving from a to b along the path marked as 1, we take the backward journey from b to a along the path marked as 2, we get a closed path (Fig. 1·27). Thus the line integral over this closed path i.e.

$$\oint \vec{A} \cdot \vec{dr} = \int \vec{A} \cdot \vec{dr} + \int \vec{A} \cdot \vec{dr}$$

over path 1 over path 2

$$\int_a^b \vec{A} \cdot \vec{dr} + \int_b^a \vec{A} \cdot \vec{dr}$$

$$\int_a^b \nabla\phi \cdot \vec{dr} + \int_b^a \nabla\phi \cdot \vec{dr} = \int_a^b d\phi + \int_b^a d\phi$$

$$= \phi(b) - \phi(a) + \phi(a) - \phi(b) = 0$$

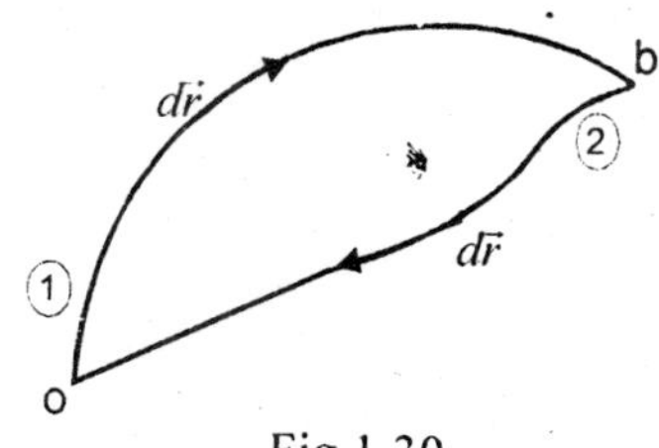

Fig 1.30

Thus the line integral of a vector field, represented by the gradient of scalar function, over a closed path vanished.

The vector field A, which can be expressed as gradient of a scalar, is called a scalar *potential* field or *Lamellar vector* field or conservative field. This is also called non-curl or irrotational field due to the reasons we shall explain later.

Example 6. If $\phi(x, y, z) = 3x^2y - yz^2$, *find grad* ϕ *at the point* (1, 2, –1),

Solution. grad $\phi = \nabla\phi = \hat{i}\dfrac{\partial\phi}{\partial x} + \hat{j}\dfrac{\partial\phi}{\partial y} + \hat{k}\dfrac{\partial\phi}{\partial z}$

$$= \hat{i}\frac{\partial}{\partial x}\left(3x^2y - yz^2\right) + \hat{j}\frac{\partial}{\partial y}\left(3x^2y - yz^2\right) + \hat{k}\frac{\partial}{\partial z}\left(3x^2y - yz^2\right)$$

$$= 6xy\hat{i} + \left(3x^2 - z^2\right)\hat{j} - 2yz\hat{k}$$

$$= 6(1)(2)\hat{i} + \left\{3(1)^2 - (-1)^2\right\}\hat{j} - 2(2)(-1)\hat{k}$$

$$= 12\hat{i} + 2\hat{j} + 4\hat{k}$$

Example 17. Prove that $\nabla(\vec{a} \cdot \vec{r}) = \vec{a}$ where $\vec{a}$ is a constant vector and $\vec{r}$ the position vector. (k.U. 1986)

Solution. Here $\vec{a} \cdot \vec{r} = \left(a_x\hat{i} + a_y\hat{j} + a_z\hat{k}\right) \cdot \left(x\hat{i} + y\hat{j} + z\hat{k}\right)$

$$= a_x x + a_y y + a_z z$$

where a_x, a_y, a_z represent the rectangular components of $\vec{a}$

or
$$\nabla(\vec{a} \cdot \vec{r}) = \left(\hat{i}\frac{\partial}{\partial x} + \hat{j}\frac{\partial}{\partial y} + \hat{k}\frac{\partial}{\partial z}\right)\left(a_x x + a_y y + a_z z\right)$$

$$= \hat{i}\frac{\partial}{\partial x}\left(a_x x + a_y y + a_z z\right) + \hat{j}\frac{\partial}{\partial y}\left(a_x x + a_y y + a_z z\right) + \hat{k}\frac{\partial}{\partial x}\left(a_x x + a_y y + a_z z\right)$$

$$= \hat{i}\, a_x\frac{\partial x}{\partial x} + \hat{j}\, ay\frac{ay}{ay} + k\, az\frac{\partial z}{\partial z}$$

$$= ax\,\hat{i} + ay\,\hat{j} + az\,\hat{k}$$

$$= \vec{a}.$$

Example 18. What is the meaning of grad r where $\vec{r}$ is position vector?

Solution. As $\vec{r}$ is position vector os $\vec{r} = x\hat{i} + y\hat{j} + z\hat{k}$.

Magnitude of $\vec{r} = r = (x^2+y^2+z^2)^{1/2}$

$\therefore$ grad r = grad $(x^2+y^2+z^2)^{1/2}$

$$= \left(\hat{i}\frac{\partial}{\partial x} + \hat{j}\frac{\partial}{\partial y} + \hat{k}\frac{\partial}{\partial z}\right)\left(x^2+y^2+z^2\right)^{1/2}$$

$$= \hat{i}\frac{\partial}{\partial x}\left(x^2+y^2+z^2\right)^{1/2} + \hat{j}\frac{\partial}{\partial y}\left(x^2+y^2+z^2\right)^{1/2} + \hat{k}\frac{\partial}{\partial z}\left(x^2+y^2+z^2\right)^{1/2}$$

$$= \hat{i}\frac{1}{2}\left(x^2+y^2+z^2\right)^{-1/2} \times 2x + \hat{j}\frac{1}{2}\left(x^2+y^2+z^2\right)^{-1/2} \times 2y + \hat{k}\frac{1}{2}\left(x^2+y^2+z^2\right)^{-1/2} 2z$$

$$= \left(x\hat{i} + y\hat{j} + z\hat{k}\right)\left(x^2+y^2+z^2\right)^{-1/2}$$

$$= \frac{\vec{r}}{\sqrt{r^2}} = \frac{\vec{r}}{r} = \hat{r}$$ is a unit vector in the direction

of $\vec{r}$.

Example 19. Prove that grad (uv)= u grad v+v grad u where u and v both are scalar:

Solution. grad (uv) $= \nabla(uv) = \hat{i}\frac{\partial}{\partial x}(uv) + \hat{j}\frac{\partial}{\partial y}(uv) + \hat{k}\frac{\partial}{\partial z}$

$$= \hat{i}\left(u\frac{\partial v}{\partial x} + v\frac{\partial v}{\partial x}\right) + \hat{j}\left(u\frac{\partial v}{\partial y} + v\frac{\partial v}{\partial y}\right) + \hat{k}\left(u\frac{\partial v}{\partial z} + v\frac{\partial u}{\partial z}\right)$$

Arranging these we get

$$= u\left(\hat{i}\frac{\partial v}{\partial x} + \hat{j}\frac{\partial v}{\partial y} + \hat{k}\frac{\partial v}{\partial z}\right) + v\left(\hat{i}\frac{\partial u}{\partial x} + \hat{j}\frac{\partial u}{\partial y} + \hat{k}\frac{\partial u}{\partial z}\right)$$

$$= u(\nabla u) + u(\nabla u)$$

$$= u\ grad\ v + v\ grad\ u.$$

Example 20. *Show that* $\nabla r^n = nr^{n-1}\,\hat{r}$ *where* $\hat{r}$ *is a unit vector in the direction of position vector* $\vec{r}$.

Solution. As $\vec{r}$ is position vector so $\vec{r} = x\hat{i} + y\hat{j} + z\hat{k}$

and $$r = \sqrt{x^2 + y^2 + z^2}$$

$\because$ $$\nabla r^n = \nabla\left(x^2 + y^2 + z^2\right)^{n/2}$$

$$= \hat{i}\frac{\partial}{\partial x}\left(x^2 + y^2 + z^2\right)^{n/2} + \hat{j}\frac{\partial}{\partial y}\left(x^2 + y^2 + z^2\right)^{n/2} + \hat{k}\frac{\partial}{\partial z}\left(x^2 + y^2 + z^2\right)^{n/2}$$

$$= \hat{i}\left[\frac{n}{2}\left(x^2 + y^2 + z^2\right)^{n/2-1}.2x\right] + \hat{j}\left[\frac{n}{2}\left(x^2 + y^2 + z^2\right)^{n/2-1}.2y\right]$$

$$+\hat{k}\left[\frac{n}{2}\left(x^2 + y^2 + z^2\right)^{n/2-1}.2z\right]$$

$$= n\left(x^2 + y^2 + z^2\right)^{n/2-1}\left[x\hat{i} + y\hat{j} + z\hat{k}\right]$$

$$= n\left(r^2\right)^{n/2-1}\vec{r} = n\left(r^2\right)^{n-2/2}\vec{r} = nr^{n-2}\vec{r}$$

$$= nr^{n-2}r\frac{r}{r} = nr^{n-1}\hat{r}$$

Example 21. *Find* $\nabla\phi$ *if* $\phi = In\,|\vec{r}|$ *where* $\vec{r}$ *is position vector.*

Solution. As $\vec{r}$ is position vector so $\vec{r} = x\hat{i} + y\hat{i} + z\hat{k}$

Then $$|\vec{r}| = \sqrt{x^2 + y^2 + z^2}\; and\; \phi = In\,|\vec{r}| = \frac{1}{2}In\left(x^2 + y^2 + z^2\right)$$

$\therefore$ $$\nabla\phi = \frac{1}{2}\nabla In\left(x^2 + y^2 + z^2\right)$$

$$= \frac{1}{2}\left[\hat{i}\frac{\partial}{\partial x}in\left(x^2 + y^2 + z^2\right) + \hat{j}\frac{\partial}{\partial y}In\left(x^2 + y^2 + z^2\right) + \hat{k}\frac{\partial}{\partial z}In\left(x^2 + y^2 + z^2\right) -\right]$$

$$= \frac{1}{2}\left[\hat{i}\frac{2x}{x^2 + y^2 + z^2} + \hat{j}\frac{2y}{x^2 + y^2 + z^2} + \hat{k}\frac{2z}{x^2 + y^2 + z^2}\right]$$

$$= \hat{i}\frac{x}{x^2+y^2+z^2} + \hat{j}\frac{y}{x^2+y^2+z^2} + \hat{k}\frac{z}{x^2+y^2+z^2}$$

$$= \frac{x\hat{i}+y\hat{j}+z\hat{k}}{x^2+y^2+z^2} = \frac{\vec{r}}{r^2} = \frac{\hat{r}}{r}$$

Example 22. *Find* $\nabla\phi$ *if* $\phi = \frac{1}{r}$ *where* $r = \sqrt{x^2+y^2+z^2}$.

Solution. $\nabla\phi = \nabla\left(\frac{1}{r}\right) = \nabla\left(\frac{1}{\sqrt{x^2+y^2+z^2}}\right)$

$$= \nabla\left(x^2+y^2+z^2\right)^{-1/2}$$

$$= \hat{i}\frac{\partial}{\partial x}\left(x^2+y^2+z^2\right)^{-1/2} + \hat{j}\frac{\partial}{\partial y}\left(x^2+y^2+z^2\right)^{-1/2} + \hat{k}\frac{\partial}{\partial z}\left(x^2+y^2+z^2\right)^{-1/2}$$

$$= \hat{i}\left[-\frac{1}{2}\left(x^2+y^2+z^2\right)^{-3/2}2x\right] + \hat{j}\left[-\frac{1}{2}\left(x^2+y^2+z^2\right)^{-3/2}2y\right]$$

$$+\hat{k}\left[-\frac{1}{2}\left(x^2+y^2+z^2\right)^{-3/2}2x\right]$$

$$= \left(-x\hat{i} - y\hat{j} - z\hat{k}\right)\left(x^2+y^2+z^2\right)^{-3/2}$$

$$= \frac{-\left(x\hat{i}+y\hat{j}+z\hat{k}\right)}{\left(x^2+y^2+z^2\right)^{-3/2}} = \frac{-\vec{r}}{\left(r^2\right)^{3/2}} = \frac{\vec{r}}{r^3}$$

Example 23. If u is a function of scalar point function v (x, y, z) prove that grad $u = \frac{du}{dv}$ grad v.

Solution. Grad $u = \nabla u = \hat{i}\frac{\partial u}{\partial x} + \hat{j}\frac{\partial u}{\partial y} + \hat{k}\frac{\partial u}{\partial z}$

Further u is a function of v and v is a function of (x, y, z)

$$\therefore \quad \text{grad } u = \hat{i}\frac{\partial u}{\partial v}\frac{\partial v}{\partial x} + \hat{j}\frac{\partial u}{\partial v}\frac{\partial v}{\partial y} + \hat{k}\frac{\partial u}{\partial v}\cdot\frac{\partial v}{\partial z}$$

$$= \frac{\partial u}{\partial v}\left[\hat{i}\frac{\partial v}{\partial x} + \hat{j}\frac{\partial v}{\partial y} + \hat{k}\frac{\partial v}{\partial z}\right]$$

$$= \frac{\partial u}{\partial v}(\nabla u)$$

($\because$ partial derivative of u w.r.t v is complete derivative.)

Example 24. find a unit normal to the surface $x^2y+2xz=4$ at the point (2,–1, 3)

Solution. It is proved in article (1.20), the grad ϕ is normal to the level surface ϕ = constant.

So normal to the given surface $x^2y+ 2xz=4$ is grad $(x^2y + 2xz)$

$$\therefore \quad \text{grad } (x^2y + 2xz) = \hat{i}\frac{\partial}{\partial x}\left(x^2y + 2xz\right) + \hat{j}\frac{\partial}{\partial y}(x2y + 2xz) + \hat{k}\frac{\partial}{\partial z}\left(x^2y + 2xz\right)$$

$$= (2xy + 2z)\hat{i} + x^2\hat{j} + 2x\hat{k}$$

$$= [2(2)(-1) + 2 \times 3]\hat{i} + 2^2\hat{j} + 2(2)\hat{k}$$

$$= 2\hat{i} + 4\hat{j} + 4\hat{k} \quad \text{at the point } (2, -1, 3).$$

$$\text{Then unit normal to the surface} = \frac{2\hat{i} + 4\hat{j} + 4\hat{k}}{\sqrt{(2)^2 + (4)^2 + (4)^2}}$$

$$= \frac{2\hat{i} + 4\hat{j} + 4\hat{k}}{6}$$

$$= \frac{1}{3}\hat{i} + \frac{2}{3}\hat{j} + \frac{2}{3}\hat{k}$$

Another unit vector is $-\frac{1}{3}\hat{i} - \frac{2}{3}\hat{j} - \frac{2}{3}\hat{k}$ directed opposite to that above.

Divergence of A Vector and Divergence Theorem

We have noticed that the net outflow of the flux of a vector field A from a closed surface S is obtained from the integral $\oint$ A. dS. We now define the divergence of A as the net outward flow of flux per unit vloume over a closed incremental surface.

The outward flux per unit volume as the volume shrinks about P is the divergence of A at a given point P.

Hence,

$$div\, A = \nabla \, . \, A = \lim_{\Delta v \to 0} \frac{\oint_s A \, . \, ds}{\Delta v} \tag{1.36}$$

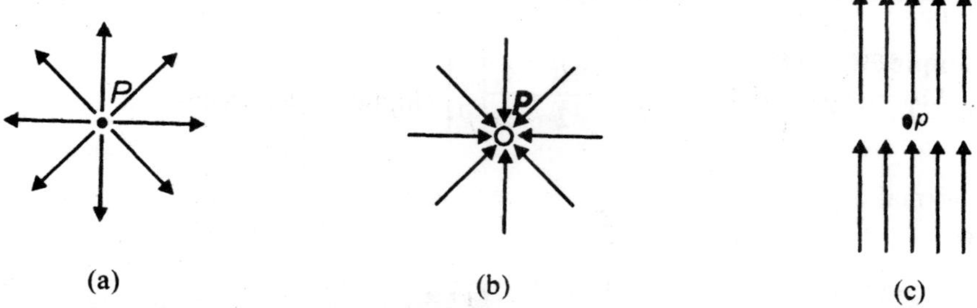

Figure.1·31 Illustration of the divergene of a vector field at P; (a) positive divergence. (b) negative divergence, (c) zero divergence.

Where Δv is the volume enclosed by the closed surface S in which P is located. Physically, we may regrad the divergence of the vector field A at a given point.as a measure of how much the field deverges or emanates from that point. Figure 1.31(a) shows that the divergence of a vector field at point P is positive because the vector diverges (or spreads out) at P. In figure 1.31(b) a vector field has negative divergence (or convergence) at p, and in Figure 1.31(c) a vector field has zero divergence at P. The divergence of a vector field can also be viewed as simply the limit of the limit of the field's source strength per unit volume (or source density); it is positive at a *source* point in the field, and negative at a sink point, or zero where there is neither sink nor source.

We can obtain an expression for $\nabla \, . \, A$ in Cartesian coordinates from the definition in eq. (1.32). Suppose we wish to evaluate the divergence of vector field A at point $P(x_o, y_o, z_o)$; we let the point be enclosed by a differential volume as in Figure 1. 33. The surface integral in eq. (1.34) is obtained from

$$\int_s A \, . \, ds = \left(\int_{front} + \int_{back} + \int_{left} + \int_{right} + \int_{top} + \int_{bottom} \right) A \, . \, ds \quad ...(1.37)$$

A three-dimensional Taylor series expansion of A_x about P is

$$A_x(x,y,z) = A_x(x_o, y_o, z_o) + (x - x_o)\left.\frac{\partial A}{\partial x}\right|_P + (y - y_o)\left.\frac{\partial A_x}{\partial y}\right|_P$$

$$+ (z - z_o)\left.\frac{\partial A_x}{\partial z}\right| + \text{higher-order terms} \quad ...(1.38)$$

For the front side, $x = x_o + dx/2$ *and* $ds = dy\, dz\, a_x$. Then,

$$\int_{front} A \, . \, ds = dy\, dz\left[A_x(x_o, y_o, z_o) + \frac{dx}{2}\frac{\partial A_x}{\partial x}\bigg|_P \right] + \text{higher-order terms}$$

For the back side, $x = x_o - dx/2, ds = dy\, dz(-a_x)$. Then,

$$\int_{back} A.ds = -dydz\left[A_x(x_o, y_o, z_o) - \frac{dx}{2}\frac{\partial A_x}{\partial x}\bigg|_P \right] + \text{higher-order terms}$$

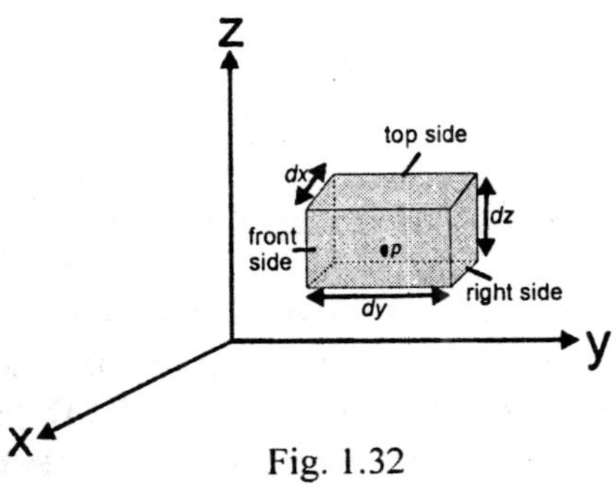

Fig. 1.32

Hence,

$$\int_{front} A \, . \, ds + \int_{back} A \, . \, ds = dx\, dy\, dz \frac{\partial A_x}{\partial x}\bigg|_P + \text{ higher-order terms} \qquad ...(1.39)$$

By taking similar steps, we obtain.

$$\int_{left} A \, . \, ds + \int_{right} A \, . \, ds = dx\, dy\, dz \frac{\partial A_y}{\partial y}\bigg|_P + \text{ higher-order terms} \qquad (1.40)$$

and

$$\int_{top} A \, . \, ds + \int_{bottom} A \, . \, ds = dx\, dy\, dz \frac{\partial A_z}{\partial z}\bigg|_P + \text{ higher-order terms } (1{\cdot}41)$$

Substituting eqs. (1·38) to (1·41)into eq. (1·37) and noting that Δv =dx dy dz, we get

$$\lim_{\Delta v \to 0} \frac{\oint_s A \, . \, ds}{\Delta v} = \left(\frac{\partial A_x}{\partial_x} + \frac{\partial A_y}{\partial y} + \frac{\partial A_z}{\partial z} \right)\bigg|_{at\ P} \qquad (1{\cdot}42)$$

because the higher-order terms will vanish as $\Delta v \to 0$. Thus, the divergence of A at point $P(x_o, y_o, z_o)$ in a Cartesian system is given by

$$\nabla . A = \frac{\partial A_x}{\partial_x} + \frac{\partial A_y}{\partial y} + \frac{\partial A_z}{\partial z} \qquad (1{\cdot}43)$$

Similar expressions for $\nabla . A$ in other coordinate systems can be obtained directly from eq. (1·36) or by trandforming eq. (1·43) into the appropriate coordnate system. In cylindrical, substituting.

$$\begin{bmatrix} A_x \\ A_y \\ A_z \end{bmatrix} = \begin{bmatrix} \cos\theta - \sin\theta & 0 \\ \sin\theta - \cos\theta & 0 \\ 0 \quad\quad 0 & 1 \end{bmatrix} \begin{bmatrix} A_q \\ A_0 \\ A_z \end{bmatrix}$$

and $\frac{\partial}{\partial x} = \cos\theta \frac{\partial}{\partial p} - \frac{\sin\theta}{p}\frac{\partial}{\partial\theta}$ and also

$\frac{\partial}{\partial y} = \sin\phi \frac{\partial}{\partial p} + \frac{\cos\theta}{p}\frac{\partial}{\partial\phi}$ in to eq. (1·43) yields

$$\nabla . A = \frac{1}{\rho}\frac{\partial}{\partial\rho}(\rho A\rho) + \frac{1}{\rho}\frac{\partial A_\phi}{\partial_\phi} + \frac{\partial A}{\partial z} \qquad ...(1.44)$$

Substituting eqs. (2·28) and (3·20) to (3·22) into eq. (1·43) we obtain be divergence of A in spherical coordinates as,

$$\nabla . A = \frac{1}{r^2}\frac{\partial}{\partial r}\left(r^2 A_r\right) + \frac{1}{r\sin\theta}\frac{\partial}{\partial\theta}\left(A_\theta \sin_\theta\right) + \frac{1}{r\sin\theta}\frac{\partial A_\phi}{\partial\phi} \qquad ...(1.45)$$

Note : the following properties of the divergence of a vector field:

1. It produces a scalar field (because scalar product is involved).
2. The divergence of a scalar V, div V, makes no Sense.
3. $\nabla . (A + B) = \nabla . A + \nabla . B$
4. $\nabla . (VA) = V\nabla . A + A . \nabla V$

From the definition of the divergence of A in eq. (1·36), it is not difficult to expect that

$$\int_s A . ds = \int_v \nabla . A \, dv \qquad (1{\cdot}46)$$

This is called the *divergence theorem*, otherwise known as the *Gauss-Ostrogradsky theorem*.

The **divergence theorem** states that the total out ward flux of a vector field **A** through the **closed** surface S is the same as the volume integral of the divergence of **A**.

To prove the divergence theorem, subdivide volume v into a large number of small cells. If the kth cell has volume Δv_k and is bounded by surface S_k.

$$\oint_s A \,.\, ds = \sum_k \int_{s_k} A \,.\, ds = \sum_k \frac{\int_{s_k} A \,.\, ds}{\Delta v_k} \Delta v_k \qquad ...(1.47)$$

Since the outward flux to one cell is inward to some neighboring cells, there is cancellation on every interior surface, so the sum of the surface integrals over Sk's is the same as the surface integral over the surface S. Taking the limit of the right-hand side of eq. (1·47) and incorporating eq. (1·36) gives.

$$\oint_s A \,.\, ds = \int_v \nabla \,.\, A \, dv \qquad (1.48)$$

Which is the divergencetheorem. The theorem applies to any volume v bounded by the closed surface S such as that shown in Figure (1·34) provided that A and ∇ . A are continuous in the region. With a little experience, it will soon become apparent that volume integrals are easier to evaluate than surface integrals. For this reason, to determine the flux of A through a closed surface we simply find the right-hand side of eq. (1·46) instead of the left-hand side of the equation.

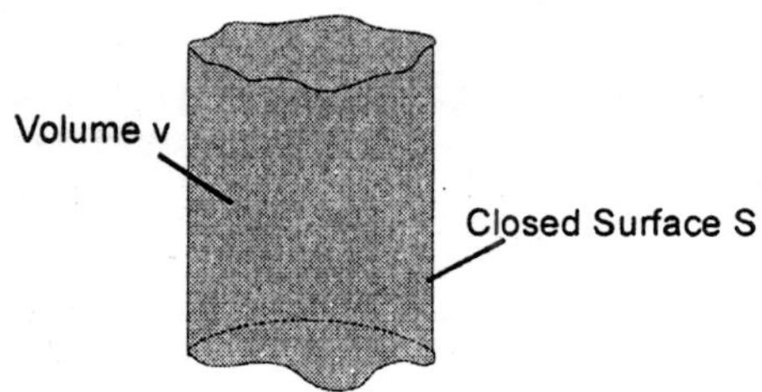

Figure (1.34) Volume V enclosed by surface S.

Example 25. Determine the divergence of these vector field:

(a) $P = x^2yz a_x + xz a_z$

(b) $Q = \rho \sin\phi \, a_\rho + \rho^2 z \, a_\phi + z \cos\phi \, a_z$

(c) $T = \frac{1}{r^2} \cos\theta \, a_r + r \sin\theta \cos\phi \, a_\theta + \cos\theta \, a\phi$

Solution:

(a) $$\nabla \,.\, P = \frac{\partial}{\partial x} P_x + \frac{\partial}{\partial y} P_y + \frac{\partial}{\partial z} P_z$$

$$= \frac{\partial}{\partial x}\left(x^2 yz\right) + \frac{\partial}{\partial y}(0) + \frac{\partial}{\partial z}(xz)$$

$$= 2xyz + x$$

(b) $$\nabla \cdot Q = \frac{1}{\rho}\frac{\partial}{\partial \rho}\left(_{\rho}Q_{\rho}\right) + \frac{1}{\rho}\frac{\partial}{\partial \phi}Q_{\phi} + \frac{\partial}{\partial z}Q_z$$

$$= \frac{1}{\rho}\frac{\partial}{\partial \rho}\left(\rho^2 \sin\phi\right) + \frac{1}{\rho}\frac{\partial}{\partial \phi}\left(_{\rho}2_{z}\right) + \frac{\partial}{\partial z}(z\cos\phi)$$

$$= 2\sin\phi + \cos\phi$$

(c) $$\nabla \cdot T = \frac{1}{r^2}\frac{\partial}{\partial r}\left(r^2 T_r\right) + \frac{1}{r\sin\theta}\frac{\partial}{\partial \theta}(T_\theta \sin\theta) + \frac{1}{r\sin\theta}\frac{\partial}{\partial \phi}(T\phi)$$

$$= \frac{1}{r^2}\frac{\partial}{\partial r}(\cos\theta) + \frac{1}{r\sin\theta}\frac{\partial}{\partial\,\theta}\left(r\sin^2\theta\cos\phi\right) + \frac{1}{r\sin\theta\,\partial\phi}(\cos\theta)$$

$$= 0 + \frac{1}{r\sin\theta}2r\sin\theta\cos\theta\cos\phi + 0$$

$$= 2\cos\theta\cos\phi$$

Exercise 7

Determine the devergence of the following vector fields and evaluate them at the specified points.

(a) A = yzax + yaz at (1, –2, 3)

(b) B = ρz sin ϕ a_ρ + cos ϕ a ϕ

(c) C = $2r\cos\theta\cos\phi a_r + r^{1/2} a_\phi at(1, \pi/6, \pi/3)$

Answer: (a) 4x, 4, (b) (2–3z)z sinϕ, –1. (c) 6 cosθcosϕ, 2.598.

Example 26. If G(r)=10_e^{-2z} $\left(\rho a_\rho + a_z\right)$, determine the flux of G out of the entire surface of the cylinder ρ=1,0 $\le z \le$ 1. Confirm the result using the divergence theorem.

Solution: If ψ is the flux of G through the given surface, shown in Figure 3.17, then

$$\psi = \oint G.ds = \psi_t + \psi_b + \psi_s$$

where ψ_t, ψ_b, and ψ_s are the fluzes through the top, bottom, and sides(curved surface) of the cylinder as in Figure3·17.

For $\psi_t, z = 1,\ dS = \rho\, dp\, d\phi\, a_z$. Hence,

$$\psi_t = \int G.dS = \int_{\rho=0}^{1} \int_{\phi=0}^{2\pi} 10e^{-2} \rho\, dp\, d\phi = 10e^{-2}(2\pi)\frac{\rho^2}{2}\Bigg|_0^1$$

$$= 10\pi e^{-2}$$

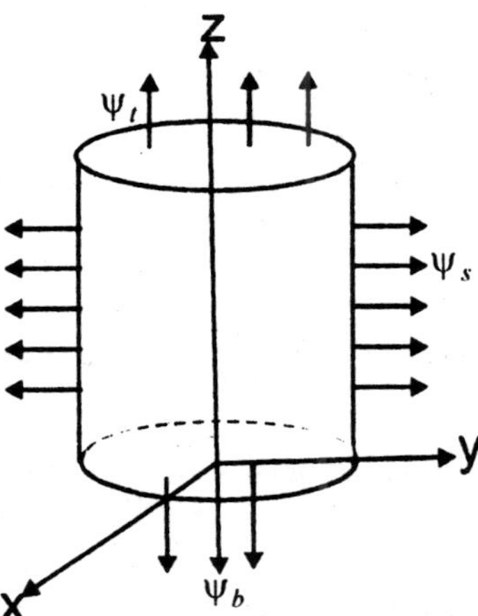

Figure 1·25 For Example 26

For $\psi_b, z = 0$ *and* $dS = \rho\, dp\, d\phi(-a_z)$. Hence,

$$\psi_b = \int_b G.\, dS = \int_{p=0}^{1} \int_{\phi=0}^{2\pi} 10e^{0} \rho\, dp\, d\phi = -10(2\pi)\frac{\rho^2}{2}\Bigg|_0^1$$

$$= -10\pi$$

For $\psi_s, \rho = 1,\ dS = \rho\, dz\, d\phi\, a\rho$. Hence,

$$\psi_s = \int_s G.dS = \int_{z=0}^{1} \int_{\phi=0}^{2\pi} 10e^{-2z} \rho^2 dz d\phi = 10(1)(2\pi)\frac{e^{-2z}}{-2}\Bigg|_0^1$$

Thus,

$$\psi = \psi_t + \psi_b + \psi_s = 10\pi e^{-2} - 10\pi + 10\pi\left(1 - e^{-2}\right) = 0$$

Alternatively, since S is a closed surface, we can apply the divergence theorem:

$$\psi = \oint_s G\, .\, dS = \int_v (\nabla . G)\, dv$$

But

$$\nabla . G = \frac{1}{\rho}\frac{\partial}{\partial \rho}(\rho G\rho) + \frac{1}{\rho}\frac{\partial}{\partial \phi}G\phi + \frac{\partial}{\partial z}Gz$$

$$= \frac{1}{\rho}\frac{\partial}{\partial \rho}\left(\rho^2 10e^{-2z}\right) - 20e^{-2z} = 0$$

Showing that G has no source. Hence,

$$\psi = \int_v (\nabla . G)\, dv = 0$$

Exercise 8

Determine the flux of $D = \rho^2 \cos^2 \phi\, a_\rho + z \sin \phi\, a_\phi$ over the closed surface of the cylinder $0 \le z \le 1, \rho = 4$. Verify the divergence theorem for this case.

Answer: 64π.

Curl of A Vector and Stokes's Theorem

We have already defined the circulation of a vector field A around a closed path L as the integral $\oint_L A . dI$.

The curl of A is an axial (or rotational vector whose magnitude is the maximum circulation of A per unit area as the tends to zero and whose direction is the normal direction of the area when the area is oriented so as to make the circulation maximum.[2]

That is,

$$\text{Curl } A = \nabla \times A = \left(\lim_{\Delta S \to 0} \frac{\oint_L A . dI}{\Delta S} \right)_{max} a_n \qquad (1{\cdot}49)$$

Where the area ΔS is bounded by the curveL and a_n is the unit vector normal to the surface ΔS and is determined using the right-hand rule.

$$\oint_L A . dI = \left(\int_{ab} + \int_{bc} + \int_{cd} + \int_{da} \right) A . dI \qquad (1.50)$$

We expand the field components in a Taylor series expanision about the center point $P(x_o, y_o, z_o)$ as in eq. (1·38) and evaluate eq. (1·50). on side ab, dI =dy a_y and $z=z_o-dz/2$, so

$$\int_{ab} A . dI = dy \left[A_y(x_o, y_o, z_o) - \frac{dz}{2} \frac{\partial A_y}{\partial z}\bigg|_p \right] \qquad (1{\cdot}51)$$

On side *bc*, dl=dz a_z and y =y_o +dy/2, so

$$\int_{bc} A \,.\, dI = d_z\left[A_z\left(x_o, y_o, z_o\right) + \frac{dy}{2}\frac{\partial A_z}{\partial y}\bigg|_p\right] \tag{1·52}$$

On side *cd*, *d*l = dy a_y and z=z_o + dz/2, so

$$\int_{cd} A \,.\, dI = -d_y\left[A_y\left(x_o, y_o, z_o\right) + \frac{dz}{2}\frac{\partial A_y}{\partial z}\bigg|_p\right] \tag{1·53}$$

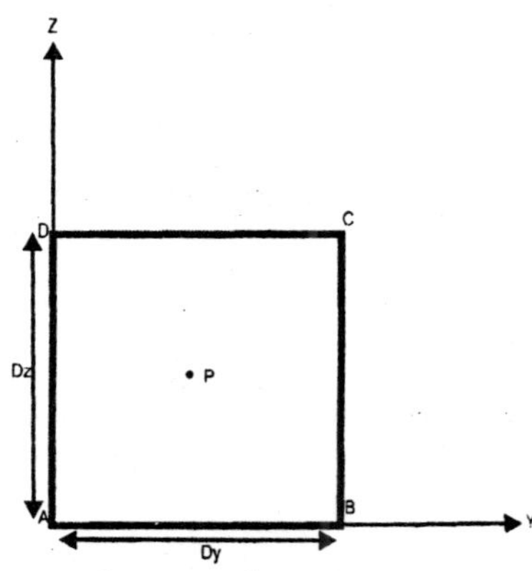

Figure 1·36 Contour used in evaluating the x-component of $\nabla \times A$ at point $P(x_o, y_o, z_o)$.

On side *da*, *d*l= dz a_z and y=y_o–dy/2, so

$$\int_{da} A \,.\, dI = -dz\left[Az\left(x_o, y_o, z_o\right) - \frac{dy}{2}\frac{\partial A_z}{\partial y}\bigg|_p\right] \tag{1·54}$$

Substituting eqs. (1·51) to (1·54) into eq. (1·50) and noting that ΔS =dy dz, we have

$$\lim_{\Delta S \to 0} \oint_L \frac{A \,.\, dI}{\Delta S} = \frac{\partial A_z}{\partial y} - \frac{\partial A_y}{\partial z}$$

or

$$\left(Curl\, A\right)_x = \frac{\partial A_z}{\partial y} - \frac{\partial A_y}{\partial z} \tag{1·55}$$

The y-and x-component of the curl of A can be found in the same way. We obtain

$$(\text{Curl A})_y = \frac{\partial A_x}{\partial z} - \frac{\partial A_z}{\partial x} \tag{1·56a}$$

$$(\text{Curl A})_z = \frac{\partial A_y}{\partial x} - \frac{\partial A_x}{\partial y} \tag{1·56 b}$$

The definition of Δ×A in eq. (1·49) is independent of the coordinate system. In cartesian coordinates the curl of A is easily found using

$$\Delta \times \text{A} = \begin{vmatrix} a_x & a_y & a_z \\ \frac{\partial}{\partial x} & \frac{\partial}{\partial y} & \frac{\partial}{\partial z} \\ A_x & A_y & A_z \end{vmatrix} \tag{1·57}$$

or

$$\nabla \times A = \left[\frac{\partial A_z}{\partial y} - \frac{\partial A_y}{\partial z}\right] a_x + \left[\frac{\partial A_x}{\partial z} - \frac{\partial A_z}{\partial x}\right] a_y + \left[\frac{\partial A_y}{\partial x} - \frac{\partial A_x}{\partial y}\right] a_z$$

By trans forming eq. (1·58) using point and vector transformation techniques. We obtain the curl of A in cylindrical coordnates as

$$\Delta \times \text{A} = \frac{1}{\rho} \begin{vmatrix} a_\rho & \rho a_y & a_z \\ \frac{\partial}{\partial \rho} & \frac{\partial}{\partial \phi} & \frac{\partial}{\partial z} \\ A_\rho & {}_\rho A_\phi & A_z \end{vmatrix}$$

or

$$\nabla \times A = \left[\frac{1}{\rho}\frac{\partial A_z}{\partial \phi} - \frac{\partial A_\phi}{\partial z}\right] a_\rho + \left[\frac{\partial A_\rho}{\partial z} - \frac{\partial A_z}{\partial \rho}\right] a_\phi + \frac{1}{\rho}\left[\frac{\partial({}_\rho A_\phi)}{\partial \rho} - \frac{\partial A_\rho}{\partial \phi}\right] a_z \tag{1·59}$$

and in spherical coordinates as

$$\nabla \times A = \frac{1}{r^2 \sin\theta} \begin{vmatrix} a_r & r a_\theta & r \sin\theta\, a\phi \\ \frac{\partial}{\partial r} & \frac{\partial}{\partial \theta} & \frac{\partial}{\partial \phi} \\ A_r & {}_r A_\theta & r \sin\theta\, A\phi \end{vmatrix}$$

or

$$\nabla \times A = \frac{1}{r \sin\theta}\left[\frac{\partial(A_\phi \sin\theta)}{\partial \theta} - \frac{\partial A_\theta}{\partial \phi}\right] a_r$$

$$+\frac{1}{r}\left[\frac{1}{\sin\theta}\frac{\partial A_r}{\partial\phi}-\frac{\partial(_r A_\phi)}{\partial r}\right]a_\theta+\frac{1}{r}\left[\frac{\partial(rA_\phi)}{\partial r}-\frac{\partial A_r}{\partial\theta}\right]a_\phi \quad (1{\cdot}60)$$

Note the Following Properties of the Curl

1. The curl of a vector field is another vector field.
2. The curl of a scalar field V; $\nabla \times V$, makes no sense.
3. $\nabla \times (A + B) = \nabla \times A + D \times B$
4. $\nabla \times (A \times B) = A\,(\nabla.B) - B(\nabla.A) + (B.\nabla A - (A.\nabla)B$
5. $\nabla \times (VA) = V\nabla \times A + \nabla V \times A$
6. The divergence of the curl of a vector field vanishes, that is, $\nabla.(\nabla \times A)=0$.
7. The curl of the gradient of a scalar field vanishes, that is, $\nabla \times \nabla V = 0$.

Other Properties of the Curl are in Appendix A.

The physical significance of the curl of a vector field is evident in eq. (1·49); the curl provides the maximum value of the circulation of the field per unit area (or circulation density) and indicates the direction along which this maximum value occurs.The curl of a vector field A at a point *P* may be regarded as a measure of the circulation or how much the field curls around *P*. For example, Figure 1·37(a)shows that the curl of a vector field around *P* is directed out of the page. Figure 1·37 (b) shows a vector field with zero curl.

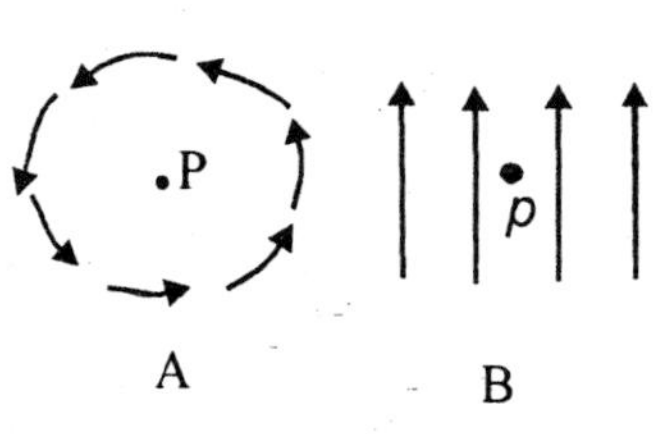

Figure (1·37) Illustration of a curl: (a) curl at *P* point out of the page; (b) curl at *P* is zero.

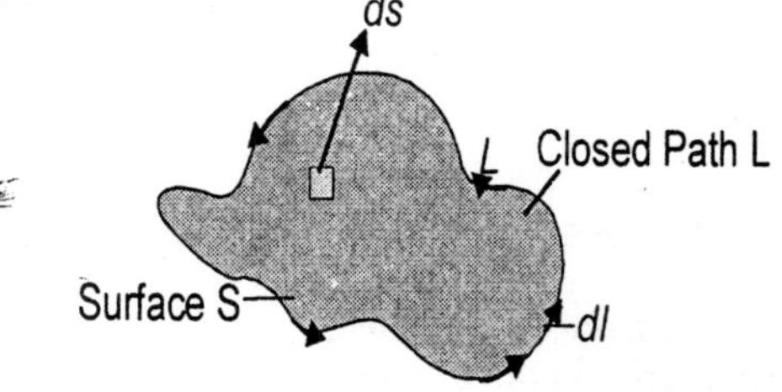

Figure (1·38) Determ-ining the sense of *d*l and *d*S involved in stokes's theorem.

Also, from the definition of the curl of A in eq. (1·49) we may expect that

$$\int_L A \,.\, dl = \int_S (\nabla \times A)\,.\,dS \quad ...(1.61)$$

***This is* Stokes's theorem.**

Stokes's theorem states that the circulation of a vector of a vect9or field A around a (closed) path L is equal to the surface integral of the curl of A over the open surface S bounded by L (see Figure 3.20) provided that A and $\nabla\times$A are continuous on S.

The proof of Stokes's theorem is similar to that of the divergence theorem. The surface S is subdivided into a large number of cells as in Figure 3·21. If the *kth* cell has surface area ∇S_k and is bounded by path L_k.

$$\int_L A \cdot dl = \sum_k \oint_{L_k} A \cdot dl = \sum_k \frac{\oint_{L_K} A \cdot dl}{\Delta S_k} \Delta S_k \qquad ...(1.62)$$

As shown in Figure 3·21, there is cancellation on every interior path, so the sum of the line integrals around Lk's is the same as the line integral around the bounding curve L. Therefore, taking the limit of the right-hand side of eq. (1·62) as $\Delta S_k \to 0$ and incorporating eq. (1·49) leads to

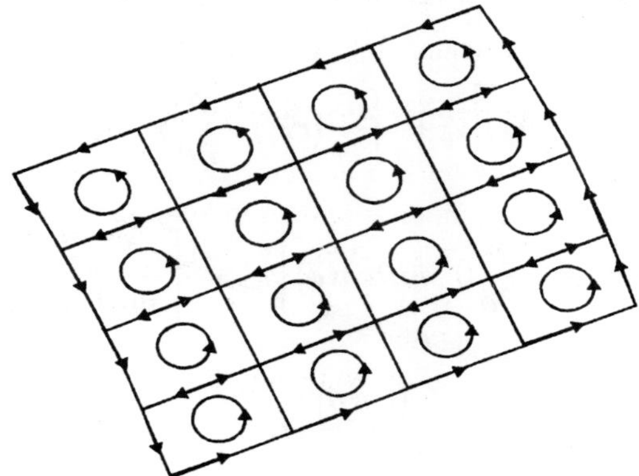

Figure 1·39 Illustration of Stokes's theorem.

$$\oint_L A \cdot dl = \int_S (\nabla \times A) \cdot dS$$

Which is Stokes's Theorem.

The direction of dl and dS in eq. (1·61) must be chosen using the right-hand rule or right-handed screw rule. Using the right-hand rule, if we let the fingers point in the direction of dl, the thumb will indicate the direction of dS (see Fig. 3·20). Note that whereas the divergence theorem relates a surface integral to a volume integral, Stokes's theorem relates a line integral (circulation) to a surface in integral.

Example 27. Determine the curl of the vector fields in Example.

Solution:

(a) $$\nabla \times P = \left(\frac{\partial P_z}{\partial y} - \frac{\partial P_y}{\partial z}\right)a_x + \left(\frac{\partial P_x}{\partial z} - \frac{\partial P_z}{\partial x}\right)a_y + \left(\frac{\partial P_y}{\partial x} - \frac{\partial P_x}{\partial y}\right)a_z$$

$$= (0-0)a_x + (x^2y - z)a_y + (0 - x^2z)a_z = (x^2y - z)a_y - x^2za_z$$

(b) $$\nabla\times Q=\left[\frac{1}{\rho}\frac{\partial Q_z}{\partial\phi}-\frac{\partial Q_\phi}{\partial z}\right]a_\rho+\left[\frac{\partial Q_\rho}{\partial z}-\frac{\partial Q_z}{\partial\rho}\right]a_\phi+\frac{1}{\rho}\left[\frac{\partial}{\partial\rho}\left({}_\rho Q_\phi\right)-\frac{\partial Q_\rho}{\partial\phi}\right]a_z$$

$$=\left(\frac{-z}{\rho}\sin\phi-\rho^2\right)a_\rho+(0-0)a_\phi+\frac{1}{\rho}\left(3\rho^2 z-\rho\cos\phi\right)a_z$$

$$=-\frac{1}{\rho}\left(z\sin\phi+\rho^3\right)a_\rho+(3\rho z-\cos\phi)a_z$$

(c) $$\nabla\times T=\frac{1}{r\sin\theta}\left[\frac{\partial}{\partial\theta}\left(T_\phi\sin\theta\right)-\frac{\partial}{\partial\phi}T_\theta\right]a_r$$

$$+\frac{1}{r}\left[\frac{1}{\sin\theta}\frac{\partial}{\partial\phi}Tr-\frac{\partial}{\partial r}\left(rT_\phi\right)\right]a_\theta+\frac{1}{r}\left[\frac{\partial}{\partial r}(rT_\theta)-\frac{\partial}{\partial\theta}Tr\right]a\phi$$

$$=\frac{1}{r\sin\theta}\left[\frac{\partial}{\partial\theta}(\cos\theta\sin\phi)-\frac{\partial}{\partial\phi}(r\sin\theta\cos\phi)\right]a_r+\frac{1}{r}\left[\frac{1}{\sin\theta}\frac{\partial}{\partial\phi}\frac{(\cos\theta)}{r^2}-\frac{\partial}{\partial r}(r\cos\theta)\right]a_\theta$$

$$+\frac{1}{r}\left[\frac{\partial}{\partial r}\left(r^2\sin\theta\cos\phi\right)-\frac{\partial}{\partial\theta}\frac{(\cos\theta)}{r^2}\right]a_\phi$$

$$=\frac{1}{r\sin\theta}(\cos 2\theta+r\sin\theta\sin\phi)a_r+\frac{1}{r}(0-\cos\theta)a_\theta+\frac{1}{r}\left(2r\sin\theta\cos\phi+\frac{\sin\theta}{r^2}\right)a_\phi$$

$$=\left(\frac{\cos 2\theta}{r\sin\theta}+\sin\phi\right)a_r-\frac{\cos\theta}{r}a_\theta+\left(2\cos\phi+\frac{1}{r^3}\right)\sin\theta\, a_\phi$$

Exercise 9

Determine the curl of the vector fields in Practice Exercise 3·6 and evaluate them at the specified points.

Answer: (a) $a_x+ya_y+(4y-z)a_z$, $a_x-2a_y-11a_z$

(b) $-6\rho z\cos\phi\, a_\rho+\rho\sin\phi\, a_\phi+(6z-1)z\cos\phi\, a_z$, $5a_\phi$

(c) $$\frac{\cot\theta}{r^{1/2}}a_r-\left(2\cot\theta\sin\phi+\frac{3}{2r^{1/2}}\right)a_\theta+2\sin\theta\cos\phi\, a\phi$$

$1.732a_r-4.5a_\theta+0.5\,a\phi$.

Example 28. If A=ρ cosϕ $a\rho$+sinϕ $a\phi$, evaluate $\oint$ A . dl around the path shown in Fingure 3·22. Confirm this using Stokes's theorem.

Solution:

Let

$$\oint_L A \,.\, dI = \left[\int_a^b + \int_b^c + \int_c^d + \int_d^a\right] A \,.\, dI$$

when path L has been divided into segments ab, bc, cd and da as in Figure 1·40.

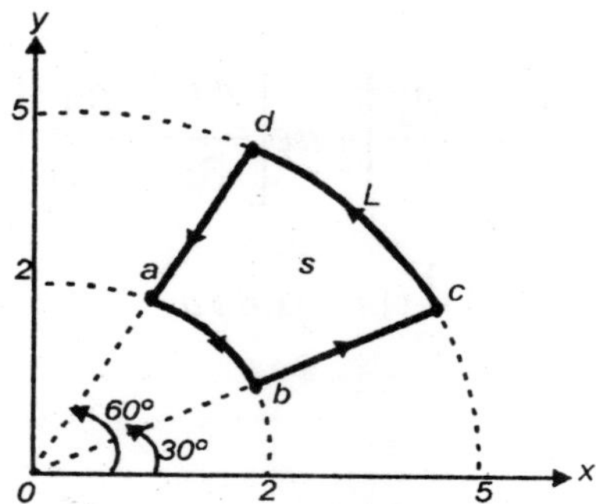

Figure (1·40). For Example 28.

Along ab, ρ=2 and dl = ρdϕ $a\phi$. Hence,

$$\int_a^b A.dI = \int_{\phi=60^\circ}^{30^\circ} \rho \sin\phi \, d\phi = 2(-\cos\phi)\Big|_{60^\circ}^{30^\circ} = -\left(\sqrt{3}-1\right)$$

Along bc, ϕ =30° and dl = dρ aρ. Hence

$$\int_b^c A \,.\, dI = \int_{\rho=2}^{5} \rho \cos\phi \, d\rho = \cos 30^\circ \frac{\rho^2}{2}\Big|_2^5 = \frac{21\sqrt{3}}{4}$$

Along cd, ρ=5 and dl = ρdϕ a_ρ. Hence.

$$\int_c^d A.dI = \int_{\phi=30^\circ}^{60^\circ} \rho \sin\phi \, d\phi = 5(-\cos\phi)\Big|_{30^\circ}^{60^\circ} = \frac{5}{2}\left(\sqrt{3}-1\right)$$

Along da, ϕ =60° and dl =dρ a_ρ. Hence,

$$\int_d^a A.dI = \int_{\rho=}^{2} \rho \cos\phi \, d\rho = \cos 60^\circ \frac{\rho^2}{2}\Big|_5^2 = -\frac{21}{4}$$

Putting all these together results in

$$\oint_L A \, . \, dl = -\sqrt{3} + 1 + \frac{21\sqrt{3}}{4} + \frac{5\sqrt{3}}{2} - \frac{5}{2} - \frac{21}{4}$$

$$= \frac{27}{4}\left(\sqrt{3} - 1\right) = 4.941$$

Using Stokes's theorem (because L is a closed path)

$$\oint_L A \, . \, dl = \int_S (\nabla \times A) \, . \, dS$$

But dS=ρ dϕ dρ•a_z and

$$\nabla \times A = a\rho\left[\frac{1}{\rho}\frac{\partial A_\phi}{\partial \phi} - \frac{\partial A_\phi}{\partial z}\right] + a\phi\left[\frac{\partial A_\rho}{\partial z} - \frac{\partial A_z}{\partial \rho}\right] + a_z \frac{1}{\rho}\left[\frac{\partial}{\partial \rho}\left({}_\rho A_\phi\right) - \frac{\partial A_\rho}{\partial \phi}\right]$$

$$= (0-0)a_\rho + (0-0)a_\phi + \frac{1}{\rho}(1+\rho)\sin\phi \, a_z$$

Hence,

$$\int_S (\nabla \times A).dS = \int_{\phi=30^\circ}^{60^\circ}\int_{\rho=2}^{5} \frac{1}{\rho}(1+\rho)\sin\phi \, \rho \, d\rho \, d\phi$$

$$= \int_{30^\circ}^{60^\circ} \sin\phi \, d\phi \int_2^5 (1+\rho) d\rho$$

$$= -\cos\phi \, \Big|_{30^\circ}^{60^\circ} \left(\rho + \frac{\rho^2}{2}\right)\Bigg|_2^5$$

$$= \frac{27}{4}\left(\sqrt{3} - 1\right) = 4.941$$

Exercise 10

Use Stokes's theorem to confirm your result in Practice Exercise 3.2

Answer: 1.

Example 29. For a vector field A, show explicitly that $\nabla \, . \, \nabla \times A=0$; that is, the divergence of the curl of any vector field is zero.

Solution: This vector identity along with the one in Oractice Exercise 3·10 is very useful in EM. For simplicity, assume that A is in Cartesian coordinates.

$$\nabla \, . \, \nabla \times A = \left(\frac{\partial}{\partial x}, \frac{\partial}{\partial y}, \frac{\partial}{\partial z}\right) . \begin{vmatrix} \partial & \partial & \partial \\ \hline \partial x & \partial y & \partial z \\ A_x & A_y & A_z \end{vmatrix}$$

$$= \left(\frac{\partial}{\partial x}, \frac{\partial}{\partial y}, \frac{\partial}{\partial z}\right) \cdot \left[\left(\frac{\partial A_z}{\partial y} - \frac{\partial A_y}{\partial z}\right), -\left(\frac{\partial A_z}{\partial x} - \frac{\partial A_x}{\partial z}\right), \left(\frac{\partial A_y}{\partial x} - \frac{\partial A_x}{\partial y}\right)\right]$$

$$= \frac{\partial}{\partial x}\left(\frac{\partial A_z}{\partial y} - \frac{\partial A_y}{\partial z}\right) - \frac{\partial}{\partial y}\left(\frac{\partial A_z}{\partial x} - \frac{\partial A_x}{\partial z}\right) + \frac{\partial}{\partial x}\left(\frac{\partial A_y}{\partial x} - \frac{\partial A_x}{\partial y}\right)$$

$$= \frac{\partial^2 A_z}{\partial x\, \partial y} - \frac{\partial^2 A_y}{\partial x\, \partial z} - \frac{\partial^2 A_z}{\partial y\, \partial x} + \frac{\partial^2 A_x}{\partial y\, \partial z} + \frac{\partial^2 A_y}{\partial z\, \partial x} - \frac{\partial^2 A_x}{\partial z\, \partial y}$$

$=0$

because $\frac{\partial^2 A_z}{\partial x\, \partial y} = \frac{\partial^2 A_z}{\partial y\, \partial x}$ and so on.

Exercise 11

For a scalar field V, show that $\nabla \times \nabla V = 0$; that is, the curl of the gradient of any scalar field vanishes.

Answer: Proof.

Laplacian of A Scalar

A single operator which is the composite of gradient and divergence operators is known as the Laplacian. It is expedient to be introduced for practical reasons.

The Laplacian of a scalar V, written as $\nabla^2 V$, is the divergence of the gradient of V. Thus, in Cartesian coordinates,

Laplacian $V = \nabla . \nabla V = \nabla^2 V$

$$= \left[\frac{\partial}{\partial x} a_x + \frac{\partial}{\partial y} a_y + \frac{\partial}{\partial z} a_z\right] \cdot \left[\frac{\partial V}{\partial x} a_x + \frac{\partial V}{\partial y} a_y + \frac{\partial V}{\partial z} a_z\right] \quad (1{\cdot}63)$$

That is,

$$\nabla^2 V = \frac{\partial^2 V}{\partial x^2} + \frac{\partial^2 V}{\partial y^2} + \frac{\partial^2 V}{\partial z^2} \quad (1{\cdot}64)$$

Notice that the Laplacian of a scalar field is another scalar field.

The Laplacian of V in other coordinate systems can be obtained from eq. (1·64) by transformation. In cylindrical coordinates,

$$\nabla^2 V = \frac{1}{\rho}\frac{\partial}{\partial \rho}\left(\rho\frac{\partial V}{\partial \rho}\right) + \frac{1}{\rho^2}\frac{\partial^2 V}{\partial \phi^2} + \frac{\partial^2 V}{\partial z^2} \tag{1·65}$$

and in spherical coordinates.

$$\nabla^2 V = \frac{1}{r^2}\frac{\partial}{\partial r}\left(r^2\frac{\partial V}{\partial r}\right) + \frac{1}{r^2 \sin\theta}\frac{\partial}{\partial \theta}$$

$$\left(\sin\theta\frac{\partial V}{\partial \theta}\right) + \frac{1}{r^2 \sin^2\theta}\frac{\partial^2 V}{\partial \phi^2} \tag{1·66}$$

A scalar field V is said to be Harmonic in a given region if its Laplacian vanishes in that region. In other words, if

$$\nabla^2 V = 0 \quad ...(1.67)$$

is satisfied in the region, the solution for V in eq. (1·67) is harmonic (it is of the form of sine or cosine). Equation (1·67) is called *Laplace's equation.*

We have only considered the Laplacian of a scalar. since the Laplacian operator ∇^2 is a scalar operator, it is also possible to define the Laplacian of vectorA. In this context, ∇^2Ashould not be viewed as the divergence of the gradient of A. which makes no sense. Rather, ∇^2A is defined as the gradient of the devergence of A minus the curl of the curl of A. That is,

$$\nabla^2 A = \nabla(\nabla . A) - \nabla \times \nabla \times A \tag{1.68}$$

This equation can be applied in finding ∇^2A in any coordicate system. In the Cartesian system (and only in that system), eq. (1·68) becomes.

$$\nabla^2 A = \nabla^2 A_x a_x + \nabla^2 A_y a_y + \nabla^2 A_z a_z \tag{1·69}$$

Example 30. Find the Laplacian of the scalar fields of Example 30: that is,

(a) V = e^{-z} sin2x cosy

(b) U = ρ^2z cos2ϕ

(c) W = 10r $\sin^2\theta$ cosϕ

Solution: The Laplacian in the Cartesian system can be found by taking the first derivative and later the second derivative.

$$\textbf{(a)}\quad \nabla^2 V = \frac{\partial^2 V}{\partial x^2} + \frac{\partial^2 V}{\partial y^2} + \frac{\partial^{2V}}{\partial z^2}$$

$$= \frac{\partial}{\partial x}\left(2e^{-z} \cos 2x \cos y\right) + \frac{\partial}{\partial y}\left(e^{-z} \cos 2x \sin y\right) + \frac{\partial}{\partial z}\left(-e^{-z} \sin 2x \cos y\right)$$

$$= -4e^{-z}\sin 2x\cos y + e^{-z}\sin 2x\cos y + e^{-z}\sin 2x\cos y$$

$$= -2e^{-z}\sin 2x\cos y$$

(b) $$\nabla^2 U = \frac{1}{\rho}\frac{\partial}{\partial\rho}\left(\rho\frac{\partial U}{\partial\rho}\right) + \frac{1}{\rho^2}\frac{\partial^2 U}{\partial\phi^2} + \frac{\partial^2 U}{\partial z^2}$$

$$= \frac{1}{\rho}\frac{\partial}{\partial\rho}\left(2\rho^2 z\cos 2\phi\right) - \frac{1}{2}4\rho^2 z\cos 2\phi + 0$$

$$= 4z\cos 2\phi - 4z\cos 2\phi$$

$$=0$$

(c) $$\nabla^2 W = \frac{1}{r^2}\frac{\partial}{\partial r}\left(r^2\frac{\partial W}{\partial r}\right) + \frac{1}{r^2\sin\theta}\frac{\partial}{\partial\theta}\cdot\left(\sin\theta\frac{\partial W}{\partial\theta}\right) + \frac{1}{r^2\sin^2\theta}\frac{\partial^2 W}{\partial\phi^2}$$

$$= \frac{1}{r^2}\frac{\partial}{\partial r}\left(10r^2\sin^2\theta\cos\phi\right) + \frac{1}{r^2\sin\theta}\cdot\frac{\partial}{\partial\theta}\left(10r\sin 2\theta\sin\theta\cos\phi\right)$$

$$-\frac{10r\sin^2\theta\cos\phi}{r^2\sin^2\theta}$$

$$= \frac{20\sin^2\theta\cos\phi}{r} + \frac{20r\cos 2\theta\sin\theta\cos\phi}{r^2\sin\theta} + \frac{10r\sin 2\theta\cos\theta\cos\phi}{r^2\sin\theta} - \frac{10\cos\phi}{r}$$

$$= \frac{10\cos\phi}{r}\left(2\sin^2\theta + 2\cos 2\theta + 2\cos^2\theta - 1\right)$$

$$= \frac{10\cos\phi}{r}\left(1 + 2\cos 2\theta\right)$$

Exercise 12

Determine the Laplacian of the scalar fields of Practice Exercise 3·3, that is,

(a) $U = x^2y + xyz$

(b) $V = \rho z\sin\phi + z^2\cos^2\phi + \rho^2$

(c) $f = \cos\theta\sin\phi \text{ in } r + r^2\phi$

Answer: (a) 2y, (b) $4+2\cos 2\phi - \frac{2z^2}{\rho^2}\cos 2\phi$, (c) $\frac{1}{r^2}\cos\theta\sin\phi(1-2\ln r\,\text{cosec}^2\theta\ln r) + 6\phi$.

Classification of Vector Fields

A vector field is unitquely characterized by its divergence and curl. Neither the divergence nor curl of a vector field is sufficient to completely describe the field. All vector fields can be classified in terms of their vanishing or nonvanishing divergence or curl as follows:

(a) $\nabla . A = 0, \nabla \times A = 0$

(b) $\nabla . A \neq 0, \nabla \times A = 0$

(c) $\nabla . A = 0, \nabla \times A \neq 0$

(d) $\nabla . A \neq 0, \nabla \times A \neq 0$

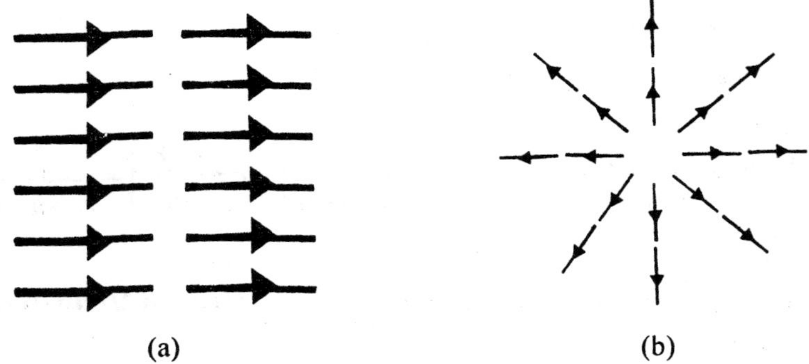

Figure 1·41 Illustrates typical fields in these categories.

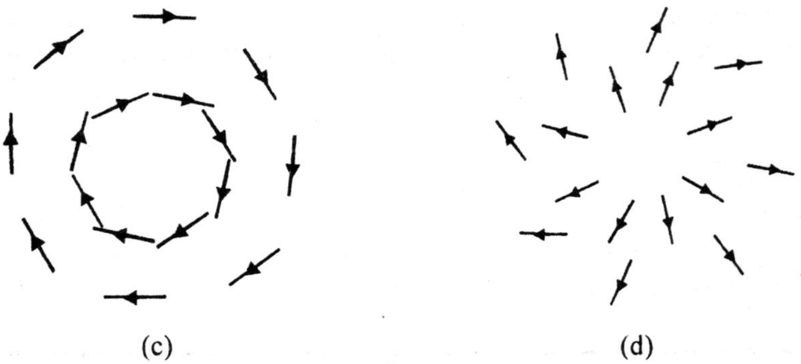

Figure 1·41. Typical fields with vanishing and nonvanishing devergence or curl.

(a) $A = ka_x$, $\nabla . A = 0$, $\nabla \times A = 0$,

(b) $A = kr$, $\nabla . A = 3k$, $\nabla \times A = 0$,

(c) $A = k \times r$, $\nabla . A = 0$, $\nabla \times A = 2k$,

(d) $A = k \times r + cr$, $\nabla . A = 3c$, $\nabla \times A$ 2k.

A vector field A is said to be solenoidal (or devergencless) if

$$\nabla . A = 0$$

Such a field has neither source nor sink of flux. From the divergence theorem,

$$\oint_S A \cdot dS = \int_v \nabla \cdot A\, dv = 0 \qquad (1{\cdot}70)$$

Hence, flux lines of A entering any closed surface must also leave it. Examples of solenoidal fields are incompressible fluids, magnetic fields, and conduction current density under steady state condition. In general, the field of curl F (for any F) is purely solenoidal because $\nabla.(\nabla \times F)=0$. Thus, a solenoidal field A can always be expressed in terms of another vector F; that is,

if $$\nabla . A = 0$$

Than $$\oint_S A.dS = 0 \quad and \quad F = \nabla \times A \qquad ...(1.71)$$

A vector field A is said to be irrotational (or potential) if

$$\nabla \times A=0.$$

That is, a curl-free vector is irrotational[3]. From Stokes's theorem

$$\oint_S (\nabla \times A) \cdot dS = \oint_L A \cdot dl = 0 \qquad (1{\cdot}72)$$

Thus in an irrotational field A, the circulation of A around a closed path is identically zero. This implies that the line integral of A is independent of the chosen path. Therefore, an irrotational field is also known as a conservative field. Examples of irrotational fields include the electrostatic field and the gravitational field. in general, the field of gradient V (for any scalar V) is purely irrotational since

$$\nabla \times (\nabla V) = 0 \qquad (1{\cdot}73)$$

Thus, an irrotational field A can always be expressed in terms of a scalar field V; that is

if

$$\nabla \times A = 0$$

then

$$\oint_L A \cdot dl = 0 \quad and\, A = -\nabla \qquad (1{\cdot}74)$$

For this reason, A may be called a potential field and V the scalar potential of A. The negative sign in eq. (1·74) has been inserted for physical reason.

A vector A is uniquely prescribed within a region by its divergence and its curl. If we let.

$$\nabla . A = \rho_v \qquad (1.75a)$$

and $$\nabla \times A = \rho_S \qquad (1{\cdot}75\ b)$$

ρ_v can be regarded as the source density of A and ρ_s its circulation density. Any vector A satisfying eq. (1·75) with both ρ_v and ρ_s vanishing at infinity can be written as the sum of two vectors: one irrotational (zero curl), the other solenoidal (zero divergence). This is called Helmholltz's theorem. Thus we may write.

$$A = -\nabla V + \nabla \times B \tag{1·76}$$

If we lwt $A_i = -\nabla V$ and $A_s = \nabla \times B$, it is evident from Example and Practice Exercise that $\nabla \times A_i = 0$ and $\nabla \times A_s = 0$, showing that A_i is irrotational and A_s is solenoidal. Finally, it is evident from eqs. (1·68) and (1·75) that any vector field has a Laplacian that satisfies.

$$\nabla^2 A = \nabla_{\rho v} - \nabla \times \rho_s \tag{1·77}$$

Example 31. Show that the vector field A is conservative if A possesses one of these two properties:

(a) The line integral of the tangential component of A along a path extending from a point P to a point Q is independent of the path.

(b) The line integral of the tangential component of A around any closed path is zero.

Solution:

(a) If A is conservative, $\nabla \times A = 0$, so there exists a potential V such that

$$A = -\nabla V = -\left[\frac{\partial V}{\partial x}a_x + \frac{\partial V}{\partial y}a_y + \frac{\partial V}{\partial z}a_z\right]$$

Hence,

$$\int_P^Q A \,.\, dI = -\int_P^Q\left[\frac{\partial V}{\partial x}dx + \frac{\partial V}{\partial y}dy + \frac{\partial V}{\partial z}dz\right]$$

$$= -\int_P^Q\left[\frac{\partial V}{\partial x}\frac{dx}{ds} + \frac{\partial V}{\partial y}\frac{dy}{ds} + \frac{\partial V}{\partial z}\frac{dz}{ds}\right]ds$$

$$= -\int_P^Q\frac{dV}{ds}ds = -\int_P^Q dV$$

or

$$\int_P^Q A \,.\, di = V(P) - V(Q)$$

showing that the line integral depends only on the end point of the curve. Thus, for a conservative field $\int_P^Q A \,.\, dI$ is simply the difference in potential at the end points.

(b) If the path is closed, that is, if *P* and *Q* coincide, then

$$\oint A \, dI = V(P) - V(P) = 0$$

Exercise 13

Show that $B=(y + z \cos xz)a_x + xa_y + x\cos xz\, a_z$ is conservative, without computing any integrals.

Answer: Proof.

Check yourself

1. The differential displacements in the Cartesian, cylindrical, and spherical systems are respectively

   $dl = dx\, a_x + dy\, a_y + dz\, a_z$

   $dl = d\rho\, a\rho + \rho\, d\phi\, a\phi + dz\, az$

   $dl = dr\, a_r + r\, d\theta\, a\theta + r \sin\theta\, d\phi\, a\phi$

   Note that dl si always taken to be in the positive direction; the direction of the displacement is taken care of by the limits of integration.

2. The differential normal areas in the three systems are respectively

   $\mathbf{dS} = dy\, dz\, a_x$

   $dx\, dz\, a_y$

   $dx\, dy\, a_z$

   $\mathbf{dS} = \rho\, d\phi\, dz\, a_\rho$

   $d\rho\, dz\, a_\phi$

   $\rho\, d\rho\, d\phi\, a_z$

   $\mathbf{dS} = r2 \sin\theta\, d\theta\, d\phi\;\; a_r$

   $r \sin\theta\, dr\, d\phi\;\; a_\theta$

   $r\, dr\, d\theta\, a_\phi$

   Note that *dS* can be in the positive or negative direction depending on the surface under consideration.

3. The differential volumes in the three system are

   $dv = dx\, dy\, dz$

   $dv = \rho\, d\rho\, d\phi\, dz$

   $dv = r^2 \sin\theta\, dr\, d\theta\, d\phi$

4. The line integral of vector A along a path L is given by $\int_L$ A. dl. If the path is closed, the line integral becomes the circulation of A around L; that is $\oint_L A \cdot dl$.

5. The flux or surface integral of a vector A across a surface s is defined as $\int_S A \cdot dS$. When the surface S is closed, the surface integral becomes the net outward flux of A across S; that is, $\oint A \cdot dS$.

6. The volume integral of a scalar ρ_v over a volume v is defined as $\int_v \rho_v \, dv$.

7. Vector differentiation is performed using the vector differential operator ∇. The gradient of a scalar field V is devoted by ∇V, the divergence of a vector field A by $\nabla . A$, the curl of A by $\nabla \times A$, and the Laplacian of V by $\nabla^2 V$.

8. The divergence theorem, $\oint_S A \cdot dS = \int_v \nabla . A \, dv$, relates a surfae integral over a closed surface to a volume integral.

9. Stokes's theorem, $\oint_L A \cdot dI = \int_S (\nabla \times A) . dS$, relates a line integral over a closed path to a surface integral.

10. If Laplace's equation, $\nabla^2 V=0$, is satisfied by a scalar field V in a given region, V is said to be harmonic in that region.

11. A vector field is solenoidal if $\nabla . A= 0$; it is irrotational or conservative if $\nabla \times A=0$.

12. A summary of the vector calculus operations in the three coordinate systems is provided on the inside back cover of the text.

13. The vector identities. $\nabla . \nabla \times A=0$ and $\nabla \times \nabla V=0$ are very useful in EM. Other vector identities are in Appendix A.10.

Example 32. Find div $\vec{r}$ where $\vec{r}$ is positive vector in three dimensional space. (K.U. 1989)

Solution. $\vec{r}$ being position vector in three dimensional space, so in cartesian co-ordinates $\vec{r} - x\hat{i} + y\hat{i} + z\hat{k}$

$$\therefore \quad Div\, \vec{r} = \nabla . \vec{r} = \left(\hat{i}\frac{\partial}{\partial x} + \hat{j}\frac{\partial}{\partial y} + \hat{k}\frac{\partial}{\partial z}\right) . \left(\hat{i}x + \hat{j}y + \hat{k}x\right)$$

$$= \frac{\partial x}{\partial x} + \frac{\partial y}{\partial y} + \frac{\partial z}{\partial z} = 1 + 1 + 1 = 3$$

Example 33. If $\vec{A} = x^2 z\hat{i} - 2y^2z^2\hat{j} + x^2y^2z\hat{k}$, Find $\nabla . \vec{A}$ at the point (1,1,1). (K . U . 1993S)

Solution:

$$\nabla . \vec{A} = \left(\hat{i}\frac{\partial}{\partial x} + \hat{j}\frac{\partial}{\partial Y} + \hat{k}\frac{\partial}{\partial z}\right) . \left(x^2 z\hat{i} - 2y^2z^2 j + x^2 z\, \hat{k}\right)$$

$$= 2xz - 4yz^2 + x^2$$

$$= 2(1)(1) - 4(1)(1)^2 + 1(1)^2 \text{ at the point } (1, 1, 1)$$

$$= 2 - 4 + 1 = -1 \text{ at } (1,1,1)$$

Example 34. Prove $\nabla.\left(\vec{A}+\vec{B}\right) = \nabla.\vec{A} + \nabla.\vec{B}$

(M.D.U. 1986)

Solution: $\nabla.\ \left(\vec{A}+\vec{B}\right) = \left(\hat{i}\frac{\partial}{\partial x} + \hat{j}\frac{\partial}{\partial y} + \hat{k}\frac{\partial}{\partial z}\right)$

$$\times\left[(Ax + Bx)\hat{i} + (Ay + By)\hat{j} + (Az + Bz)\hat{k}\right]$$

$$= \frac{\partial}{\partial x}(A_x + B_x) + \frac{\partial}{\partial y}\left(A_y + B_y\right) + \frac{\partial}{\partial z}(A_z + B_z)$$

$$= \frac{\partial A_x}{\partial x} + \frac{\partial B_x}{\partial x} + \frac{\partial A_y}{\partial y} + \frac{\partial B_y}{\partial y} + \frac{\partial A_1}{\partial z} + \frac{\partial B_z}{\partial z}$$

Rearranging

$$= \left(\frac{\partial A_x}{\partial x} + \frac{\partial A_y}{\partial y} + \frac{\partial A_z}{\partial z}\right) + \left(\frac{\partial B_x}{\partial x} + \frac{\partial B_y}{\partial y} + \frac{\partial B_z}{\partial z}\right)$$

$$= \nabla\ .\ A + \nabla\ .\ B.$$

Example 35. Show that $= \vec{\nabla}\ .\ \left(\phi\vec{A}\right) = (\nabla\phi)\ .\ \vec{A} + \phi\left(\nabla.\vec{A}\right)$

(K.U. 1985)

Solution.

$$\nabla.\left(\phi\vec{A}\right) = \left(\hat{i}\frac{\partial}{\partial x} + \hat{j}\frac{\partial}{\partial y} + \hat{k}\frac{\partial}{\partial z}\right).\left(\phi A_x\hat{i} + \phi A_y\hat{j} + \phi A_z\hat{k}\right)$$

$$= \frac{\partial}{\partial x}(\phi A_x) + \frac{\partial}{\partial y}\left(\phi A_y\right) + \frac{\partial}{\partial z}(\phi A_z)$$

$$= \phi\frac{\partial A_x}{\partial x} + Ax\frac{\partial \phi}{\partial x} + \rho\frac{\partial A_y}{\partial y} + Ay\frac{\partial \phi}{\partial y} + \phi\frac{\partial A_z}{\partial z} + A_z\frac{\partial \phi}{\partial z}$$

Rearranging

$$=\left(\frac{\partial A_x}{\partial x}+\frac{\partial A_y}{\partial y}+\frac{\partial A_z}{\partial z}\right)\phi+\left(A_x\frac{\partial \phi}{\partial x}+A_y\frac{\partial \phi}{\partial y}A_z\frac{\partial \phi}{\partial z}\right)$$

$$=\left(\nabla\,.\,\vec{A}\right)\phi+\vec{A}.$$

Example 36. If $\vec{r}$ and r have their ussual meaning, show that div ($r^n\ \vec{r}$)=(n+3)^{rn}.

Solution: Div ($r^n\ \vec{r}$) = ∇ . ($r^n\ \vec{r}$)

$$=\left(\hat{i}\frac{\partial}{\partial x}+\hat{j}\frac{\partial}{\partial y}+\hat{k}\frac{\partial}{\partial z}\right).r^n\left(x\hat{i}+y\hat{j}+z\hat{k}\right)$$

$$=\frac{\partial}{\partial x}\left(r^n x\right)+\frac{\partial}{\partial y}\left(r^n y\right)+\frac{\partial}{\partial z}\left(r^n z\right)$$

$$=\frac{\partial}{\partial x}\left[\left(x^2+y^2+z^2\right)^{n/2}x\right]+\frac{\partial}{\partial y}\left[\left(x^2+y^2+z^2\right)^{n/2}y\right]+\frac{\partial}{\partial z}\left[\left(x^2+y^2+z^2\right)^{n/2}z\right]$$

$$\left(\therefore\ x^2+y^2+z^2=r^2\right)$$

$$=x\frac{n}{2}\left(x^2+y^2+z^2\right)^{n/2-1}2x+\left(x^2+y^2+z^2\right)^{n/2}$$

$$+y\frac{n}{2}\left(x^2+y^2+z^2\right)^{n/2-1}2y+\left(x^2+y^2+z^2\right)^{n/2}$$

$$+z\frac{n}{2}\left(x^2+y^2+z^2\right)^{n/2-1}2z+\left(x^2+y^2+z^2\right)^{n/2}$$

Rearranging the term, we get

$$=3\left(x^2+y^2+z^2\right)^{n/2}+n\left(x^2+y^2+z^2\right)^{n/2-1}\cdot\left(x^2+y^2+z^2\right)$$

$$=3\left(x^2+y^2+z^2\right)^{n/2}+n\left(x^2+y^2+z^2\right)^{n/2}$$

$$=(n+3)\left(x^2+y^2+z^2\right)^{n/2}=(n+3)\left(r^2\right)^{n/2}$$

$$=(n+3)^{r^n}$$

Expample 37. Prove that divergence $\vec{r} = \frac{2}{r}$

where $\vec{r} = x\hat{i} + y\hat{j} + z\hat{k}.$

$$\hat{r} = \frac{\vec{r}}{r} = \frac{x\hat{i} + y\hat{j} + z\hat{k}}{\left(x^2 + y^2 + z^2\right)^{1/2}}$$

$$\therefore \nabla . \hat{r} = \nabla . \left(\frac{\vec{r}}{r}\right) = \left(\hat{i}\frac{\partial}{\partial x} + \hat{j}\frac{\partial}{\partial y} + \hat{k}\frac{\partial}{\partial z}\right) . \left[\frac{x\hat{i} + y\hat{j} + z\hat{k}}{\left(x^2 + y^2 + z^2\right)^{1/2}}\right]$$

$$= \frac{\partial}{\partial x}\left[x\left(x^2 + y^2 + z^2\right)^{-\frac{1}{2}}\right] + \frac{\partial}{\partial y}\left[y\left(x^2 + y^2 + z^2\right)^{-\frac{1}{2}}\right] + \frac{\partial}{\partial z}\left[z\left(x^2 + y^2 + z^2\right)^{-\frac{1}{2}}\right]$$

$$= \left(x^2 + y^2 + z^2\right)^{-\frac{1}{2}} - \frac{1}{2}x\left(x^2 + y^2 + z^2\right)^{-3/2} 2x$$

$$+ \left(x^2 + y^2 + z^2\right)^{-\frac{1}{2}} - \frac{1}{2}y\left(x^2 + y^2 + z^2\right)^{-3/2} 2y$$

$$+ \left(x^2 + y^2 + z^2\right)^{-\frac{1}{2}} - \frac{1}{2}z\left(x^2 + y^2 + z^2\right)^{-3/2} 2z$$

Rearranging the terms

$$= 3 + \left(x^2 + y^2 + z^2\right)^{-1/2} - \left(x^2 + y^2 + z^2\right)^{-3/2}\left(x^2 + y^2 + z^2\right)$$

$$= 3 + \left(x^2 + y^2 + z^2\right)^{-1/2} - \left(x^2 + y^2 + z^2\right)^{-1/2} = \left(x^2 + y^2 + z^2\right)^{-1/2}$$

$$= \frac{2}{\sqrt{x^2 + y^2 + z^2}} = \frac{2}{r}$$

Example 38. *The electric field due to a point charge q is expressed as* $\vec{E} = \frac{q\hat{r}}{4\pi\varepsilon_0 r^2}$, *prove that the divergence of electric field due to that point change is zero (* ε_0 *is the permitivity of free space).*

(M.D.U. 1982 A)

Solution: Electric field due to point charge $\vec{E} = \dfrac{q\hat{r}}{4\pi\varepsilon_0 r^2}$,

So $$\nabla.\vec{E} = \nabla.\left(\frac{q\hat{r}}{4\pi\varepsilon_0 r^2}\right) = \frac{q}{4\pi\varepsilon_0}\nabla.\left(\frac{\hat{r}}{r^2}\right)$$

$$\left\{\because \frac{q}{4\pi\varepsilon_0} \text{ is constant}\right\}$$

$$= \frac{q}{4\pi\varepsilon_0}\nabla.\left(\frac{\vec{r}}{r^3}\right)$$

But $$\nabla.\left(\frac{\vec{r}}{r^3}\right) = 0$$

(By putting n=-3 in Example 27)

$\therefore$ $$\nabla.\vec{E} = \frac{q}{4\pi\varepsilon_0}(0)\ 0.$$

Example 39. *if* $\vec{A} = \dfrac{\vec{r}}{r}$, *Find grad div* $\vec{A}$.

Solution. grad div $\vec{A} = \nabla(\nabla.\vec{A}) = \nabla\left(\nabla.\dfrac{\vec{r}}{r}\right)$

$= \nabla\left(\dfrac{2}{r}\right)$ $\left(\therefore \nabla.\dfrac{\vec{r}}{r} = \dfrac{2}{r}\right)$ as proved in Example 28.

$= 2\nabla\left(\dfrac{1}{r}\right)$

$= 2\left(\dfrac{-\vec{r}}{r^3}\right)$ $\therefore \nabla.\dfrac{1}{r} = \dfrac{-\vec{r}}{r^3}$ as proved in Example 19)

$= \dfrac{-2\vec{r}}{r^3} = -2r^{-3}\vec{r}$

Example 40. If $\vec{e}$ is a unit vector, prove div $(\vec{e}\,.\,\vec{r})\vec{e} = 1$

Solution: Let e_1, e_2, e_3, are the component of $\vec{e}$ along x, y, and z axis. Then $e_1^2+e_2^2+e_3^2=1$ (Since $\vec{e}$ is a unit vector)

Therefore div $(\vec{e}\,.\,\vec{r})\vec{e} = (\vec{e}\,.\,\vec{r})$ div $\vec{e} + \vec{e}\,.$ grad $(\vec{e}\,.\,\vec{r})$

[from Example 26)

But $\quad div\,\vec{e} = 0 \quad (\because \vec{e}$ is a constant vector)

$\therefore \quad$ div $(\vec{e}\,.\,\vec{r})\,\vec{e} = \vec{e}.grad\,(\vec{e}\,.\vec{r})$

$= e.e$ ($\because$ grad $(\vec{e}\,.\,\vec{r}) = e$ [from Example 14]

$=1.$

Example 41. Evaluate $\iint_S \vec{r}.\hat{n}$ ds *where s is a close surface.*

Solution: By divergence theorem

$\iint_S \vec{r}.\hat{n}\,ds. = \iiint_v (\vec{\nabla}.\,\vec{r})\,dv$ where v is the volume enclosed by the surface s.

$$= \iiint_v \left(\hat{i}\frac{\partial}{\partial x} + \hat{j}\frac{\partial}{\partial y} + \hat{k}\frac{\partial}{\partial z}\right).(x\hat{i} + y\hat{j} + z\,\hat{k})\,dv$$

$$= \iiint_v \left(\frac{\partial x}{\partial x} + \frac{\partial y}{\partial y} + \frac{\partial z}{\partial z}\right) dv$$

$$= \iiint_v 3dv = 3\iiint_v dv = 3v$$

Example 42. Verify gauss theorem for the field $\vec{A} = x^2\hat{i} + y^2\hat{j} + z^2\hat{k}$ where v is the volume of the cubical box bounded by the planes x=0, x=1, y=0, y=1, z=0, z=1.

Solution. Consider a cube ABCDOFGH as shown in Fig. 1·30 taking the adjacent edges OA, OF, OH as axis of reference with O as origin.

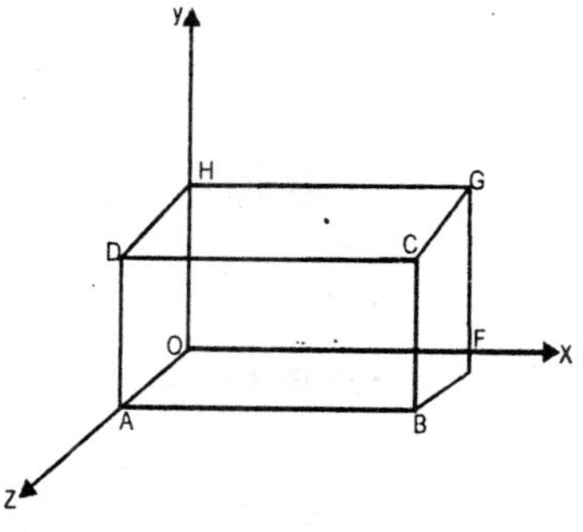

As $\vec{A} = x^2\hat{i} + y^2\hat{j} + z^2\hat{k}$

$$\because \quad \nabla.\vec{A} = \left(\hat{i}\frac{\partial}{\partial x} + \hat{j}\frac{\partial}{\partial y} + \hat{k}\frac{\partial}{\partial z}\right)\cdot\left(x^2\hat{i} + y^2\hat{j} + z^2\hat{k}\right)$$

=2y+2y+2z.

Figure. 1.42

Now the values of x,y,z very from 0 to 1, and therefore for the volume of cube, we have

$$\iiint(\nabla\cdot\vec{A})dv=\int_0^1\int_0^1\int_0^1 2(x+y+z)dx\,dy\,dz$$

Interating w.r.t.z

$$=2\int_0^1\int_0^1\left(xz+yz+\frac{z^2}{2}\right)\Bigg|_0^1 dx\,dy$$

$$=2\int_0^1\int_0^1\left(x+y+\frac{1}{2}\right)dy\,dx$$

Integrating w.r.t.y

$$=2\int_0^1\left(xy+\frac{y^2}{2}+\frac{y}{2}\right)\Bigg|_0^1 dx=2\int_0^1\left(x+\frac{1}{2}+\frac{1}{2}\right)dx$$

$$=2\int_0^1(x+1)dx=2\left(\frac{x^2}{2}+x\right)\Bigg|_0^1=2\left(\frac{1}{2}+1\right)=3$$

For the surface of the cube, we have

$=\iint_s \vec{A}\cdot\vec{ds}=\iint_s \vec{A}\cdot\hat{n}\,ds$ where s is area of all the six faces of the cube. For the surface ABCD, $\hat{n}=\hat{k}, z=1$

$$\therefore\quad \iint_{ABCD}\vec{A}\cdot k\,dx\,dy=\int_0^1\int_0^1\left(x^2\hat{i}+y^2\hat{j}+\hat{k}\right).kdxdy=\int_0^1$$

$$\int_0^1 dxdy=x\Big|_0^1\,y\Big|_0^1=1$$

For the surface BFGC $\hat{n}=\hat{i}, x=1$

$$\therefore\quad \iint_{BFGC}\vec{A}.\hat{i}\,dydz\int_0^1\int_0^1\left(\hat{i}+y^2\hat{j}+z^2\hat{k}\right).$$

$$\hat{i}dydz = \int_0^1\int_0^1 dydz = y\Big|_0^1 z\Big|_0^1 = 1$$

For the surface FGHO, $\hat{n} = -\hat{k},\ z = 0;\ \int_0^1\int_0^1 -\left(x^2\hat{i} + y^2\hat{j}\right).\hat{k}\,dxdy = 0$

For the surface OHDA $\hat{n} = -\hat{i},\ x = 0; \int_0^1\int_0^1 -\hat{A}\,.\,\hat{i}dydz$

$$= -\int_0^1\int_0^1 \left(y^2\hat{j} + z^2\hat{k}\right).\,\hat{i}\,dydz = 0$$

For the surface CDHG $\hat{n} = \hat{j}, y = 1$

$$\therefore \qquad \int\int_{CDHG} \vec{A}\,.\,\hat{j}dxdz = \int_0^1\int_0^1 \left(x^2\hat{i} + \hat{j} + z^2\hat{k}\right).$$

$$\hat{j}dxdz = \int_0^1\int_0^1 dxdz$$

$$= x\Big|_0^1 z\Big|_0^1 = 0$$

For the surface OABF, $\hat{n} = -\hat{j}, y = 0;\ \int_0^1\int_0^1 -\vec{A}\,.\,\hat{j}\,dxdz$

$$= -\int_0^1\int_0^1 \left(x^2\hat{i} + z^2\hat{k}\right),\hat{j}\,dydz = 0$$

Adding these, we get

$$\int\int \vec{A}\,.\,\vec{ds} = 1 + 1 + 0 + 0 + 1 + 0 = 3$$

Hence Gauss's theorem is verified.

Example 43. *Determine the constant 'a', so that the vector*

$\vec{A} = (x + 3y)\hat{i} + (2y + 3z)\hat{j} + (x + az)\hat{k}$ *is solenoidal.*

Solution: A vector $\vec{A}$ is solenoidal if its divergence vanishes

i.e.,

$$\vec{\nabla} . \vec{A} = 0$$

or
$$\frac{\partial}{\partial x}(x+3y)+\frac{\partial}{\partial y}(2y+3z)+\frac{\partial}{\partial z}(x+az)=0$$

$$1+2+a=0$$

or
$$3+a = 0 \text{ or } a = -3.$$

Example 44. *If* $\phi=x^2+y^2+2z$, find div gradϕ.

(K.U. 1988 A)

Solution: Div grad $\phi = \nabla.(\nabla \phi)$

$\because$
$$\nabla\phi = \hat{i}\frac{\partial}{\partial x}(x^2+y^2+2z)+\hat{j}\frac{\partial}{\partial y}(x^2+y^2+2z)+\hat{k}\frac{\partial}{\partial y}(x^2+y^2+2z)$$

$$= 2x\hat{i}+2y\hat{j}+2\hat{k}$$

$$\nabla . \nabla\phi = \left[\hat{i}\frac{\partial}{\partial x}+\hat{j}\frac{\partial}{\partial y}+\hat{k}\frac{\partial}{\partial z}\right] . [2x\hat{i}+2y\hat{j}+2\hat{k}\}$$

$$=\frac{\partial}{\partial x}(2x)+\frac{\partial}{\partial y}(2y)+\frac{\partial}{\partial z}(2)$$

$$=2+2+0=4.$$

Example 45. Prove that $\nabla^2\left(\frac{1}{r}\right)=0$ where $\vec{r}$ is position vector.

Solution: As $\vec{r}$ position vector, so $r=\sqrt{x^2+y^2+z^2}$

$\therefore$
$$\nabla^2\left(\frac{1}{r}\right)=\left(\frac{\partial^2}{\partial x^2}+\frac{\partial^2}{\partial y^2}+\frac{\partial^2}{\partial z^2}\right)\left(\frac{1}{\sqrt{x^2+y^2+z^2}}\right)$$

$$=\frac{\partial^2}{\partial x^2}\left(\frac{1}{\sqrt{x^2+y^2+z^2}}\right)+\frac{\partial^2}{\partial y^2}\left(\frac{1}{\sqrt{x^2+y^2+z^2}}\right)+\frac{\partial^2}{\partial z^2}\left(\frac{1}{\sqrt{x^2+y^2+z^2}}\right)$$

Now
$$\frac{\partial}{\partial x}\left(\frac{1}{\sqrt{x^2+y^2+z^2}}\right)=\frac{\partial}{\partial x}(x^2+y^2+z^2)^{-1/2}$$

$$= -\frac{1}{2}\left(x^2 + y^2 + z^2\right)^{-3/2} 2x$$

$$= -x\left(x^2 + y^2 + z^2\right)^{-3/2}$$

$$\frac{\partial^2}{\partial x^2}\left(\frac{1}{\sqrt{x^2 + y^2 + z^2}}\right) = \frac{\partial}{\partial x}[-x\left(x^2 + y^2 + z^2\right)^{3/2}$$

$$= -x\left\{-\frac{3}{2}\right\}\left(x^2 + y^2 + z^2\right)^{-5/2} 2x + \left(x^2 + y^2 + z^2\right)^{-3/2}(-1)$$

$$=3x^2\ (x^2+y^2+z^2)^{-5/2} - (x^2+y^2+z^2)^{-3/2}$$

$$=(x^2+y^2+z^2)^{-5/2}\ [3x^2-(x^2+y^2+z^2)]$$

$$= \frac{2x^2 - y^2 - z^2}{\left(x^2 + y^2 + z^2\right)^{5/2}}$$

Similarly $\frac{\partial^2}{\partial y^2}\left(\frac{1}{\sqrt{x^2 + y^2 + z^2}}\right) = \frac{2y^2 - z^2 - x^2}{\left(x^2 + y^2 + z^2\right)^{5/2}}$

and $\frac{\partial^2}{\partial z^2}\left(\frac{1}{\sqrt{x^2 + y^2 + z^2}}\right) = \frac{2z^2 - x^2 - y^2}{\left(x^2 + y^2 + z^2\right)^{5/2}}$

By adding, we have

$$\nabla^2\left(\frac{1}{r}\right) = \left(\frac{\partial^2}{\partial x^2} + \frac{\partial^2}{\partial y^2} + \frac{\partial^2}{\partial z^2}\right)\left(\frac{1}{\sqrt{x^2 + y^2 + z^2}}\right)$$

$$= \frac{\left(2x^2 + y^2 + z^2\right) + \left(2y^2 + z^2 + x^2\right) + \left(2z^2 + x^2 + y^2\right)}{\sqrt{\left(x^2 + y^2 + z^2\right)^5}}$$

$$= \frac{0}{\sqrt{\left(x^2 + y^2 + z^2\right)^5}} = 0.$$

Example 46. *Prove Div curl* $\vec{A}$=0.

(M.D.U. 1984, K.U.K. 1990 A)

Solution

Div Curl $\vec{A} = \nabla \cdot \nabla \times \vec{A},\ as\ \vec{A}.\left(\vec{B}\times\vec{C}\right) = \begin{vmatrix} A_x & A_y & A_z \\ B_x & B_y & B_z \\ C_x & C_y & C_z \end{vmatrix}$

So $\nabla \cdot \nabla \times \vec{A} = \begin{vmatrix} \frac{\partial}{\partial x} & \frac{\partial}{\partial y} & \frac{\partial}{\partial z} \\ \frac{\partial}{\partial x} & \frac{\partial}{\partial y} & \frac{\partial}{\partial z} \\ A_x & A_y & A_z \end{vmatrix} = 0$ (as two rows of the determinant are identical)

Thus curl of any vector field is itself a field where the lines of flow from closed curves. Here the divergence of that field is obviously be zero.

Example 47. *If* $\vec{A} = zx^3\hat{i} - 2x^3yz\hat{j} + 2yz^4\hat{k}$, *Find curl* $\vec{A}$ *at the point (1,1,1)*

(M.D.U. 1985 S)

Solution: Curl $\vec{A} = \nabla \times A = \begin{vmatrix} \hat{i} & \hat{j} & \hat{k} \\ \frac{\partial}{\partial x} & \frac{\partial}{\partial y} & \frac{\partial}{\partial z} \\ zx^3 & -2x^2yz & 2yz^4 \end{vmatrix}$

$$\left[\frac{\partial}{\partial y}\left(2yz^4\right) - \frac{\partial}{\partial z}(2x^2yz\right]\hat{i} + \left[\frac{\partial}{\partial z}\left(zx^3\right)\right] - \frac{\partial}{\partial y}2yz^4\Big]\hat{j} + \left[\frac{\partial}{\partial x}\left(-2x^2yz\right) - \frac{\partial}{\partial y}\left(zx^3\right)\right]\hat{k}$$

$$= \left(2z^4 + 2x^3y\right)\hat{i} + x^3\hat{j} - 4xyz\,\hat{k}$$

$= (2\,(1)^4+2\,(1)^2(1)\,)\hat{i} + (1)^3 - \hat{j}\,4\,(1)\,(1)\ \ (1)\,\hat{k}$ at the point (1,1,1)

$= 4\hat{i} + \hat{j} - 4\hat{k}$.

Example 48. Prove Curl curl $\vec{V}$ =grad div $\vec{V} - \nabla^2\vec{V}$.

(K.U. 1977, 1989S)

Solution. Curl curl $\vec{V} = \nabla \times \left(\nabla \times \vec{V}\right)$

$$=(\nabla \,.\, \vec{V})\nabla-(\nabla.\nabla)\,\vec{V}$$

$\left[\because \text{ From vector triple product } \vec{A}\times(\vec{B}\times\vec{C})=(\vec{A}\,.\,\vec{C})\vec{B}-(\vec{A}\,.\,\vec{B})\vec{C}\right]$

$$=\nabla(\nabla\,.\,\vec{V})-\nabla^2\vec{V}$$

$$=\text{grad div } \vec{V}-\nabla^2\vec{V}.$$

Example 49. *If u is scalar field at a point and $\vec{V}$ is a vector field at that, prove* $\nabla\times(u\,\vec{V})=(\nabla u)\times\vec{V}+u(\nabla\times\vec{V})$. ***(M.D.U. 1985)***

Solution. $\nabla\times(u\vec{V})=\begin{vmatrix}\hat{i} & \hat{j} & \hat{k}\\ \dfrac{\partial}{\partial x} & \dfrac{\partial}{\partial y} & \dfrac{\partial}{\partial z}\\ uV_x & uV_y & uV_z\end{vmatrix}$

$$=\hat{i}\left\{\frac{\partial}{\partial y}(uV_z)-\frac{\partial}{\partial z}(uV_y)\right\}+\hat{j}\left\{\frac{\partial}{\partial z}(uV_x)-\frac{\partial}{\partial x}(uV_z)\right\}+\hat{k}\left\{\frac{\partial}{\partial x}(uV_y)-\frac{\partial}{\partial y}(uV_x)\right\}$$

$$=\hat{i}\left(V_z\frac{\partial u}{\partial y}+u\frac{\partial V_z}{\partial y}-V_y\frac{\partial u}{\partial z}-u\frac{\partial V_y}{\partial z}\right)+\hat{j}\left(V_x\frac{\partial u}{\partial z}+u\frac{\partial V_x}{\partial z}-V_z\frac{\partial u}{\partial x}-u\frac{\partial V_z}{\partial x}\right)$$

$$+\hat{k}\left(V_y\frac{\partial u}{\partial x}+u\frac{\partial V_y}{\partial x}-V_x\frac{\partial u}{\partial y}-u\frac{\partial V_x}{\partial y}\right)$$

Re-arranging

$$=\hat{i}\left(V_z\frac{\partial u}{\partial y}-V_y\frac{\partial u}{\partial z}\right)+\hat{j}\left(V_x\frac{\partial u}{\partial z}-V_z\frac{\partial u}{\partial x}\right)+\hat{k}\left(V_y\frac{\partial u}{\partial x}-V_x\frac{\partial u}{\partial y}\right)$$

$$+\hat{i}u\left(\frac{\partial V_z}{\partial y}-\frac{\partial V_y}{\partial z}\right)+\hat{j}u\left(\frac{\partial V_x}{\partial z}-\frac{\partial V_z}{\partial x}\right)+\hat{k}u\left(\frac{\partial V_y}{\partial x}-\frac{\partial V_x}{\partial y}\right)$$

$$=\begin{vmatrix}\hat{i} & \hat{j} & \hat{k}\\ \dfrac{\partial u}{\partial x} & \dfrac{\partial u}{\partial y} & \dfrac{\partial u}{\partial z}\\ V_x & V_y & V_z\end{vmatrix}+u(\nabla\times\vec{V})$$

$$=(\nabla u)\times\vec{V}-u(\nabla\times\vec{V}).$$

Example 50. Prove that $\nabla \times (\vec{A} + \vec{B}) = \nabla \times \vec{A} + \nabla \times \vec{B}$

Solution.

Let $\vec{A} = A_x\hat{i} + A_y\hat{j} + A_s\hat{k}$ and $\vec{B} = B_x\hat{i} + B_y\hat{j} + B_s\hat{k}$ then

$$\nabla \times (\vec{A} + \vec{B}) = \left(\hat{i}\frac{\partial}{\partial x} + \hat{j}\frac{\partial}{\partial y} - \hat{k}\frac{\partial}{\partial z}\right) \times [(A_x + B_x)\hat{i} + (A_y + B_y)\hat{j} + (A_z + B_z)\hat{k}]$$

$$= \begin{vmatrix} \hat{i} & \hat{j} & \hat{k} \\ \frac{\partial}{\partial x} & \frac{\partial}{\partial y} & \frac{\partial}{\partial z} \\ A_x + B_x & A_y + B_y & A_z + B_z \end{vmatrix}$$

$$= \left[\frac{\partial}{\partial z}(A_z + B_z) - \frac{\partial}{\partial z}(A_y + B_y)\right]\hat{i} + \left[\frac{\partial}{\partial z}(A_x + B_x) - \frac{\partial}{\partial x}(A_z + B_z)\right]\hat{j}$$

$$+ \left[\frac{\partial}{\partial x}(A_y + B_y) - \frac{\partial}{\partial y}(A_x + B_x)\right]\hat{k}$$

Re-arranging the terms

$$= \left(\frac{\partial A_z}{\partial y} - \frac{\partial A_y}{\partial z}\right)\hat{i} + \left(\frac{\partial A_x}{\partial z} - \frac{\partial A_z}{\partial x}\right)\hat{j} + \left(\frac{\partial A_y}{\partial x} - \frac{\partial A_x}{\partial y}\right)\hat{k} + \left(\frac{\partial B_z}{\partial y} - \frac{\partial B_y}{\partial z}\right)\hat{i}$$

$$+ \left(\frac{\partial B_x}{\partial z} - \frac{\partial B_z}{\partial x}\right)\hat{j} + \left(\frac{\partial B_y}{\partial x} - \frac{\partial B_x}{\partial y}\right)\hat{k}$$

$$= \nabla \times \vec{A} + \nabla \times \vec{B}$$

Example 51. *If* $\vec{A} = x^2y\,\hat{i} - 2xz\,\hat{j} + 2yz\,\hat{k}$, *find curl curl* $\vec{A}$.

Solution. Curl curl $\vec{A} = \nabla \times (\nabla \times \vec{A})$

$$= \nabla \times \begin{vmatrix} \hat{i} & \hat{j} & \hat{k} \\ \frac{\partial}{\partial x} & \frac{\partial}{\partial y} & \frac{\partial}{\partial z} \\ x^2y & -2xz & 2yz \end{vmatrix}$$

$$= \nabla \times \left[\left\{ \frac{\partial}{\partial y}(2yz) - \frac{\partial}{\partial z}(-2xz) \right\} \hat{i} + \left\{ \frac{\partial}{\partial z}(x^2 y) - \frac{\partial}{\partial x}(2yz) \hat{j} + \left\{ \frac{\partial}{\partial x}(-2xz) - \frac{\partial}{\partial y}(x^2 y) \right\} \hat{k} \right]$$

$$= \nabla \times \left[(2z + 2x)\hat{i} + 0\hat{j} + \left(-2z - x^2\right)\hat{k} \right]$$

$$= \begin{vmatrix} \hat{i} & \hat{j} & \hat{k} \\ \dfrac{\partial}{\partial x} & \dfrac{\partial}{\partial y} & \dfrac{\partial}{\partial z} \\ 2z + 2x & 0 & -x^2 - 2z \end{vmatrix} = (2 + 2x)\hat{j}$$

Example 52. A rigid body is rotating with a uniform angular velocity about an axis passing through it. Show that curl of linear velocity of the particles of body is equal to twice the angular velocity.

(M.D.U. 1983)

Solution. Let the rigid body be rotating with angular velocity $\vec{\omega}$ about y-axis. O is taken as origin. Then any particle at *P* (Fig. 1·35) moves in a circular path about the axis with a tangential linear velocity $v = \omega r$ where $\vec{r}$ is position vector for *P*. In vector form.

$$\vec{v} = \vec{\omega} \times \vec{r}$$

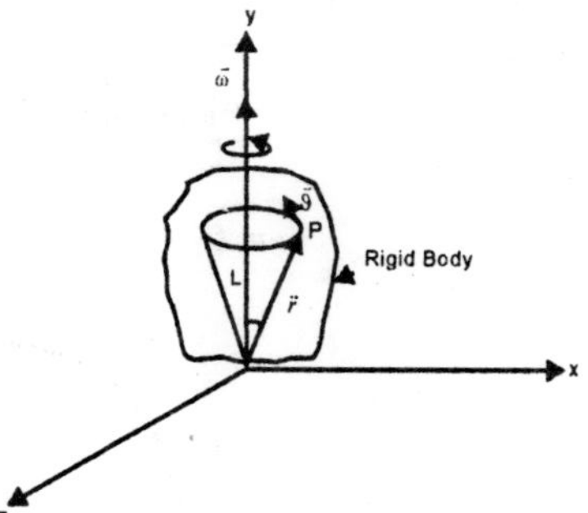

Now curl $\vec{v} = \nabla \times \vec{v}$

$$= \nabla \times (\vec{\omega} \times \vec{r})$$

$$= \nabla \times \begin{vmatrix} \hat{i} & \hat{j} & \hat{k} \\ \varpi_x & \varpi_y & \varpi_z \\ x & y & z \end{vmatrix}$$

$$= \nabla \times \left[\omega_y z - \omega_x y)\hat{i} + (\omega_z x - \omega_x z)\hat{j} + (\omega_x y - \omega_y x)\hat{k} \right]$$

$$= \begin{bmatrix} \hat{i} & \hat{j} & \hat{k} \\ \dfrac{\partial}{\partial x} & \dfrac{\partial}{\partial y} & \dfrac{\partial}{\partial z} \\ \omega_y z - \omega_z y & \omega_z x - \omega_x z & \omega_x y - \omega_y x \end{bmatrix}$$

$$= \left[\frac{\partial}{\partial y}(\omega_x y - \omega_y x) - \frac{\partial}{\partial z}(\omega_z x - \omega_x z) \right] \hat{i}$$

$$+\left[\frac{\partial}{\partial z}(\omega_y z-\omega_z y)-\frac{\partial}{\partial x}(\omega_x y-\omega_y x)\right]\hat{j}+\left[\frac{\partial}{\partial x}(\omega_z x-\omega_x y)-\frac{\partial}{\partial x}(\omega_z y-\omega_x x)\right]\hat{k}$$

$$=(\omega_x+\omega_x)\hat{i}+(\omega_y+\omega_y)\hat{j}+(\omega_z+\omega_z)\hat{k}$$

($\because$ $\vec{\omega}$ is constant, so $\varpi_x \varpi_y \varpi_z$ are contants)

$$=2(\omega_x\hat{i}+\omega_y\hat{j}+\omega_z\hat{k})$$

$$=2\vec{\omega}$$

This problem also indicates that curl of a vector field has something to do with rotational properties of the field.

Example 53. *Find the surface integral of a curl of a vector field when the surface is closed.*

Solution. Let S be any closed surface and let it cut open by any plane into portions S_1 and S_2 as shown in Fig. 1·36. If 'C' be the common curve bounding both the surfaces, then

$$\int_S curl\,\vec{A}\,.\,\vec{dS}=\int_{S1} curl\,\vec{A}\,.\,\vec{dS}+\int_{S2} curl\,\vec{A}\,.\,\vec{dS}$$

$$=\int_C \vec{A}\,.\,\vec{dl}-\int_C \vec{A}\,.\,\vec{dl} \text{ (By stoke's Theorem)}$$

$$=0.$$

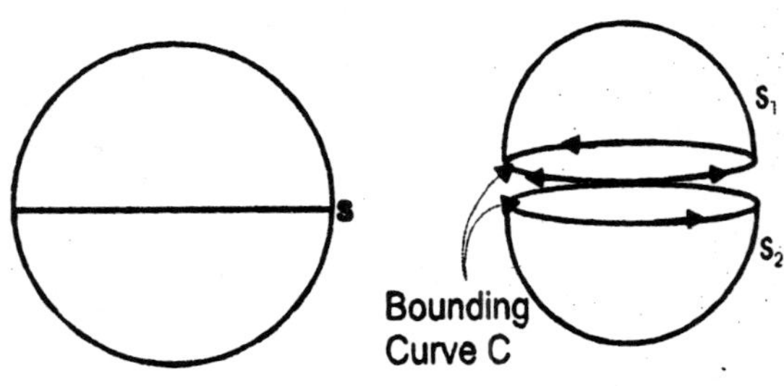

Fig 1.45

The second integral is taken negative because it is traced in the opposite direction as shown in Fig. 1·45.

Thus if S is closed surface, then $\int_S curl\,\vec{A}\,.\,\vec{dS}=0.$

Example 54. *Verify Stoke's theorem for the vector field $\vec{A}=y\hat{i}-x\hat{j}$ where C is boundary of the circle of radius 1 unit with centre at the origin in xy -plane.*

Solution. As C is the boundary of the circle in xy plane of radius one, and centre at the origin O, then the parametric equations of C are

$$x = \cos\theta$$

$$y = \sin\theta$$

as shown in Fig. 1·46.

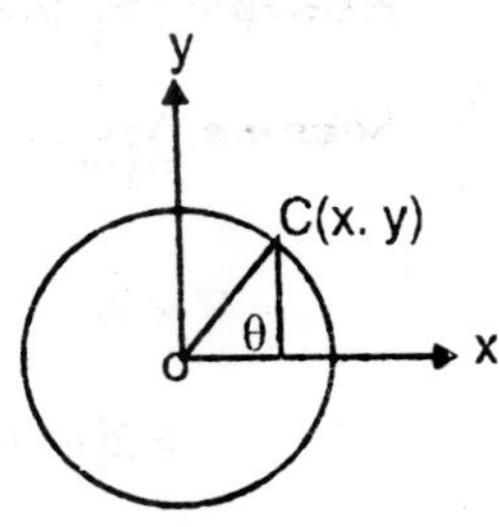

Fig 1.46

Then $dy=\cos\theta$, $dx=-\sin\theta$

where θ varies from 0 to 2π.

Accrding to Stoke's theorem $\oint_C \vec{A} \cdot \vec{dl}$

$$= \iint_S (\nabla \times \vec{A}) \cdot \vec{dS}$$

Then $$= \oint_c \vec{A} \cdot \vec{dr} = \oint_c (y\hat{i} - x\hat{j}).(\hat{i}\,dx + \hat{j}\,dy)$$

$$\oint_c ydx - xdy = \int_0^{2\pi} (\sin\theta(-\sin\theta)d\theta \cos\theta\cos\theta d\theta)$$

$$= \int_0^{2\pi} (-\sin^2\theta - \cos^2\theta) d\theta = -\int_0^{2\pi} d\theta = -2\pi$$

Now

$$\nabla \times \vec{A} = \begin{vmatrix} \hat{i} & \hat{j} & \hat{k} \\ \frac{\partial}{\partial x} & \frac{\partial}{\partial z} & \frac{\partial}{\partial z} \\ y & -x & 0 \end{vmatrix} = -2\,\hat{k}$$

Also $$\iint_s (\nabla \times \vec{A}) .\vec{ds} = \iint \nabla \times (y\hat{i} - x\hat{j}).\vec{ds}$$

$$= \iint_S -2\hat{k} \cdot \vec{ds}$$

$= \iint -2\hat{k} \cdot \hat{k}\, ds$ (as the area element $\vec{ds}$ of a circle is traced counter-clockwise in xy plane is along z-direction, so $\vec{ds} = \hat{k}\,ds$

$$= -2\iint_S ds$$

$$= -2\pi (\because \iint ds = \pi = \text{area of circle of unit radius})$$

Hence Stoke's theorem is verified.

Example 55. *Prove that vector* $\vec{A} = \frac{\vec{r}}{r}$ is irrotational, where $\vec{r}$ is position vector.

Solution. Vector $\vec{A}$ is irrotational if $\nabla \times \vec{A} = 0$

$$\therefore \quad \nabla \times \vec{A} = \nabla \times \frac{\vec{r}}{r} = \nabla \times \left(\frac{\hat{i}x + \hat{j}y + \hat{k}z}{\sqrt{x^2 + y^2 + z^2}} \right)$$

$$= \begin{vmatrix} \hat{i} & \hat{j} & \hat{k} \\ \frac{\partial}{\partial x} & \frac{\partial}{\partial z} & \frac{\partial}{\partial z} \\ \frac{x}{r} & \frac{y}{r} & \frac{z}{r} \end{vmatrix}$$

$$= \hat{i}\left[\left(\frac{\partial}{\partial y}\left(\frac{z}{r}\right) - \frac{\partial}{\partial z}\left(\frac{y}{r}\right)\right)\right] + \hat{j}\left[\frac{\partial}{\partial z}\left(\frac{x}{r}\right) - \frac{\partial}{\partial y}\left(\frac{z}{r}\right)\right] + \hat{k}\left[\frac{\partial}{\partial x}\left(\frac{y}{r}\right) - \frac{\partial}{\partial y}\left(\frac{x}{y}\right)\right]$$

Considering the first term

$$= \frac{\partial}{\partial y}\left(\frac{z}{r}\right) - \frac{\partial}{\partial z}\left(\frac{y}{r}\right)$$

$$= \frac{\partial}{\partial y}\left(\frac{z}{\sqrt{x^2 + y^2 + z^2}}\right) - \frac{\partial}{\partial z}\left(\frac{y}{\sqrt{x^2 + y^2 + z^2}}\right)$$

$$= z\left(-\frac{1}{2}\right)\left(x^2 + y^2 + z^2\right)^{-3/2} 2y - y\left(-\frac{1}{2}\right) \times \left(x^2 + y^2 + z^2\right)^{-3/2} 2z$$

$$= -yz\left(x^2 + y^2 + z^2\right)^{-3/2} + yz\left(x^2 + y^2 + z^2\right)^{-3/2} = 0$$

Similarly, second term = 0 and third term = 0.

So $\nabla \times \vec{A} = 0$

Hence, $\vec{A}$ is irrotational vector.

Example 56. *if $\vec{A}$ and $\vec{B}$ are irrotational vectors, prove that $\vec{A} \times \vec{B}$ is solenoidal.*

Solution. As $\vec{A}$ and $\vec{B}$ are irrotational, so

$$\nabla \times \vec{A} = 0 \quad and \quad \nabla \times \vec{B} = 0$$

$\vec{A} \times \vec{B}$ will be solenooidal if $\nabla.(\vec{A} \times \vec{B}) = 0$

$$\nabla.(\vec{A} \times \vec{B}) = \nabla.\begin{vmatrix} \hat{i} & \hat{j} & \hat{k} \\ A_x & A_y & A_z \\ B_x & B_y & B_z \end{vmatrix} = \nabla.\ \left[(A_y B_z - A_z B_y)\hat{i}\right.$$

$$\left. +(A_z B_x - A_x B_z)j + (A_x B_y - A_y B_x)\hat{k}\right]$$

$$= \left(\hat{i}\frac{\partial}{\partial x} + j\frac{\partial}{\partial y} + k\frac{\partial}{\partial z}\right)\left\{(A_y B_z - A_z B_y)\hat{i} + (A_z B_x - A_x B_z)\hat{j} + (A_x B_y - A_y B_x)\hat{k}\right\}$$

$$= \frac{\partial}{\partial x}(A_y B_z - A_z B_y) + \frac{\partial}{\partial y}(A_z B_x - A_x B_z) + \frac{\partial}{\partial z}(A_x B_y - A_y B_x)$$

$$= A_y \frac{\partial B_z}{\partial x} + \frac{\partial A_x}{\partial x} B_z - A_z \frac{\partial B_y}{\partial x} - \frac{\partial A_z}{\partial x} B_y + A_z \frac{\partial B_z}{\partial y}$$

$$+ \frac{\partial A_z}{\partial y} B_x - A_x \frac{\partial B_z}{\partial y} - \frac{\partial A_x}{\partial y} B_z + A_x \frac{\partial B_y}{\partial z} + \frac{\partial A_x}{\partial z} B_y - A_y \frac{\partial B_x}{\partial z} - \frac{\partial A_y}{\partial z} B_x$$

Re-arranging,

$$= B_x\left(\frac{\partial A_z}{\partial y} - \frac{\partial A_y}{\partial x}\right) + B_y\left(\frac{\partial A_x}{\partial z} - \frac{\partial A_z}{\partial x}\right) + B_z\left(\frac{\partial A_y}{\partial x} - \frac{\partial A_x}{\partial z}\right) + A_x\left(\frac{\partial B_y}{\partial z} - \frac{\partial B_z}{\partial y}\right)$$

$$+ A_y\left(\frac{\partial B_z}{\partial x} - \frac{\partial B_x}{\partial z}\right) + A_x\left(\frac{\partial B_x}{\partial y} - \frac{\partial B_y}{\partial x}\right)$$

$$= \vec{B}\,.\,(\nabla \times \vec{A}) - \vec{A}.\,(\nabla \times \vec{B})$$

$=0-0=0$

($\therefore$ $\vec{A}$ and $\vec{B}$ are irrotational vectors)

Hence $\vec{A} \times \vec{B}$ is soienoidal.

Example 57. *Find the constants a , b and c so that*

$$\vec{A} = (x + 2y + cz)\hat{i} + (ax + 3y - z)\hat{j} + (4x + by + 2z)\hat{k}$$

is irrotational

Solution. Vector $\vec{A}$ is Irrotational if $\nabla \times \vec{A} = 0$

$$\nabla \times \vec{A} = \begin{vmatrix} \hat{i} & \hat{j} & \hat{k} \\ \frac{\partial}{\partial x} & \frac{\partial}{\partial y} & \frac{\partial}{\partial z} \\ x+2y+cz & ax-3y-z & 4x+by+2z \end{vmatrix} = 0$$

$$= \left\{\frac{\partial}{\partial y}(4x+by+2z) - \frac{\partial}{\partial z}(ax-3y-z)\right\}\hat{i} + \left\{\frac{\partial}{\partial z}(x+2y+cz) - \frac{\partial}{\partial x}(4x+by+2z)\right\}\hat{j}$$

$$+\left\{\frac{\partial}{\partial x}(ax-3y-z) - \frac{\partial}{\partial y}(x+2y+cz)\right\}\hat{k}$$

$$= (b+1)\hat{i} + (c-4)\hat{j} + (a-2)\hat{k} = 0$$

Putting each term equal to zero, we have b=–1, c=4, a=2.

Green's Theorem

Green's Theorem is a special case of Gauss's Theorem, so by the use of Gauss's Theorem, we can derive Green's theorem as follow: Taking the vector field $\vec{A}$ as the product of one scalar *u* and the gradient of another scalar w, i,e.,

$$\vec{A} = u \text{ grad } w = u\nabla w = \hat{i}u\frac{\partial w}{\partial x} + \hat{j}u\frac{\partial w}{\partial y} + \hat{k}u\frac{\partial w}{\partial z}$$

Here we assume that *u* and *w* are continuous scalar functions with continuous derivative of second order at least.

Now $$\nabla . \vec{A} = \frac{\partial A_x}{\partial x} + \frac{\partial A_y}{\partial y} + \frac{\partial A_z}{\partial z}$$

$$= \frac{\partial}{\partial x}\left[u\frac{\partial w}{\partial x}\right] + \frac{\partial}{\partial y}\left[u\frac{\partial w}{\partial y}\right] + \frac{\partial}{\partial z}\left[u\frac{\partial w}{\partial z}\right]$$

$$= \frac{\partial u}{\partial x}.\frac{\partial w}{\partial x} + u\frac{\partial^2 w}{\partial x^2} + \frac{\partial u}{\partial y}.\frac{\partial w}{\partial y} + u\frac{\partial^2 w}{\partial y^2} + \frac{\partial u}{\partial z}.\frac{\partial w}{\partial z} + u\frac{\partial^2 w}{\partial z^2}$$

Re-arranging the terms

$$= u\left[\frac{\partial^2 w}{\partial x^2} + \frac{\partial^2 w}{\partial y^2} + \frac{\partial^2 w}{\partial z^2}\right] + \frac{\partial u}{\partial x}.\frac{\partial w}{\partial x} + \frac{\partial u}{\partial y}.\frac{\partial w}{\partial y} + \frac{\partial u}{\partial z}.\frac{\partial w}{\partial z}$$

$$= u\nabla^2 w + \nabla u.\nabla w$$

Putting these values for $\vec{A}$ and $\nabla.\vec{A}$ in the equation of Gauss's theorem, i.e.,

$$\iiint_V (\nabla.\vec{A})dv = \iint_s \vec{A}.d\vec{s}$$

we get

$$\iiint_v \left(u\nabla^2 w + \nabla u.\nabla w\right)dv = \iint_s (u\nabla w).d\vec{s} \tag{1·78}$$

This transformation is called first of Green's theorem. In equation (1·78) if we interchange the function u and w, we get

$$\iiint_v \left(w\nabla^2 u + \nabla w.\nabla u\right)dv = \iint_s (w\nabla u).d\vec{s}$$

Subtracting (1·79) from (1·78), we get (1·79)

$$\iiint_v \left(u\nabla^2 w - w\nabla^2 u\right)dv = \iint_s (u\nabla w - w\nabla u).d\vec{s}$$

This trandfomation is referred to as the second form of Green's theorem.

Equations (1·78) and (1·80) are of great importance in the field of electrodyamins.

Green's Theorem in the plane

Statement. If 'S' is a closed region ogf the x-y plane bounded by a simple closed curve 'C' and if M and N are continuous scalar functions of x and y having continuous derivative in S, then

$$\oint_c (Mdx + Ndy) = \iint_s \left[\frac{\partial N}{\partial x} - \frac{\partial M}{\partial y}\right]dx\,dy$$

where C is transversed in the positive (anti-clockwise direction.

Green's theorem in the plane is a special case of Stoke's theorem, so by its use we can derive Green's theorm in the plane as follows:

Stoke's theorem States that

$$\oint_c \vec{A}\,.\,\vec{dr} = \iint_s \left(\nabla \times \vec{A}\right).\hat{n}\,ds$$

If the curve C is plotted in xy plane then $\hat{n} = \hat{k}$ and $\vec{A} = A_x\hat{i} + Ay\hat{j}$ for this plane.

So
$$\oint_c \vec{A}\,\vec{dr} = \iint_s \left(\nabla \times \vec{A}\right).\hat{k}\,ds$$

Let A_x=M and A_y=N so that $\vec{A} = M\hat{i} + N\hat{j}$, then

$$\nabla \times \vec{A} = \nabla \times \left(M\hat{i} + N\hat{j}\right)$$

$$= \begin{vmatrix} \hat{i} & \hat{j} & \hat{k} \\ \dfrac{\partial}{\partial x} & \dfrac{\partial}{\partial y} & \dfrac{\partial}{\partial z} \\ M & N & 0 \end{vmatrix}$$

$$= -\frac{\partial N}{\partial z}\hat{i} + \frac{\partial M}{\partial z}\hat{j} + \left[\frac{\partial N}{\partial x} - \frac{\partial M}{\partial y}\right]\hat{k}$$

So $$\left(\nabla \times \vec{A}\right).\hat{k}\right) = \frac{\partial N}{\partial x} - \frac{\partial M}{\partial y}$$

and $$\vec{A}\,.\,\vec{dr} = \left(M\hat{i} + N\hat{j}\right).(\hat{i}dx + \hat{j}dy + \hat{k}dz)$$

$$= Mdx + Ndy$$

So Stoke's theorem states that

$$= \oint_c \left(Mdx + Ndy\right) = \iint_s \left[\frac{\partial N}{\partial x}\frac{\partial M}{\partial y}\right] dx\, dy$$

which is Green's theorem in a plane.

LONG ANSWER QUESTIONS

1. The resultant of two vector p and q is a vector r. After reversing to direction of q their resultant is another vector D. prove that

   $(R^2 + D^2 = 2\ (P^2\ + Q^2)$.

2. The sum and difference of two vectors are perpendicular to each other. Show that their magnitude are equal. Draw a diagram showing these vectors.
3. What do you understand by the scalar or dot product of two vectors. Prove that the scalar product of two vectors is distributive.
4. Define a unit vector. If $\hat{i}, \hat{j}, \hat{k}$ are unit vectors along x, y and z axes repectively, prove that

   $\hat{i}.\hat{i} = 1,\ \hat{i}.\hat{j} = 0,\ \hat{i} \times \hat{i} = 0$ and $\hat{i} \times \hat{j} = \hat{k}$.

5. If $\vec{A}$ and $\vec{B}$ two vectors given by

   $\vec{A} = A_x\hat{i} +| A_y\hat{j} + A_z\hat{k}$ and $\vec{B} = B_x\hat{i} +| B_y\hat{j} + B_z\hat{k}$

   Prove that $\vec{A}\,.\,\vec{B} = A_xB_x + A_yB_y + A_zB_z$ and $\vec{A} \times \vec{B} = -\vec{B} \times \vec{A}$.

6. What do you mean by the vector product of two vectors? Show that magnitude of this product gives the area of the parallelogram fprmed by the two vectors as sides.

(K.U. 1975, 79)

7. Express vector product of two vectors in terms of certesian co-ordinates,

(K.U. 1976)

8. Prove that $\vec{A}.\left(\vec{B}\times\vec{C}\right)$ represents the volume of a parallelopiped formed with $\vec{A}$, $\vec{B}$ and $\vec{C}$ as sides.

9. What do you mean by scalar and vector fields. Give at least two examples of each.

10. What do you understand by the gradient of a scalar function? physical significance.

(M.D.U 1982; K.U. 1995S; 1993A)

11. Explain clearly what do you understand by divergence and curl of a vector field $\vec{F}$. Obtain expressions in certesian co-ordinates for (i) div $\vec{F}$. (ii) curl $\vec{F}$.

(K.U. 1989A, 1994S; M.D.U 1985)

12. Discuss physical significance of $\nabla\times\vec{V}$ for linear velocity vector $\vec{V}$.

(K.U. 1979, 79S)

13. Prove that if the triple product of these vectors vanishes, the vectors are co-planar.

14. Discuss the physical significance of curl of a vector field.

(K.U. 1994A, 1995A)

15. Discuss the physical significance of divergence of a vector field.

(K.U. 1994A, 1995A)

16. Explain line integral, surface integral and volume integral.

17. State and prove Gauss's Divergence theorem.

(K.U. 1995A,1994S, 1992A)

18. State and prove Stoke's theorem. Give its practical applications.

(M.D.U 1980; K.U. 1993S, 1994A, 1995S)

19. State and prove Green's theorem in the plane.

(K.U.1992 A)

20. If $\vec{F}$ is a continuous differentible vector point function and S is a closed surface enclosed the volume $\vec{V}$, then

$$\iint_S \vec{F}.\hat{n}\,ds = \int \iint_V \left(\nabla.\vec{F}\right)dv.$$

21. If $\vec{V}$ (x, y, z) is a continuous differentiable vector point function show that $\nabla.\vec{V}$ is a scalar quantity.

(Jammu Univ. 1986)

22. The line integral of a vector equal to grad of a scalar over a closed path is zero.

(M.D.U. 1986; K.U. 1981, 1992 A)

23. Show that the curl of gradient of a scalar is always zero.

24. Show that 'Line integral' of conservative vector over a closed path is zero.

(K. U. 1988 A)

25. What are special features of a vector obtained as gradient of scalar function?

(M.D.U. 1985)

26. If the line integral of a vector $\vec{A}$ around a closed curve is equal to surface integral of the vector $\vec{B}$ taken over the surface bounded by the given closed curve, show that $\vec{B} = \text{curl}\,\vec{A}$.

27. Explain the situations

(i) div $\vec{F} = 0$ but curl $\vec{F} \neq 0$ (ii) div $\vec{F} \neq 0$ but curl $\vec{F} = 0$

(iii) curl $\vec{F} = 0$ and div $\vec{F} = 0$ (vi) curl $\vec{F} \neq 0$ and div $\vec{F} \neq 0$.

(K.U.1986)

28. Explain what do you understand by solenoidal and irrotational fields.

(K.U. 1988)

30. If $\vec{A} = A_x\hat{i} + A_y\hat{j} + A_z\hat{k},$ prove that $A = \sqrt{\vec{A}\,.\,\vec{A}}$

$$A = \sqrt{A_x^2 + A_y^2 + A_z^2}.$$

31. Find the angle between

$\vec{A} = 2\hat{i} + 2\hat{j} - 2\hat{k}$ and $\vec{B} = 6\hat{i} + 3\hat{j} - 2\hat{k}.$ **[Ans. 79°1']**

32. Find the work done in moving an object along a vector $\vec{r} = 3\hat{i} + 2\hat{j} - 5\hat{k}$ if the applied force is $\vec{F} = 2\hat{i} + \hat{j} - \hat{k}.$

[Ans. 9 units]

33. If $\vec{A} = 2\hat{i} - 3\hat{j} - \hat{k}$ and $\vec{B} = \hat{i} - 4\hat{j} - 2\hat{k},$ find

(a) $(\vec{A} \times \vec{B})$ (b) $\vec{B} \times \vec{A}$ (c) $(\vec{A} + \vec{B}) \times (\vec{A} - \vec{B}).$

[Ans. (a) $10\hat{i} + 3\hat{j} + 11\hat{k}$ **(b)** $-10\hat{i} - 3\hat{j} + 11\hat{k}$ **(c)** $-20\hat{i} - 5\hat{j} - 22\hat{k}$ **]**

34. If $\vec{A} = 3\hat{i} - \hat{j} + 2\hat{k}$, $\vec{B} = 2\hat{i} + \hat{j} + \hat{k}$ and $\vec{C} = \hat{i} - 2\hat{j} + 2\hat{k}$,

(a) $(\vec{A} \times \vec{B}) \times \vec{C}$ (b) $\vec{A} \times (\vec{B} \times \vec{C})$.

[Ans. (a) $12\hat{i} + 11\hat{j} + 5\hat{k}$, **(b)** $15\hat{i} + 15\hat{j} - 15\hat{k}$ **]**

35. For what value of a are $\vec{A} = a\hat{i} - 2\hat{j} + \hat{k}$ and $\vec{B} = 2a\hat{i} - a\hat{j} + 4\hat{k}$ Perpendicular?

[Ans. a = 2, – 1]

36. If $\vec{A} = 4\hat{i} - \hat{j} + 3\hat{k}$ and $\vec{B} = 2\hat{i} + \hat{j} - 3\hat{k}$, find a unit vector perpendicular to both $\vec{A}$ and $\vec{B}$.

$$\left[\textit{Ans.} \pm \frac{(0\hat{i} + 3\hat{j} + \hat{k})}{\sqrt{12}} \right]$$

Hint. $\vec{A} \times \vec{B}$ is normal to both $\vec{A}$ and $\vec{B}$. A unit vector parallel to $\vec{A} \times \vec{B}$ is $\dfrac{\vec{A} \times \vec{B}}{|\vec{A} \times \vec{B}|}$.

37. Find the condition that vectors $\vec{A} = a_1\hat{i} + b_1\hat{j} + c_1\hat{k}$ and $\vec{B} = a_2\hat{i} + b_2\hat{j} + c_2\hat{k}$ are parallel.

$$\left[\textit{Ans.} \frac{a_1}{a_2} = \frac{b_1}{b_2} = \frac{c_1}{c_2} \right]$$

38. Given are two vectors such that $\vec{A} + \vec{B} = 11\hat{i} - \hat{j} + 5\hat{k}$ and $\vec{A} + \vec{B} = -5\hat{i} + 11\hat{j} + 9\hat{k}$. Find (a) $\vec{A}$ and $\vec{B}$, (b) angle included between $\vec{A}$ and $\vec{A} + \vec{B}$.

$$\begin{bmatrix} \textbf{Ans. (a) } 3\mathbf{i} + 5\mathbf{j} + 7\mathbf{k},\ 8\mathbf{i} - 6\mathbf{j} - 2\mathbf{k} \\ \textbf{(b) } \cos^{-1}(\cdot 57) \end{bmatrix}$$

40. Show that vectors, $\vec{A} = 3\hat{i} - 2\hat{j} + \hat{k}$, $\vec{B} = \hat{i} - 3\hat{j} + 5\hat{k}$ and $\vec{C} = 2\hat{i} + \hat{j} - 4\hat{k}$ form a right angled triangle.

[Hint. Prove $|\vec{C}|^2 = |\vec{A}|^2 + |\vec{B}|^2$. **(K.U. 1992 S)**

41. If $\vec{A}$, $\vec{B}$, $\vec{C}$ are there vectors, prove that

$$(\vec{A} \times \vec{B}).(\vec{B} \times \vec{C}) \times (\vec{C} \times \vec{A}) = [\vec{A}\ \vec{B}\ \vec{C}]^2$$

42. Show that

$$\vec{A}\times\left[\vec{B}\times\left(\vec{C}\times\vec{D}\right)\right]=\left(\vec{A}\times\vec{C}\right)-\left(\vec{B}.\vec{D}\right)-\left(\vec{A}\times\vec{D}\right)\left(\vec{B}.\vec{C}\right).$$

43. If $\vec{r}$ is the position vector at a point at which the value of the gravitational potential due to mass m at origin is given by $\frac{-m}{r}$, find the gravitational field intensity at the point *P*.

$\left[\textbf{Ans.} \frac{\mathbf{m\vec{r}}}{\mathbf{r^3}}\right]$

[Hint. Position Vector $\vec{r}=x\hat{i}+y\hat{j}+z\hat{k}$ find grad $\left(\frac{-m}{r}\right)$]

44. If $\vec{A}=xz^3\hat{i}-2x^2yz\hat{j}+2yz^4\hat{k}$, find curl $\vec{A}$ at (1, –1, 1).

$\left[\textbf{Ans. } 3\hat{\mathbf{j}}+4\hat{\mathbf{k}}\right]$

45. Show that is a vector perpendicular to surface $\phi(x,y,z)=$ C where C is constant.

46. If $\vec{A}=2yz\hat{i}-x^2y\hat{j}+xz^2\hat{k},\ \vec{B}=x^2\hat{i}+yz\hat{j}+xy\hat{k}$ and $\phi = 2x^2yz^3$, find

(a) $\left(\vec{A}.\nabla\right)\phi$ (b) $\vec{A}.\nabla\phi$ (c) $\left(\vec{A}\times\nabla\right)\phi$ (d) $\vec{A}\times\nabla\phi$ (e) $\left(\vec{B}.\nabla\right)\vec{A}$.

[Ans. (a) 8xy2z4 –2x4yz3+6x3yz4

(b) 8xy2z4–2x4yz3+6x2yz4

(c) and (d)–$(6x^4y^2z^2+2x^3z^5)\,\hat{i}+(4x^2yz^5-12x^2y^2z^3)\,\hat{j}+(4x^2yz^4+4x^3y^2z^3)\,\hat{k}$

(e) $[(2yz^2-2xy^2)\,\hat{i}-(2x^3y+x^2yz)\,\hat{j}+(x^2z^2-2x^3yz)\,\hat{k}\,]$

47. Show that $r^n\,\vec{r}$ is an irrotational vector for any value of n but is solenoidal only for n = –3 where $\vec{r}$ is position vector and is its magnitude.

48. Find the angle between the surface $x^2+y^2+z^2=9$ and $z=x^3+y^2-3$ at the point (2–1, 2). **[Ans. 54°-25']**

49. Given $\phi = 2x^3y^2z^4$, find div grad ϕ.

[Ans. $12xy^2z^4+4x^3z^4+24x^3y^2z^2$]

50. If $\phi = 3x^2z-y^2z^3+4x^3y+2x-3y-5$, find $\nabla^2\phi$

[Ans. $6z+24xy-2z^3-6y^2z$]

51. Evaluate curl grad u, where u is scalar function.

(K.U. 1986 A) [Ans. 0]

52. Evaluate grad (1/r) where $\vec{r}$ is position vector.

$$\left[\text{Ans. } \frac{-\vec{1}}{r^3}\right]$$

(M.D.U. 1985; K.U. 1986S)

53. Evaluate $\nabla \cdot \left[r\nabla\left(\frac{1}{r^3}\right)\right]$

[Ans. $3r^{-4}$]

54. Evaluate $\nabla^2 r$ where $r = \sqrt{x^2 + y^2 + z^2}$.

$\left[\text{Ans. } \frac{2}{r}\right]$ **(K.U. 1988s; 1992A)**

55. If $\vec{A} = \frac{r}{r}$, find grad div $\vec{A}$.

[Ans. $-2r^{-3}\vec{r}$]

56. Evalute ∇^2(In r).

[Ans. $1/r^2$]

57. Evaluate $\int\int_s \vec{F}.d\vec{s}$ where $\vec{F} = 4xz\hat{i} - y\hat{j} + yz\hat{k}$ and 's' is the surface bounded by x = 0, x = 1, y = 0, y = 1, z = 0, z = 1. **[Ans. 1.5]**

$$\left[\text{H int.} \int\int_s \vec{F}.d\vec{s} = \int\int\int_v \text{div } \vec{F}dv = \int_{\alpha=0}^{1}\int_{y=0}^{1}\int_{z=0}^{1} \text{div } \vec{F}dx\, dy\, dz\right]$$

58. Verify Stoke's theorem for the function $\vec{F} = x^2\hat{i} + xy\hat{j}$ integrated round the square in the plane z = 0 whose sides are along the lines x=0, y = 0, x = a, y = a.

59. If $\vec{F} = 3xy\hat{i} + y^2\hat{j}$ evaluate $\int_C \vec{F}.d\vec{r}$ where C is given by curve in xy plane, $y = 2x^2$ from (0, 0) to (1, 2).

$$\left[\textit{Ans.} \frac{-7}{6}\right]$$

OBJECTIVE QUESTION

1. A differential volume in spherical coordinates is shown in Figure 3·26. For the volume element, match the items in the left column with those in the right column.

(a) dl from A to D	(i) $-r^2 \sin\theta\, d\theta\, d\phi\, a_r$
(b) dl from E to A	(ii) $-r \sin\theta\, dr\, d\phi\, a_\theta$
(c) dl from A to B	(iii) $r\, dr\, d\theta\, a\phi$
(d) dS for face EFGH	(iv) $dr\, a_r$
(e) dS for face AEHD	(v) $r\, d\theta\, a_\theta$
(f) dS for face ABFE	(vi) $r \sin\theta\, d\phi\, a_\phi$

2. If $r = xa_x + ya_y + za_z$, the position vector of point (x,y,z) and r=|r|, which of the following is incorrect?

(a) $\nabla r = r/r$ (b) $\nabla . r = 1$

(c) $\nabla^2 (r.r) = 6$ (d) $\nabla \times r = 0$

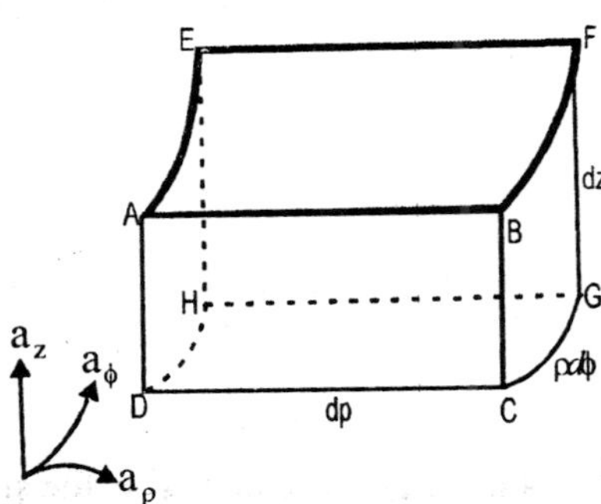

Fig 1.47. For Review Question 1

3. which of the folling is a meaningless combination?

(a) grad div (b) div curl

(c) curl grad (d) curl grad

(e) div curl

4. which of the following is zero?

(a) grad div (b) div grad

(c) curl grad (d) curl curl

5. Given field $A = 3x^2yz\, a_x + x^3z\, a_y + (x^3y - 2z)a_z$, it can be said that A is.

(a) Harmonic (b) Divergenceless

(c) Solenoidal (d) Rotational

(e) Conservative

6. The surface current density J in a rectangular waveguide is plotted in Figure 1·29. It is evident from the figure that Jdiverges at the top wall of the guide whereas it is divergenceless at the side wall.

(a) True (b) False

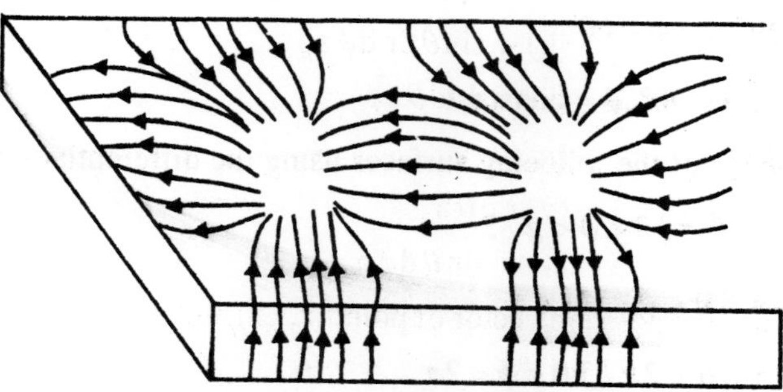

Fig 1.48. For Review Question 6.

7. Stokes's theorem is applicable only when a closed path exists and the vector field and its derivatives are continuous within the path.

(a) True

(b) False

(c) Not necessarily

8. If a vector field Q is solenoidal, which of these is true?

(a) $\oint_L Q.d\mathrm{l}=0$

(b) $\oint_S Q.d\mathrm{S}=0$

(c) $\nabla \times \mathrm{Q} =0$

(d) $\nabla \times \mathrm{Q} \neq 0$

(e) $\nabla^2 Q=0$

Answers: 1.1a-(vi), b-(vii), c-(v), d-(i), e-(ii), f-(iv), g-(ii), 1.2a-(vi), b-(v), c-(vii), d-(ii), e-(i), f-(iv), g-(ii), 1.3a-(v), b-(vi), c-(iv), d-(iii), e-(i), f-(ii), 1.4b, 1.4b, 1.5c, 1.6c, 1.7e, 1.8a, 1.9a, 1.10b.

CHECK YOURSELF

1. Using the differential length dl, find the length of each of the following curves:

   (a) $\rho = 3, \pi/4 < \phi < \pi/2, z = \text{constant}$

   (b) $r = 1, \theta = 30°, 0 < \phi < 60°$

   (c) $r = 4,\ 30° < \theta < 90°, \phi = \text{constant}$

2. Calculate the areas of the folloeing surfaces using the differential surface area ds:

   (a) $\rho = 2, 0 < z < 5, \pi/3 < \phi < \pi/2$

   (b) $z = 1, 1 < \rho < 3, 0 < \phi < \pi/4$

   (c) $r = 10, \pi/4 < \theta < 2\pi/3, 0 < \phi < 2\pi$

   (d) $0 < r, < 4, 60° < \theta < 90°, \phi = \text{constant}$

3. Use the differential volume dv to determine the volumes of the following regions:

   (a) $0 < x < 1, 1 < y < 2, -3 < z < 3$

   (b) $2 < \rho < 5, \pi/3 < \phi < \pi, -1 < z < 4$

   (c) $1 < r < 3, \pi/2 < \theta < 2\pi/3, \pi/6 < \phi < \pi/2$

4. Given that $\rho_s = x^2 + xy$, calculate $\int s\rho_s dS$ over the region $y \leq x$, $0 < x < 1$.

5. Given that $H = x^2 a_x + y^2 ay$, evalute $\int_L h.dl$, where L is along the curve $y = x^2$ from (0,0) to (1,1).

6. Find the volume cut from the sphere radius r=a by cone $\theta = \alpha$. Calculate the volume when $\alpha = \pi/3$ and $\alpha = \pi/2$.

7. If the integral $\int_A^B F.dl$ is regarded as the work done in moving a particle from A to B, find the work done by the force field.

$$F = 2xy\, a_x + (x^2 - z^2)a_y - 3xz^2\, a_z$$

   on a particle that travels from A(0, 0, 0) to B(2, 1,3) along

   (a) The segment (0, 0, 0) → (0, 1, 0) → (2, 1, 0) → (2, 1, 3).

   (b) The straight line (0, 0, 0) to (2, 1,3).

8. If

$$\vec{H} = (x - y)a_x + (x^2+zy)a_y+5yz\ a_z$$

evaluate $\int H.\ dl$ along the contour of Figure 3·28.

9. If V=(x+y)z, evaluate $\int_S VdS$, where S is the durface of the cylindrical wedge defined by $0<\phi<\pi/2$, $0<z<2$ and dS is normal to that surface.

10. Let A = $2xya_x + xza_y - ya_z$. Evaluate $\int A\ dv$ over:

(a) a rectangular region $0 \le x \le 2, 0 \le y \le 2, 0 \le z \le 2$.

(b) a cylindrical region $\rho \le 3, 0 \le z \le 5$.

(c) a spherical region $r \le 4$.

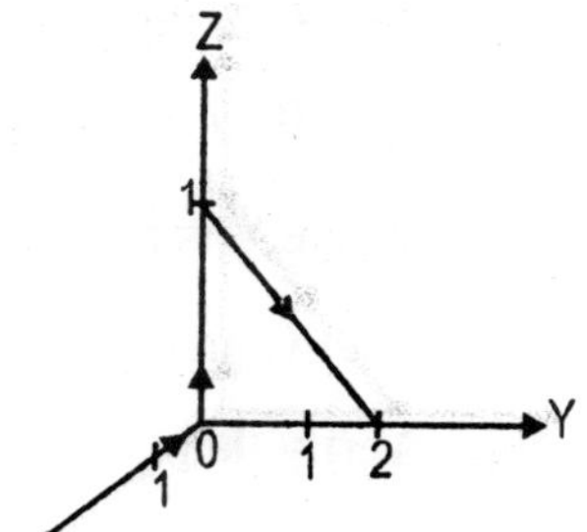

Fig 1.40. For Problem 8.

11. The acceleration of a particle is given by $a=2.4a_z$ m/s². The initial position of the particle is r=(0, 0,0), while its initial velocity is $v=-2a_x + 5a_z$ m/s. (a) Find the position of the particle at time $t = 1$. (b) Determine the velocity of the particle as a function of t.

12. Find the gradient of the these scalar fields:

(a) U = $4xz^2 + 3yz$

(b) W = $2\rho\,(z^2 + 1)\cos\phi$

(c) $H\ r^2 \cos\theta \cos\phi$

13. Determine the gradient of the followwing fields and compute its value at the specified point.

(a) V = $e^{(2x+3y)} \cos 5z$, (0.1, –0.2, 0.4)

(b) T = $5\rho e^{-2z} \sin\phi$, (2, π/3, 0)

(c) Q = $\dfrac{\sin\theta \sin\phi}{r^2}$, (1, π/6, π/2)

14. Determine the unit vector normal to $S(x, y, z) = x^2 + y^2 - z$ at point (1, 3, 0).

15. The temperature in an auditorium is given by $T=x2+y2-z$. A mosquito located at (1, 1, 2) in the auditorium desires to fly in such a direction that it will get warm as soon as possible. In what direction must it fly?

16. Find the divergence and curl of the following vectors:

(a) A = $e^{xy}\ a_x + \sin xy\ a_y + \cos^2 xz\ a_z$

(b) B = $\rho z^2 \cos\phi\ a\rho + z \sin^2\phi\ a_z$

(c) C = $r \cos\theta\ a_r\ \frac{1}{r} \sin\theta\ a_\theta + 2r^2\ son\theta\ a\phi$

17. Evaluate $\nabla \times A$ and $\nabla . \nabla \times A$ if:

(a) $A = x^2ya_x + y^2za_y - 2xza_z$

(b) $A = \rho^2 za_\rho + \rho^3 a_\phi + 3\rho z^2 a_z$

(c) $A = \dfrac{\sin\phi}{r^2} a_r - \dfrac{\cos\phi}{r^2}$

18. The heat flow vector $H = k\nabla T$, where T is the temperature and k is the thermal conductivity. Show that where

$$T = 50 \sin\frac{\pi x}{2} \cosh\frac{\pi x}{2}$$

then $\nabla^{*}. H = 0$.

19. (a) Prove that

$$\nabla . (VA) = V\nabla . A + A . \nabla V$$

where V is a scalar field and A is a vector field.

(b) Evaluate $\nabla . (VA)$ when $A = 2xa_y + 3ya_y - 4za_z$ and V=xyz.

20. Verify that

$$\nabla \times (VA) = V(\nabla \times A) + \nabla V \times A$$

where V and A are scalar and vector fields, respectively.

(b) Evaluate $\nabla \times (VA)$ when $V = \dfrac{1}{r^2}$ and $A = r\cos\theta\, a_r + r\sin\theta\, a_\theta + \sin\theta\cos\phi\, a_\phi$.

21. If $U = xz - x^2y + y^2z^2$, evaluate div grad U.

22. Show that ∇ in $\rho = \nabla \times \phi a_z$.

23. Prove that $\nabla\phi = \nabla \times \left(\dfrac{r\nabla\theta}{\sin\theta}\right)$.

24. Evaluate ∇V, $\nabla . \nabla V$, and $\nabla \times \nabla V$ if:

(a) $V = 3x^2y + xz$

(b) $V = \rho z \cos\phi$

(c) $V = 4r^2 \cos\theta$ son ϕ

25. If $r = xa_x + ya_y + za_z$ and $T = 2zya_x + xy^2a_y + x^2yza_z$ determine

(a) $(\nabla . r)T$ (b) $(r . \nabla)T$

(c) $\nabla . r (r.T)$ (d) $(r. \nabla) r^2$

26. If r = xax + yay + zaz is the position vector of point (x, y, z), r |r|, **and n is an integer,** show that:

    (a) $\nabla . r^n r = (n+3)r^n$

    (b) $\nabla \times r^n r = 0$

27. If r and r are as definded in the previous problem, prove that:

    (a) $\nabla (\text{In } r) = \frac{r}{r}$ (b) $\nabla^2 (\text{In } r) = \frac{1}{r^2}$

28. For each of the following scalar fields, find $\nabla^2 V$

    (a) $V_1 = x^3 + y3 + z^3$

    (b) $V_2 = \rho z^2 \sin^2\phi$

    (c) $V_3 = r^2(1 + \cos\theta \sin\phi)$

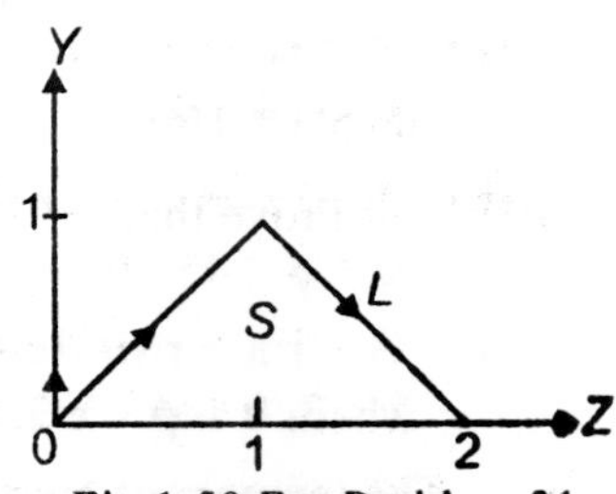

Fig 1.50 For Problem 31.

29. Find the Laplacian of the following scalar fields and compute the value at the specified point.

    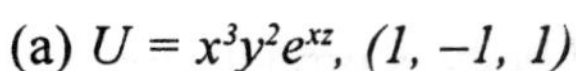
    (a) $U = x^3y^2e^{xz}$, (1, –1, 1)

    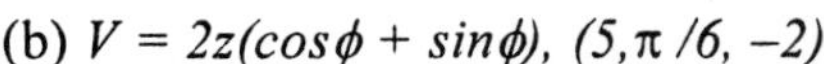
    (b) $V = 2z(\cos\phi + \sin\phi)$, (5, π /6, –2)

    (c) $W = e^{-r} \sin\theta \cos\phi$, (1, π/3, π/6)

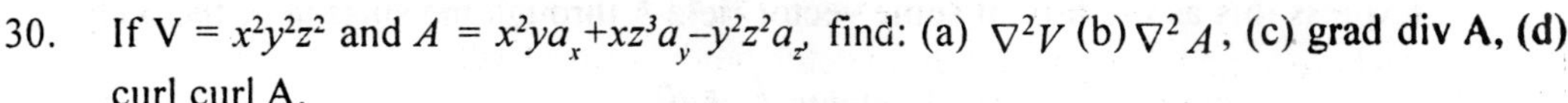
30. If $V = x^2y^2z^2$ and $A = x^2ya_x + xz^3a_y - y^2z^2a_z$ find: (a) $\nabla^2 V$ (b) $\nabla^2 A$, (c) **grad div A, (d)** curl curl A.

31. Given that $F = x^2ya_x - ya_y$, find

    (a) $\oint_L F.dl$ where L is shown in Figure 3·29.

    (b) $\int_S (\nabla \times F).dS$ where S is the area bounded by L.

    (c) Is Stokes's theorem satisfied?

32. Let $D = 2\rho z^2 a\rho + \rho \cos^2\phi \, a_z$. Evaluate

    (a) $\oint_S D.dS$

    (b) $\int_V \Delta . Ddv$

    over the region defined by $0 \le \rho \le 5, -1 \le z \le 1, 0 < \phi < 2\pi$.

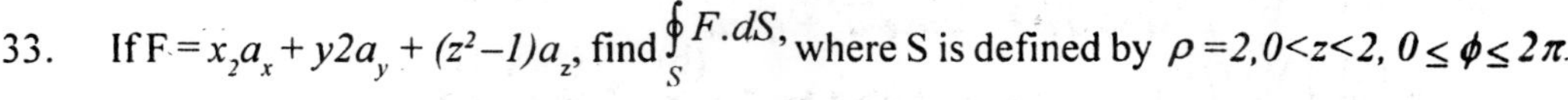
33. If $F = x_2a_x + y2a_y + (z^2 - 1)a_z$, find $\oint_S F.dS$, where S is defined by $\rho = 2, 0 < z < 2, 0 \le \phi \le 2\pi$.

34. (a) Given that A = $xya_x + yza_y + xza_z$, evaluate $\oint_S A\,.dS$, where S is the surface of thecube defined by

$$0 \le x \le 1, 0 \le y \le 1, 0 \le z \le 1.$$

(b) Repeat part (a) if S remains the same but A = $yza_x + xza_y + xya_z$.

35. Verify the divergence theorem

$$\oint_S A.dS = \int_v \nabla . A\, dv$$

for each of the following cases:

(a) $A = xy^2ax + y^3ay + y^2zaz$ and S is the surface of the cuboid defined by $0 < x < 1$, $0 < y < 1$, $0 < z < 1$.

(b) $A = 2\rho za\rho + 3z \sin\phi\, a\phi - 4\rho \cos\phi\, a_z$ and S is the surface of the wedge $0 < \rho < 2$, $0 < \phi < 45°$, $0 < z < 5$.

(c) $A = r^2a_r + r \sin\theta \cos\phi\, a_\theta$ and S is the surface of a quarter of a sphere defined by $0 < r < 3$, $0 < \phi < \pi/2$, $0 < \theta < \pi/2$.

36. The moment of inertia about the z-axis of thea rigid body is proportional to

$$\int_v \left(x^2 + y^2\right) dx\, dy\, dz$$

Express this as the flux of some vector field A through the surface of the body.

37. Let $A = \rho \sin\phi\, a\rho + \rho^2 a\phi$. Evaluate $\oint_L A.dl$

(a) L is the contour of Figure 3·30 (a)

(b) L is the contour of Figure 3·30 (b)

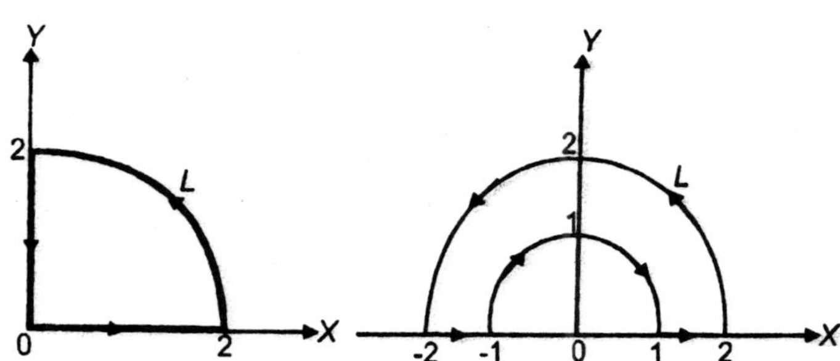

Fig. 1.51. For Problem 37.

38. Calculate the total outward flux of vector

$$F = \rho^2 \sin\phi\, a\rho + z \cos\phi\, a_\phi + \rho z a_z$$

through the hollow cylinder defind by $2 \le \rho \le 3$, $0 \le z \le 5$.

39. Find the flux of the curl of field.

$$T = \frac{1}{r^2}\cos\theta\, a_r + r\sin\theta\cos\phi\, a_\theta + \cos\theta\, a_\phi$$

Through the hemishphere r = 4, $z \leq 0$.

40. A vector field is given by

$$Q\frac{\sqrt{x^2+y^2+z^2}}{\sqrt{x^2+y^2}}\,[(x-y)a_x + (x+y)a_y$$

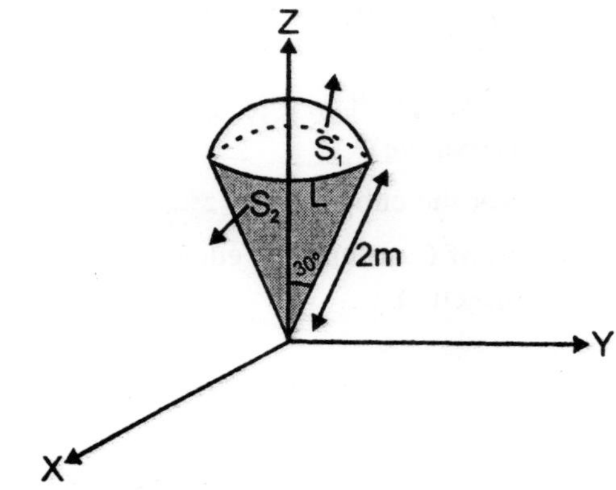

Fig. 1.52. Volume in form of ice-cream cone for Problem 40.

Evaluate the following integral:

(a) $\int_L Q \cdot dl$ where L is the circular edge of the volume in the from of an ice-cream come shown in Figure 3·31

(b) $\int_{S_1} (\nabla \times Q) \cdot dS$ where S_1 is the top surface of the volume

(c) $\int_{S_2} (\nabla \times Q) \cdot dS$ where S_2 is the slanting surface of the volume

(d) $\int_{S_1} Q \cdot dS$

(e) $\int_{S_2} Q \cdot dS$

(f) $\int_v \nabla \cdot Q\, dS$

How do your results in parts (a) to (f) compare?

41. A rigid body spins about a fixed axis through its center with angular velocity ω. If u is the velocity at any point in the body, show that $\omega = 1/2\nabla \times u$.

42. Let U and V be scalar fields, show that

$$\oint_L U\nabla V.dl = -\oint_L V\nabla U.dl$$

43. Show that

$$\int r^n dv = \frac{1}{n+3}\oint r^n r.dr$$

where r, r, and n are as defined in problem 3·26.

44. Given the vector field

$$G = (16xy - z)a_x + 8x^2a_y - xa_z$$

(a) Is G irrotational (or conservative)?

(b) Find the net flux of G over the cube $0 < x, y, z < 1$.

(c) Determine the circulation of G around the edge of the square z = 0, 0 < x, y < 1. Assume anticlockwise direction.

2

Electric Charges and Fields

General Concept : You may have experienced that when we rub our comb with hairs and put it in touch of small (light) objects such as small piece of papers, it attracts that. What does it signify?

(i) *It signifies the existence of charges in comb*: There are so many other general concepts about the existence of charges. The most basic experiment for the existence of electric charge is that when an abonite rod is rubbed with cat's fur, it acquires the property of attracting light objects like small bits of paper or straw towards it. Similarly, if a glass rod is rubbed with silk, it also reveals the same property. Bodies which show this attractive property are said to be charged or electrified. there are large number of substances in the universe, which become charged with rubbed with some material.

(ii) *Attraction and Repulsion between charges* : There are two kinds of force between charged body i.e. attraction and repulsion. The charged bodies always exert force on each other. Two identically charged bodies exert repulsive while opposite charges attract each other. For example; two ebonite rods A and B rubbed with cat's fur, repel one another when they are brought near one another (Fig. 2.1a). Two glass rods C and D both rubbed with silk and put them close to each other, are again found to repel one another (Fig. 2.1b). But a charged ebonite rod A is found to attract a charged glass rod C when former is brought near the latter. (Fig. 2.1c)

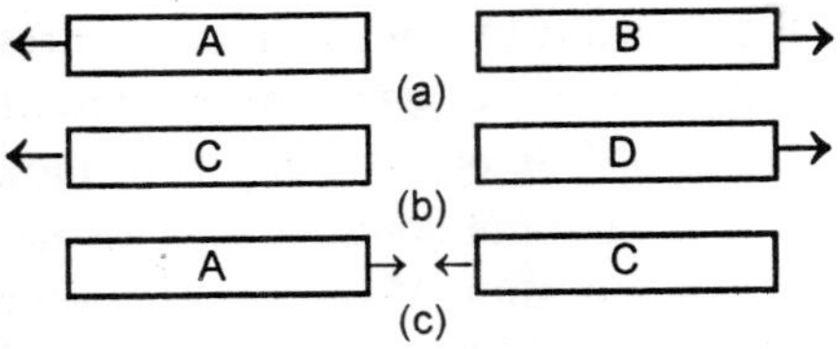

A B
(a)
C D
(b)
A C
(c)

Fig. 2.1

(iii) *It is found that there exist two and only two types of charges*: Taking a charged glass rod as a standard charged body of one type say **'A'** and the charged abonite rod as a standard charged body of another type say **'B'**. It is found that any other charged body in the universe will either repel **A** or **B** but not both. If is repels **A**, it will attract **B** and vice-versa. All bodies repelling A will also repel one another and all bodies repelling **B** will also repel one another. Therefore, we must say that there are mainly two categories of charged bodies.All bodies which repel **A** forms one category and all bodies repelling **B** form second category. Bodies belonging to the same category repel one another and those belongin to different categories attract one another. Thus two and only two types of charged bodies exist in the universe. Two bodies in the smae category are called similarly charged and in different categories are called dissimilarly charged.

(iv) Electronictheory of charges : This theory says that all the objects are made of atoms. every atom consists of positively charged particles called protons, negatively charged particles called electrons and uncharged particles called neutrons.The electrons revolve round the nucleus in various orbits. Experiments have shown that a proton carries a positive charge which is exactly equal in magnitude but opposite in sign to that on an electron. The exact equality of the magnitude of charge in proton and electron has been experimentally verified to a great accuracy 1 to 10^{20}. Matter in general is electrically neutral which means that in a neutral atom, number of protons inside the nucleus is equal to the number of electrons revolving round it. When the two bodies are brought in intimate contact (rubbing for example), there is transference of electrons in the outermost orbit from one-body to the other and the body which loses the electrons becomes positively charged and the other which gains electrons becomes negatively charged. this transference of electrons does not continue for an indefinite time because the substance with more electron affinity, accumulate electrons and the excess electrons in it repel others from coming to it. therfore, we must say that if there is lack of electrons in a body then it becomes positively charged and if there is affluency of electrons in a body then it becomes negatively charged. Electric charge consists of certain minimum charge. It is not a continuous fluid. Electron is the element which carry the minimum charge. Except electron, there is no other element which carry minimum charge.

Charge Quantisation

Electric Charge consists of certain minimum charge. It is not a continous fluid. Electron is the element which carry the minimum charge except electron, there is no other element which carry minimum charge.

This minimum charge is denoted by symbol '*e*' and its magnitude is 1.60×10^{-19} coulomb. Any physically existing charge '*q*', irrespective of its origin, will be equal to the whole number multiple of '*e*', i.e., q=ne, where *n* is either a positive or a negative integer and we cannot have a charge of 1·3e or 2·5 *e* etc. Thus the charge on a body exists in terms of discrete packets of charge '*e*' and not continuous. This fact of the existence of charge only in discrete packets is called quantisation of charge. The principle of quantisation of charge becomes important only when we deal things on atomic scale. In our study of electrostatics we shall be interested in scales much bigger than the atomic scale because any perceptible electrical effect involves very very large number of elementary charges ($n>10^6$).

Conservation and Invariance of Charge

An isolated system is one in which no matter is allowed to cross the boundary of the system. According to the principle of conservation, total net charge in an isolated system remains unchanged. This principle of conservation of charge holds good for both large scale events and events at atomic and nuclear level. This hypothesis of charge conservation implies that charge can neither be created nor destroyed. Whenever object loses charge of either kind an equal amount of charge appears somewhere else. As an instant, when an ebonite rod is rubbed with cat's fur, the negative charge produced on the ebonite rod is exactly equal to the positive charge on the fur. (All that has happened is that some of the electrons for fur have transferred to ebonite rod carrying their charge with them). Thus the total charge in the system consistingof bodies being rubbed together remains conserved, i.e. unchanged (in this case zero). At atomic level, events like pair producation and annibilation of an electron and a positron, provide examples of charge conservation. For demonstrating the example of pair production, let us consider a thin-walled box, in vacuum. Which is electriocally neutral, when this box is exposed to Y-rays or high energy photons, then under certain conditions a photon ends up its own existence and transfroms itself into an electron and positron pair. And the box still remains electrically neutral (since the electron and positron have exactly equal but opposite charge and the photon has no charge) indicating the validity of conservation of charge. Further more, the experiments have established the principle of inveriance of charge. According to this principle the total amout of charge of an isolated system does not depend upon the relative motion of the observers.

Example 1. Give it a positive charge of $6{\cdot}4\times10^{-19}$ C How many electrons must be removed from a body Given e=$1{\cdot}9\times10^{-19}$ C.

Solution. As a body is neutral initially, so to give a positive charge of 6.4×10-19 C to it, a negative charge $9{\cdot}6\times10^{-19}$ C must be removed from it. Hence

$$\text{Number of electrons removed} = \frac{\text{Total negative charge removed}}{\text{Charge on an electron}}$$

$$= \frac{6.4\times10^{-19}}{1.6\times10^{-19}} = 4$$

Example 2. How many electrons flow per second through the hot filament of a lighted lamp of 60 watt and 220 volt?

Solution.

$$VC=W$$

where, V – is potential

C – is the amout of flowing current

W – energy consumed

or $$220\times C=60$$

$\therefore$ $$C = \frac{60}{120} = \frac{3}{11}A.$$

But one ampere current is defined as the rate of flow of one coulomb charge per second.

Hence $\frac{3}{11}C$ charge will flow per second through the filament.

$\therefore$ Number of electrons flowing per second

$$=\frac{\text{Charge flowing per second}}{\text{Charge on an electron}}$$

$$=\frac{\frac{3}{11}}{1.6\times10^{-19}}=\frac{3}{11\times1.6}\times10^{-19}$$

$$=\frac{300}{11\times16}\times10^{-19}=1.704\times10^{18}$$

Coulomb's Law

French colonel, charles Augustine coulomb postulated a law in 175 after an extensive experiment which is know as coulomb's Law the Law pertains to the exiostence of force between two point charges.

By a point charge we mean a charge that is loocated on a body whose dimensions are much smaller than other relevant dimensions. For example, a collection of electric charges on a pinhead may be regarded as a point charge. Charges are generally measured in coulombs (C). One coulomb is approximately equivalent to 6×10^{18} electrons; it is a very large unit of charge because one electron charge e = -1.6019×10^{-19}C.

Coulomb's law states that the force F between two point charges Q_1 and Q_2 is:

1. Along the line joining them
2. Directly proportional to the product Q_1Q_2 of the charges
3. Inverseiy proportional to the square of the distance R between them.[3]

$$F\alpha\, Q_1Q_2$$

$$F\alpha\,\frac{1}{R^2}$$

Q1 ← R → Q2

Fig (2.2)

Now; $F\alpha\,\frac{Q_1Q_2}{R^2}$ or $F=\frac{kQ_1Q_2}{R^2}$ ...(2.1)

where k is the proportionality constant. In SI units, charges Q_1 and Q_2 are in coulombs (C), the distance R is in meters (m), and the force F is in newtons (N) so that $k = 1/4\pi\varepsilon_o$. The constant ε_o is known as the permitivity of free space (in farads per meter) and has the value.

$$\varepsilon_o = 8.854\times10^{-12}\cong\frac{10^{-9}}{36\pi}F/m$$

$$or\ k=\frac{1}{4\pi\varepsilon_o}\cong 9\times10^{9}m/F \quad ...(2.2)$$

Thus eq. (2.1) becomes

$$F = \frac{Q_1 Q_2}{4\pi\varepsilon_0 R^2} \qquad \text{...(2.3)}$$

It point charges Q_1 and Q_2 are located at points having position vectors r_1 and r_2 due to Q_1 shown in Figure 4.1, is given by

$$F_{12} = \frac{Q_1 Q_2}{4\pi\varepsilon_0 R^2} a_{R_{12}} \qquad \text{...(2.4)}$$

where

$$R_{12} = r_2 - r_1 \qquad \text{...(2.5 a)}$$

$$R = |R_{12}| \qquad \text{...(2.5b)}$$

$$a_{R_{12}} = \frac{R_{12}}{R} \qquad \text{...(2.5c)}$$

By substituting eq. (4.5) into eq. (4.4), we may write eq. (4.4) as

$$F_{12} = \frac{Q_1 Q_2}{4\pi\varepsilon_0 R^3} R_{12} \qquad \text{...(2.6a)}$$

or

$$F_{12} = \frac{Q_1 Q_2 (r_2 - r_1)}{4\pi\varepsilon_0 |r_2 - r_1|^3} \qquad \text{...(2.6b)}$$

It is worthwhile to note that

1. As shown in Figure 4.1, the force $\mathbf{F}_{21}$ on Q_1 due to Q_2 is given by

$$F_{21} = |F_{12}| a_{R_{21}} = |F_{12}|(-a_{R_{12}})$$

or

$$F_{21} = -F_{12} \qquad \text{...(2.7)}$$

since

$$a_{R_{21}} = -a_{R_{12}}$$

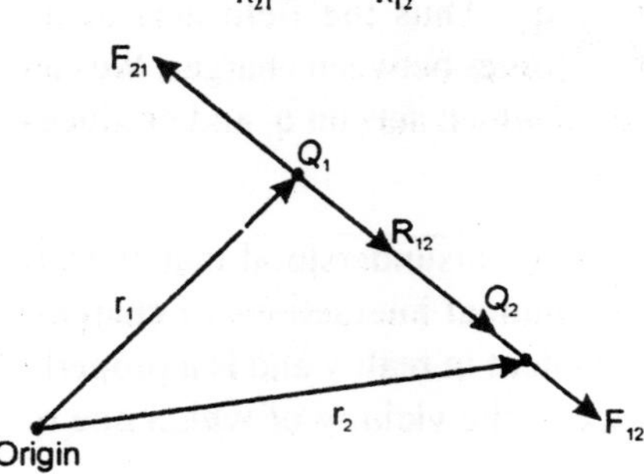

Fig. 2.3 Coulomb vector force on point changes Q_1 and Q_2.

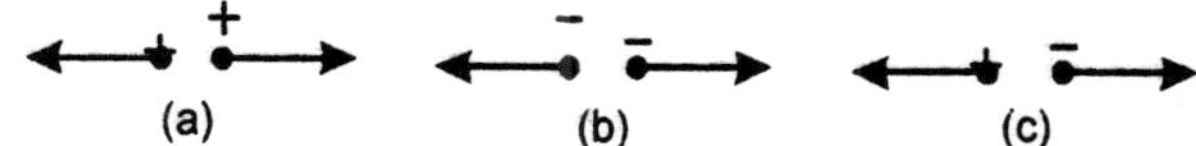

Fig. 2.4 (a), (b) Like charges repel (c) unlike charges attract

2. Like charges (charges of the same sign) repel each other while unlike charges attract. This is illustrated in Figure 2.2.
3. The distance R between the charges bodies Q_1 and Q_2 must be large compared with the linear dimensions of the bodies; that is, Q_1 and Q_2 must be point charges.
4. $Q1$ and Q_2 must be static (at rest).
5. The signs of Q_1 and Q_2 must be taken into account in eq. (2.4).

Principle of superposition is applied once there are more than two charges to determine the force on a particular charge principle states that if there are N charges Q_1, Q_2, ..., Q_N located, respectively, at points with position vectors r_1, r_2, ..., r_N, the resultant force F on a charge Q located at point r is the vector sum of the forces exerted on Q by each of the charges Q_1, Q_2. . . . Q_N. Hence:

$$F = \frac{QQ_1(r-r_1)}{4\pi\varepsilon_0|r-r1|^3} + \frac{QQ_2(r-r_2)}{4\pi\varepsilon_0|r-r_2|^3} + \ldots + \frac{QQ_N(r-r_n)}{4\pi\varepsilon_0|r-r_N|^3}$$

or

$$F = \frac{Q}{4\pi\varepsilon_0}\sum_{k=1}^{N}\frac{Q_k(r-r_k)}{|r-r_k|^3} \qquad \ldots(2.8)$$

Electric Field

The space of a field which is surrounded by a charged body and also affected by its presence is called electric field.

If a charge q_1 is placed at any point, it sets up an electric field in space surrounding it. This field is shown by doted region in Fig. 2.5. This field is shown by dotted region in Fig. 2.5. This field region. This action of the field of q_1 on q_2 results in the force F that is experienced by q_2. Thus the field acts as an intermediary for transmission of forces between charges. We can also imagine that q_2 sets up a field which acts on q_1 and produces a force- $\vec{F}$ on it.

This, however, should not be misunderstood that field is merely an aid in visualising the mutual interactions of charges. The field is taken to be a self-existent in reality and is a property assigned to every point in space in the vicinity of which one or more charged bodies are present.

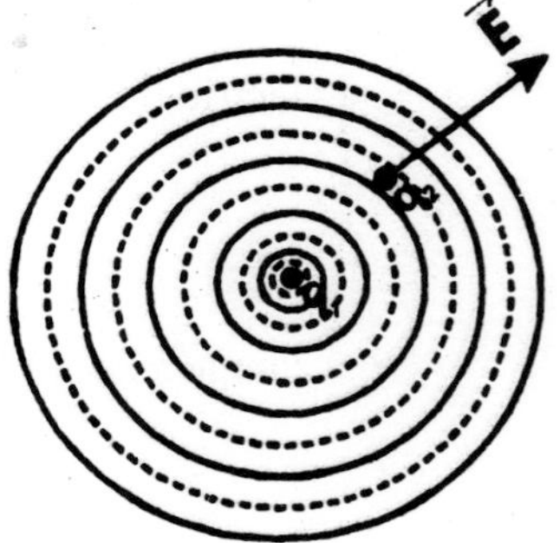

Fig 2.5

Electric Field Intensity

We place a small test charge q (assumed to be positive for convenience) at that point. Let us further suppose that the total force experienced by this test charge due to other charges (called source charge) be $\vec{F}$. Hence, the force per unit charge is called the intensity of electric field.

$$\vec{E} = \frac{\vec{F}}{q} \qquad ...(2.9)$$

The electric field strength $\vec{E}$ is a vector quantity because $\vec{F}$ is a vector and q is a scalar. Direction of $\vec{E}$ is the direction of $\vec{F}$, that is, it is the direction in which a unit positive charge placed at the point of observation would tend to move if free to do so.

For the authentic measurement of field intensity, the test charge must be vanishingly small because a large test charge might influence distribution of field due to the primary (source) charges, thus changing the very quant'ty we are trying to measure. Therefore the above equation (2.9) should be replaced by

$$\vec{E} = \lim_{q \to 0} \frac{\vec{F}}{q} \qquad ...(2.10)$$

Therefore the force per unit positive charge on a vanishingly small test charge at a point is called the intensity of electric field at that point.

Calculation of $\vec{E}$

(a) ***Electric field due to a point charge*** : Assume a test charge qo is placed at a point p which is lying in vacuum at a distance r from a point charge 'q' lyig at) (Fig. 2.6) and let a test charge q_o be placed at P. Now according to Coulomb's law, the force F acting on q_o due to q in S.I. units is given by

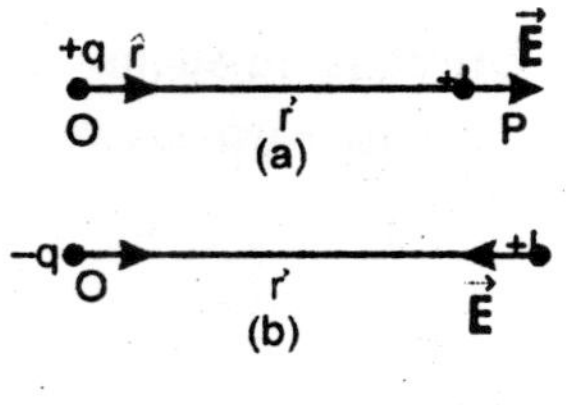

+q r̂ O r P Ē (a)

-q O r Ē (b)

Fig 2.6

$$\vec{F} = \frac{1}{4\pi\varepsilon_0} \frac{qqo}{r^2} \hat{r}$$

Hence the electric field intensity at the point P is given by

$$\vec{E} = \frac{\vec{F}}{qo} = \frac{1}{4\pi\varepsilon_0} \frac{q}{r^2} \hat{r} \qquad ...(2.11)$$

If the charge q lies in a medium of dielectric constant K, then the equation (2.12) becomes

$$\vec{E} = \frac{1}{4\pi\varepsilon_0} \frac{q}{Kr^2} \hat{r} \qquad ...(2.12)$$

In cgs units, equations (2.11) and (2.12) reduce to

$$\vec{E} = \frac{1}{r^2}\hat{r} \text{ and } \vec{E} = \frac{1}{K}\frac{q}{r^2}\hat{r}$$

The direction of $\vec{E}$ is along the line joining O and P, pointing away from q if it is poistive [Fig. 2.6(a)] and towards it if is negative [Fig. 2.6 (b)].

(b) *Field due to group of charges* : The electric field $\vec{E}$ due to each charge at the given point as if it was only charge present at that point and then adding these fields vectorially we can get the electric field at any given point due to group of point charges. If $\vec{E}_1, \vec{E}_2, \vec{E}_3$, etc. are the fields a the observation point due to indiovidual charges and $\vec{E}$ be the resultant intensity at that point, then we have

$$\vec{E} = \vec{E}_1 + \vec{E}_2 + \vec{E}_3 + \ldots$$

$$\sum_{i=1}^{n} \vec{E}_i \qquad \ldots (2.13)$$

where $\vec{E}i$ is the field at the point of observation due to the it point charge.

Example 3. A practical application of electrostatics is in electrostatic separation of solids. For example, Florida phosphate ore, consisting of small particles of quartz and phosphate rock, can be separated into its components by applying a uniform electric field as in Figure 2.7. Assuming zero initial velocity and displacement, determine the separation between the particles after falling 80cm. Take E=500kV/m and $Q/m = 9\,\mu C/kg$ for both positively and negatively charged particles.

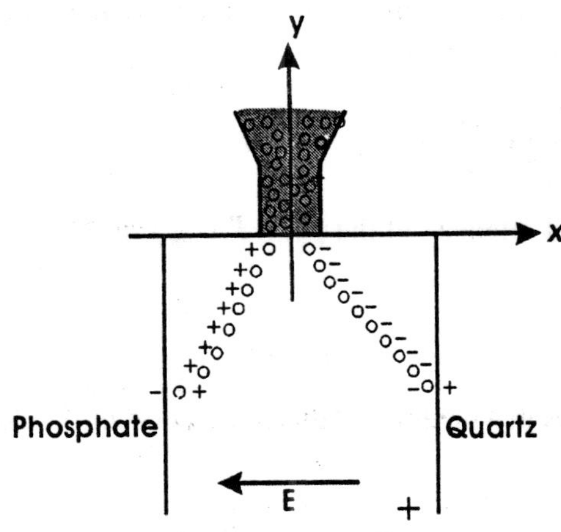

Figure 2.7 Electrostatic separation of solids; for Example 3.

Solution :

Ignoring the coulombic force between particles, the electrostatic force is acting horizontally while the gravitational force (weight) is acting vertically on the particles. Thus,

$$QE = m\frac{d^2}{dt^2}a_x$$

or

$$\frac{d^2x}{dt^2} = \frac{Q}{m}E$$

Integrating twice gives

$$x = \frac{Q}{2m}Et^2 + c_1t + c_2$$

where C_1 and C_2 are integration constants. Similarly,

$$-mg = m\frac{d^2y}{dt^2}$$

or

$$\frac{d^2y}{dt^2} = -g$$

Integrating twice, we get

$$y=1/2gt^2+c_3t+c_4$$

Since the initial displacement is zero,

$$x(t = o) \rightarrow c_2 = 0$$

$$y(t = o) \rightarrow c_4 = 0$$

Also, due to zero initial velocity,

$$\left.\frac{dx}{dt}\right|_{t=0} = 0 \rightarrow c_1 = 0$$

$$\left.\frac{dx}{dt}\right|_{t=0} = 0 \rightarrow c_3 = 0$$

Thus

$$x = \frac{QE}{2m}t^2 \quad y = \frac{1}{2}gt^2$$

When y = - 80cm = -0.8m

$$t^2 = \frac{0.8 \times 2}{9.8} = 0.1633$$

and

$$x = 1/2 \times 9 \times 10^{-6} \times 5 \times 10^{5} \times 0.1633 = 0.3673\text{m}$$

The separation between the particles is $2x$ = 73.47cm.

Exercise 1

An ion rocket emits positive cesium ions from a wedge-shape electrode into the region descibed by $x > |y|$. The electric field is E = -400a_x + 200a_y kV/m. The ions have single electronic charges e = - 1.6019 × 10^{-19} C and mass m = 2.22 × 10^{-25} kg and travel in a vacuum with zero initial velocity. If the emission is condined to -40cm <y <40 cm, find the largest value of x which can be reached.

Answer : 0.8 m.

Electric Fields Due to Continuous Charge Distributions

Till now, we have only discussed about point charges and force as well as electric fields due to that there is possibility of continuous charge distribution along a line, on a surface, or in a volume as illustrated in Figure 2.8. ρ*l* (in C/m), ρs (in C/m²), and ρv (in C/m³), respectively is used to denote the line charge density, surface charge density and volume charge density respectively. These must not be confused with ρ (with subscript) used for radial distance in cylindrical coordinates.

The charge element *dQ* and the total charge *Q* due to these charge distribution are obtaioned from Figure (2.8) as

$$dQ = \rho_L dl \rightarrow Q = \int_L \rho_L dl \quad \text{(line charg)} \qquad ...(2.14a)$$

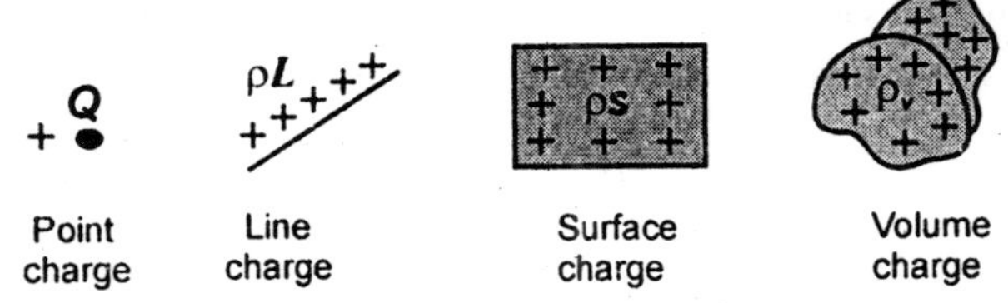

Figure 2.8 Various charge distributions and charge elements.

$$dQ = \rho_L dS \rightarrow Q = \int_L \rho_S dS \quad \text{(surface charge)} \qquad (2.14b)$$

$$dQ = \rho_v dv \rightarrow Q = \int_v \rho_v dv \quad \text{(volume charge)} \qquad (2.14c)$$

The summation of the field contributed by the numerous point charges making up the charge distribution is considered as the electric field intensity due to each of the charge distribution ρ_L, ρ_S and ρ_V Thus by replacing Q in eq. (2.13) with charge element $dQ=\rho L\ dl$, $\rho s dS$, or $\rho v\ dv$ and integrating, we get

$$E = \int \frac{\rho_L dl}{4\pi\varepsilon_0 R^2} a_R \qquad \text{(line charge)} \tag{2.15}$$

$$E = \int \frac{\rho_S dS}{4\pi\varepsilon_0 R^2} a_R \qquad \text{(surface charge)} \tag{2.16}$$

$$E = \int \frac{\rho_v dv}{4\pi\varepsilon_0 R^2} a_R \qquad \text{(volume charge)} \tag{2.17}$$

It should be noted that R^2 and a_R vary as the integrals in eqs. (2.14) to (2.17) are evaluated. We shall now apply these formulas to some specific charge distributions.

A. A Line Charge

Assume a line charge extending from A to B along the z-axis with uniform charge density ρ_L as shown in Figure 2.8. The charge element dQ associated with element $dl=dz$ of the line is

$$dQ=\rho_L\ dl=\rho_L dz$$

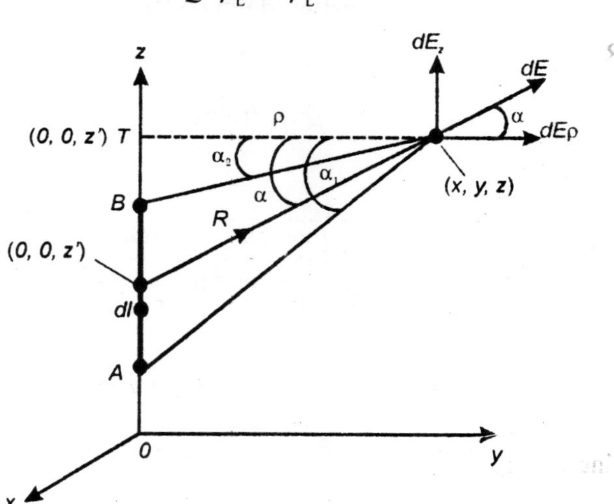

Figure 2.8 Evaluation of the E field due to a line charge and hence the total charge Q is

and hence the total charge Q is

$$Q = \int_{Z_A}^{Z_B} \rho_L dz \tag{2.18}$$

By using equa (2.15), the electric field intensity at an arbitrary point P(x, y, z) can be found. It is important that we learn to derive and substitute each term in eqs. (2.15) to (2.16) for a given charge distribution. It is custom ary to denote the field point by (x, y, z) and the source point by (x', y', z'). Thus from Figure 2.8,

$$dl = dz'$$

$$\mathbf{R} = (x, y, z)-(0, 0, z')=x\mathbf{a}_x+y\mathbf{a}_y+(z-z')\mathbf{a}_z$$

or

$$\mathbf{R}=\rho\mathbf{a}_\rho+(z-z')\mathbf{a}_z$$

$$R^2 = |R|^2 = z^2 + y^2 + (z - z')\ 2 + \rho^2 + (z-z')^2$$

$$\frac{a_2}{R^2} = \frac{R}{|R|^3} = \frac{\rho a_\rho + (z-z')a_z}{[\rho^2 + (z-z')^2]^{3/2}}$$

Substituting all this into eq. (2.15) we get

$$E = \frac{\rho_L}{4\pi\varepsilon_0}\int\frac{\rho a_\rho + (z-z')a_z}{[\rho^2+(z-z')^2]^{3/2}}dz' \tag{2.19}$$

To evaluate this it is convenient that we define α, α_1 and α_2 as in Figure 2.8.

$$R = [\rho^2+(z-z')^2]^{1/2}=\rho \sec\alpha$$

$$z' = OT-\rho\ \tan\alpha, dz' = -\rho\sec^2\alpha\, d\alpha$$

Hence, eq. (2.19) become

$$E = \frac{-\rho_L}{4\pi\varepsilon_0}\int_{\alpha_1}^{\alpha_2}\frac{\rho \sec^2\alpha[\cos\alpha\, a_\rho + \sin\alpha\, a_z]d\alpha}{\rho^2\sec^2\alpha} \tag{2.20}$$

$$= \frac{\rho_L}{4\pi\varepsilon_0\rho}\int_{\alpha_1}^{\alpha_2}[\cos\alpha\, a_\rho + \sin\alpha\, a_z]d\alpha$$

Thus for a finite line charge,

$$E = \frac{\rho_L}{4\pi\varepsilon_0\rho}[-(\sin\alpha_2 - \sin\alpha_1 + (\cos\alpha_2 - \cos\alpha_1)a_z] \tag{2.21}$$

Point A is at $(0, 0,-\infty)$ and B is at $(0, 0, \infty)$ for an infinite line charge therefore; $\alpha_1=\pi/2$, $\alpha_2=-\pi/2$:, the z – Component vanishes and eq. (4.20) becomes

$$\boxed{E = \frac{\rho_L}{2\pi\varepsilon_0\rho}a_\rho} \tag{2.22}$$

If the line is not along the z-axis then ρ is the perpendicular distance from the line to the point of interest and ar is a unit vector along that distance directed from the line charge to the field point.

B. A Surface Charge

The charge associated with an elemental area dS of an infinite sheet of charge in the xy-plane with uniform charge density ρs is given:

$$dQ = \rho s\ dS$$

and hence the total charge is

$$Q = \int \rho s dS \tag{2.23}$$

From eq. (2.16), the contribution to the E field at point $P(0, 0, h)$ by the elemental surface 1 shown in Figure 2.9 is

$$dE = \frac{dQ}{4\pi\varepsilon_0 R^2} a_R \tag{2.24}$$

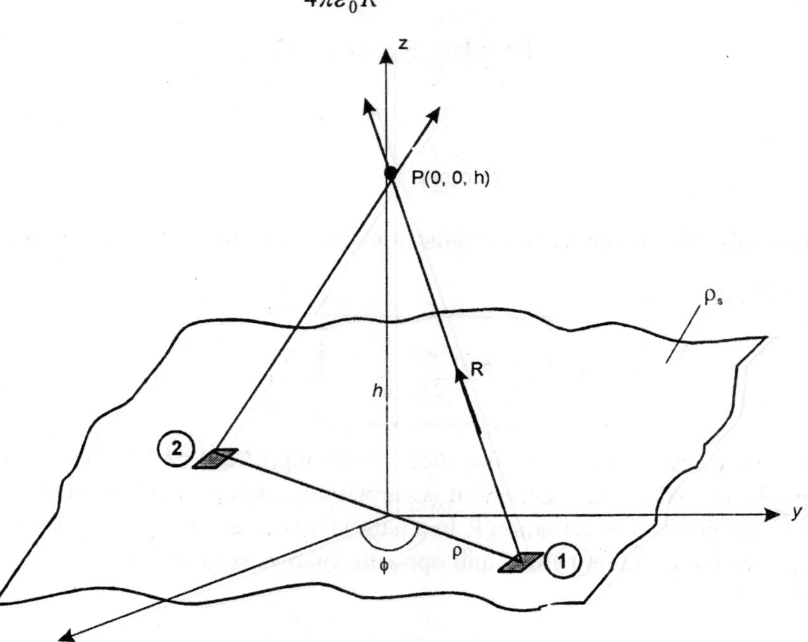

Figure 2.9 Evaluation of the E Field due to an infinite sheet of charge.

From Figure 2.9,

$$\mathbf{R} = \rho\,(\text{-}a\rho) + ha_z,\ R = |\mathbf{R}| = [\rho^2 + h^2]^{1/2}$$

$$a_R = \frac{R}{R,} dQ = \rho s dS = \rho s \rho d\phi d\rho\ ;\ a_R = \frac{R}{R}, dQ = \rho_s ds = \rho_s \rho d\phi d\rho$$

Substitution of these terms into eq. (2.24) gives

$$dE = \frac{\rho s \rho d\phi d\rho[-\rho a_\rho + h a_z]}{4\pi\varepsilon_0 [\rho^2 + h^2]^{3/2}} \tag{2.25}$$

Due to the symmetry of the charge distribution, for every element 1, there is a corresponding element 2 whose contribution along ap cancels that of element 1, as illustrated in Figure 2.9. Thus the contributins to Eρ add up to zero so that E has only *z*-component. This can also be shown mathematically by replacing ap with cos . Integration of cos or sin over 0<<2p given zero. Therefore,

$$E = \int dE_z = \frac{\rho s}{4\pi\varepsilon_0} \int_{\phi=0}^{2\pi} \int_{\rho=0}^{\infty} \frac{h\rho d\rho d\phi}{[\rho^2 + h^2]^{3/2}} a_z$$

$$= \frac{\rho s h}{4\pi\varepsilon_0} 2\pi \int_0^{\infty} [\rho^2 + h^2]^{-3/2} \frac{1}{2} d(\rho^2) a_Z$$

$$= \frac{\rho s h}{2\varepsilon_0} \left\{ -[\rho^2 + h^2]^{-1/2} \right\}_0^{\infty} a_z$$

$$E = \frac{\rho s}{2\varepsilon_0} a_z \qquad \text{...(2.26)}$$

that is, **E** has only *z*-component if the charge is in the *xy*-plane, In general, for an infinite sheet of charge

$$\boxed{E = \frac{\rho s}{2\varepsilon_0} a_n} \qquad \text{...(2.27)}$$

where a is a unit vector normal to the sheet. From eq. (2.26) or (2.27), we notice that the electric field is normal to the sheet and it is surprisingly independent of the distance between the sheet and the point of observation P. In a parallel plate capacitor, the electric field existing between the two plates having equal and opposite charges is given by

$$E = \frac{\rho s}{2\varepsilon_0} a_n + \frac{-\rho s}{2\varepsilon_0}(-a_n) = \frac{\rho s}{\varepsilon_0} a_n \qquad \text{...(2.28)}$$

C. A Volume Charge

The charge *dQ* associated with the elemental volume *dv* of the volume charge distribution with uniform charge density *Pv* is given:

$$dQ = \rho_v dv$$

and hence the total charge in a sphere of radius a is

$$Q = \int \rho_v dv = \rho_v \int dv \qquad ...(2.29)$$

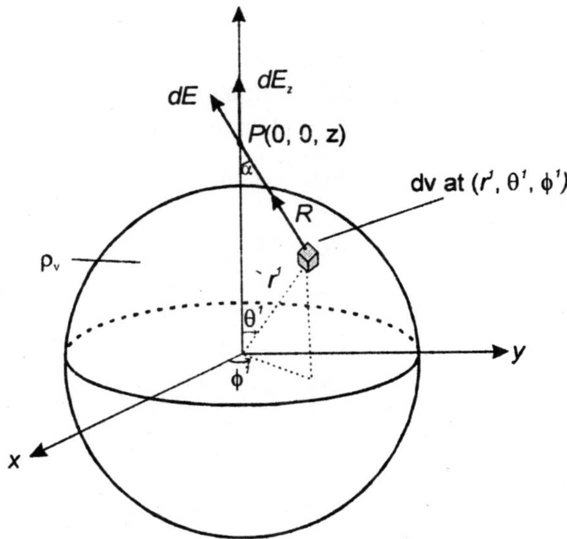

Figure (2.10) Evaluation of the E field due to a volume charge distribution.

$$= \rho_v \frac{4\pi a^3}{3}$$

The electric field $d\mathbf{E}$ at $P(0, 0, z)$ due to the elementary volume charge is

$$dE = \frac{\rho_v dv}{4\pi\varepsilon_0 R^2} a_R$$

where $\mathbf{a}_R = \cos a_z + \sin \alpha\, \mathbf{a}_p$. Due to the symmetry of the charge distribution, the contributions to E_x or E_y add up to zero. We are left with only E_z, given by

$$E_z = E \bullet a_z = \int dE \cos\alpha = \frac{\rho_v}{4\pi\varepsilon_0} \int \frac{dv \cos\alpha}{R^2} \qquad ...(2.30)$$

Again, we need to derive expressions for dv, R^2, and cos α

$$dv = r'^2 \sin\theta' dr' d\theta' d\phi' \qquad ...(2.31)$$

Applying the cosine rule to Figure (2.10), we have

$$R^2 = z^2 + r'^2 - 2zr' \cos\theta'$$

$$r'^2 = z^2 + R^2 - 2zR \cos\alpha'$$

It is convenient to evaluate the integral in eq. (2.30) in terms of R and r'. Hence we express cos θ', cos α, and sin θ' dθ' in terms of R and r', that is,

$$\cos\alpha = \frac{z^2 + R^2 - r'^2}{2zR} \qquad ...(2.32a)$$

$$\cos\theta' = \frac{z^2 + r'^2 - R^2}{2zr'} \qquad ...(2.32b)$$

Differentiating eq. (2.32b) with respect to θ' keeping z and r' fixed, we obtain

$$\sin\theta' d\theta' = \frac{RdR}{zr'} \qquad ...(2.33)$$

Substituting eqs. (2.31) to (2.33) into eq. (2.30) yields

$$E_z = \frac{\rho_v}{4\pi\varepsilon_0}\int_{\phi=0}^{2\pi} d\phi \int_{R=z-r'}^{z+r'} r'^2 \frac{RdR}{zr'} dr' \frac{z^2 + R^2 - r'^2}{2zR}\frac{1}{R^2}$$

$$= \frac{\rho_v 2\pi}{8\pi\varepsilon_0 z^2}\int_{\phi'=0}^{2\pi} d\phi' \int_{R=z-r'}^{z+r'} r'\left[1 + \frac{z^2 - r'^2}{R^2}\right] dRdr'$$

$$= \frac{\rho_v \pi}{4\pi\varepsilon_0 z^2}\int_0^a r'\left[R + \frac{z^2 - r'^2}{R}\right]_{z-r'}^{z+r'} dr'$$

$$= \frac{\rho_v \pi}{4\pi\varepsilon_0 z^2}\int_0^a 4r'^2\, dr' = \frac{1}{4\pi\varepsilon_0 z^2}\left(\frac{4}{3}\pi a^3 \rho v\right)$$

or

$$E = \frac{Q}{4\pi\varepsilon_0 x^2} a_z \qquad (2.34)$$

This result is obtained for E at P(0, 0, z). Due to the symmetry of the charge distribution. the electric field at P(r, θ, ϕ) is readily obtained from eq. (2.34)as

$$E = \frac{Q}{4\pi\varepsilon_0 x^2} a_r \qquad (2.35)$$

which is identical to the electric field at the same point due to a point charge Q located at the origin or the center of the spherical charge distribution.

Example 4. A circular ring of radius a carries a uniform charge pl C/m and is placed on the xy-plane with axis the same as the z-axis.

(a) Show that

$$E(0,0,h) = \frac{\rho_L ah}{2\varepsilon_0\left[h^2 + a^2\right]^{3/2}} a_z$$

(b) What values of h gives the maximum value of **E**?

(c) If the total charge on the ring is Q, find **E** as $a \to 0$.

Solution:

(a) Consider the system as shown in Figure (2.11). Again the trick in finding E using eq. (4.14) is deriving each term in the equation. In this case,

$$dl = a\,d\phi, \quad \mathbf{R} = a(-\mathbf{a}_\rho) + h\mathbf{a}_z$$

$$R = |R| = [a^2 + h^2]^{1/2}, \quad a_R = \frac{R}{R}$$

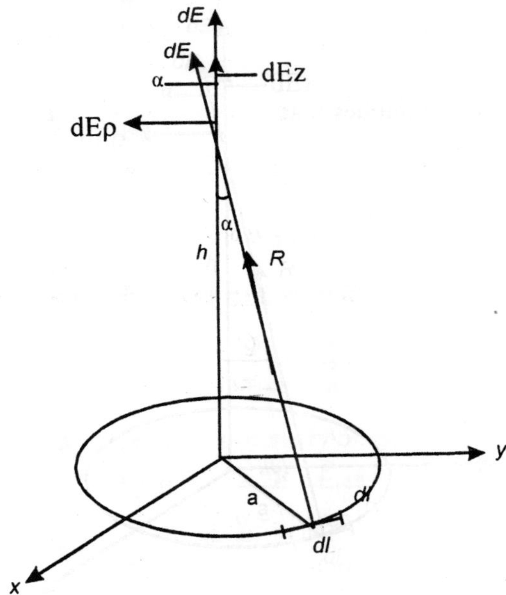

Figure 2.11 Charged ring; for Example 4.4

or

$$E = \frac{a_R}{R^2} = \frac{R}{|R|^3} = \frac{-aa_\rho + ha_z}{\left[a^2 + h^2\right]^{3/2}}$$

Hence

$$E = \frac{\rho_L}{4\pi\varepsilon_0} \int_{\phi=0}^{2\pi} \frac{(-a a_\rho + h a_z)}{[a^2 + h^2]^{3/2}} a\, d\phi$$

By symmetry, the contributions along $\mathbf{a}_\rho$ add up to zero. This is evident from the fact that for every element dl there is a corresponding element diametrically opposite it that gives an equal but opposite dE_ρ so that the two contributions cancel each other. Thus we are left with the z-component. That is,

$$E = \frac{\rho_L a h a_z}{4\pi\varepsilon_0 [h^2 + a^2]^{3/2}} \int_0^{2\pi} d\phi = \frac{\rho_L a h a_z}{2\varepsilon_0 [h^2 + a^2]^{3/2}}$$

as required.

(b)

$$\frac{d|E|}{dh} = \frac{\rho_L a}{2\varepsilon_0} \left\{ \frac{[h^2 + a^2]^{3/2}(1) - \frac{3}{2}(h)2h[h^2 + a^2]^{1/2}}{[h^2 + a^2]^3} \right\}$$

For maximum E, $\frac{d|E|}{dh} = 0$, which implies that

$$[h^2 + a^2]^{1/2}\, [h^2 + a^2 - 3h^2] = 0$$

$$a^2 - 2h^2 = 0 \text{ or } h = \pm\frac{a}{\sqrt{2}}$$

(c) Since the charge is uniformly distributed, the line charge density is

$$\rho_L = \frac{Q}{2\pi a}$$

so that

$$E = \frac{Qh}{4\pi\varepsilon_0 [h^2 + a^2]^{3/2}} a_z$$

As $a \to 0$

$$E = \frac{Q}{4\pi\varepsilon_0 h^2} a_z$$

or in general

$$E = \frac{Q}{4\pi\varepsilon_0 r^2} a_R$$

which is the same as that of a point charge as one would expect.

Exercise 2

A circular disk of radius a is uniformly charged with ps C/m2. If the disk lies on the z=0 plane with its axis along the *z*-axis,

(a) Show that at point (0, 0, *h*)

$$E = \frac{\rho_s}{2\varepsilon_0}\left\{1 - \frac{h}{\left[h^2 + a^2\right]^{1/2}}\right\}a_z$$

(b) From this, derive the E field due to an infinite sheet of charge on the z=0 plane.

(c)If $a << h$, show that **E** is similar to the field due to a point charge.

Answer : (a) Proof, (b) $E = \frac{\rho_s}{2\varepsilon_0}a_z$ (c) Proof

Example 5 : The finite sheet $0 \le x \le 1$, $0 \le y \le 1$ on the z = 0 plane has a charge density ps = $xy(x^2 + y^2 + 25)^{3/2}$ nC/m². Find

(a)The total charge on the sheet

(b) The electric field at (0, 0, 5)

(c)The force experienced by a -1mC charge located at (0, 0, 5)

Solution:

(a)

$$Q = \int \rho s dS = \int_0^1\int_0^1 xy\left(x^2 + y^2 + 25\right)^{3/2} dxdy nC$$

Since $x\,dx = 1/2\ d(x^2)$, we now integrate with respect to x^2 (or change variables: $x^2 = u$ so that $x\,dx = du/2$).

$$Q = \frac{1}{2}\int_0^1 y\int_0^1 \left(x^2 + y^2 + 25\right)^{3/2} d\left(x^2\right)dy nC$$

$$= \frac{1}{2}\int_0^1 y\frac{2}{5}\left(x^2 + y^2 + 25\right)^{5/2}\Big|_0^1 dy$$

$$= \frac{1}{5}\int_0^1 \frac{1}{2}\left[\left(y^2 + 26\right)^{5/2} - \left(y^2 + 25\right)^{5/2}\right]d\left(y^2\right)$$

$$= \frac{1}{10}\bullet\frac{2}{7}\left[\left(y^2 + 26\right)^{7/2} - \left(y^2 + 25\right)^{7/2}\right]_0^1$$

$$= \frac{1}{35}\left[(27)^{7\ 2} + (25)^{7\ 2} - 2(26)^{7\ 2}\right]$$

Q = 33.15 nC

(b) $$E = \int \frac{\rho s dS a_R}{4\pi\varepsilon_0 r^2} = \int \frac{\rho s dS(r - r')}{4\pi\varepsilon_0 |r - r'|^3}$$

where **r** – **r**'= (0, 0, 5) - (*x*, *y*, 0) = (–*x*,–*y*, 5). Hence,

$$E = \int_0^1\int_0^1 \frac{10^{-9} xy\left(x^2 + y^2 + 25\right)^{3\ 2}\left(-xa_x - ya_y + 5a_z\right)dxdy}{4\pi \bullet \frac{10^{-9}}{36\pi}\left(x^2 + y^2 + 25\right)^{3/2}}$$

$$= 9\left[-\int_0^1 x^2dx\int_0^1 ydya_x - \int_0^1 xdx\int_0^1 y^2dya_y + 5\int_0^1 xdx\int_0^1 ydya_z\right]$$

$$= 9\left(\frac{-6}{6}, \frac{-1}{6}, \frac{5}{4}\right)$$

= (–1.5, –1.5, 11.25) V/m

(c) **F** = *q***E** = (1.5, 1.5, – 11.25)mN

Exercise 3

A square plate described by -2 ≤ x ≤ 2, -2 ≤ y ≤ 2, z = 0 carries a charge 12 |y| mC/m². Find the total charge on the plate and the electric field intensity at (0, 0, 10).

Answer : 192 mC, 16.46 a_z MV/m.

Example 6 Planes *x* = 2 and *y* = –3, respectively, carry charges 10π nC/m2 and 15π nC/m2. and 15π nC/m2. If the line *x* = 0, *z* = 2 carries charge 10π nC/m, calculate E at (1, 1, –1) due to the three charge distributions.

Solution :

Let

$$\mathbf{E} = \mathbf{E}_1 + \mathbf{E}_2 + \mathbf{E}_3$$

where $\mathbf{E}_1$, $\mathbf{E}_2$, and $\mathbf{E}_3$ are, respectively, the contributions to **E** at point (1, 1,–1) due to the infinite sheet 1, infinite sheet 2, and infinite line 3 as shown in Figure 4.10(a). Applying eqs. (4.26) and (4.21) gives

$$E_1 = \frac{\rho s}{2\varepsilon_0}\left(-a_x\right) = \frac{10 \bullet 10^{-9}}{2 \bullet \frac{10^{-9}}{36\pi}} a_x = -180\pi a_x$$

$$E_2 = \frac{\rho s_2}{2\varepsilon_0} a_y = \frac{15 \cdot 10^{-9}}{2 \cdot \frac{10^{-9}}{36\pi}} a_y = -270\pi a_y$$

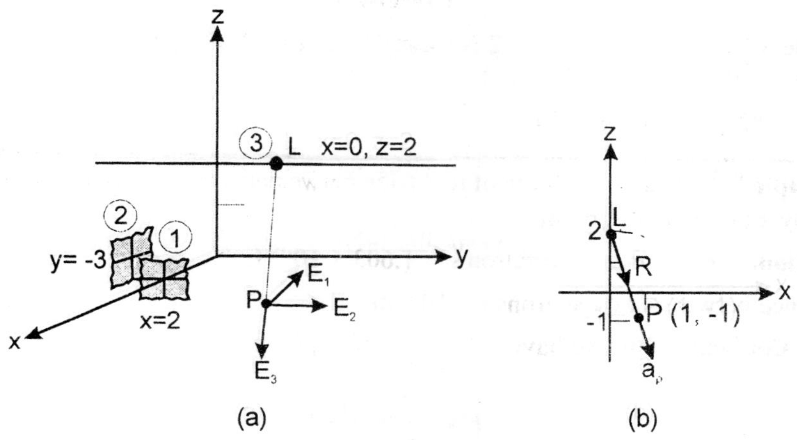

Figure 2.12 For Example 4.6: (a) three charge distributions; (b) finding ρ and $\mathbf{a}\rho$ on plane y = 1.

and

$$E_3 = \frac{\rho_L}{2\pi\varepsilon_0 \rho} a_\rho$$

where aρ (not regular aρ but with a similar meaning) is a unit vector along *LP* perpendicular to the line charge and ρ is the length *LP* to be determined from Figure 4.10(b). Figure 4.10(b) results from Figure 4.10(a) if we consider plane $y = 1$ on which $\mathbf{E}_3$ lies. From Figure 4.10(b), the distance vector from *L* to *P* is

$$\mathbf{R} = -3\mathbf{a}_z + \mathbf{a}_x$$

$$\rho = |R| = \sqrt{10}, \quad a_\rho = \frac{R}{|R|} = \frac{1}{\sqrt{10}} a_x - \frac{3}{\sqrt{10}} a_z$$

Hence,

$$E_3 = \frac{10\pi \cdot 10^{-9}}{2\pi \cdot \frac{10^{-9}}{36\pi}} \cdot \frac{1}{10}(a_x - 3a_z)$$

$$= 18\pi\,(a_x - 3a_z)$$

Thus by adding E_1, E_2, and E_3, we obtain the total field as

$$E = -162\pi a_x + 270\pi a_y - 54\pi a_x \text{ V/m}$$

Note that to obtain $\mathbf{a}_r$, $\mathbf{a}_\pi$, or $\mathbf{a}_v$, which we alwways need for finding **F** or **E**, we must go from the charge (at position vector **r'**) to the field point (at position vector **r**); hence $\mathbf{a}_r$, $\mathbf{a}_\pi$, or $\mathbf{a}_v$, is a unit vector along **r-r'**. Observe this carefully in Figures 4.6 to 4.10.

Exercise 4

In Example 4.6 if the line $x = 0$, $z = 2$ is rotated through 90° about the point (0, 2, 2) so that it becomes $x = 0$, $y = 2$, find **E** at (1, 1, -1).

Answer : $- 282.7a_x + 564.5a_y$ V/m

Example 7: Calculate the force of repulsion between two electrons in vacuum which are separated by a distance of 10^{-14} m.

Solution. Charge on each electron $e = 1.603 \times 10^{-19}$ C

Distance between two electrons $r = 10^{-16}$ m

From Coulomb's law, we have

$$F = \frac{1}{4\pi\varepsilon_0} \bullet \frac{q_1 q_2}{r_2}$$

Here $$q_1 = q_2 = e \text{ and } \frac{1}{4\pi\varepsilon_0} = 9 \times 10^9 \, Nm^2C^{-2}$$

$$F = \frac{9 \times 10^9 \, Nm^2C^{-2} \times \left(1.603 \times 10^{-19} C\right)^2}{\left(10^{-14} m\right)^2}$$

$$= 9 \times (1.603)^2 \times 10^{-1} = 2{\cdot}31 \text{ N.}$$

Example 8. Compare the magnitude of electrostatic and gravitational force in hydrogen atom in which the electron and the proton are about 5·3 × 10-11 m apart. Given electronic charge = 1·6 × 19-19-19C, electronic mass = $9{\cdot}1 \times 10^{-31}$ kg, proton mass = $1{\cdot}7 \times 10^{-27}$ kg and G = $6{\cdot}7 \times 1^{-11}$ Nm² kg⁻².

Solution. Charge on the electron = Charge on the porton

$$= 1{\cdot}6 \times 10^{-19} \text{C}$$

Distance between electron and proton

$$= r = 5{\cdot}3 \times 10^{-11} \text{m}$$

Mass of electron $m_1 = 9{\cdot}1 \times 10^{-31}$ kg

Mass of proton $m_2 = 1{\cdot}7 \times 10^{-27}$ kg

Gıavitation constant $= 6{\cdot}7 \times 10^{-11}$ Nm² kg⁻²

From Coulomb's law, the electrostatic force between an electron and proton is given by

$$F_e = \frac{1}{4\pi\varepsilon_0}\frac{q_1q_2}{r^2}$$

$$= \frac{\left(9\times10^9\,Nm^2C^{-2}\right)\left(1\bullet6\times10^{-19}C\right)^2}{\left(5\bullet3\times10^{-11}m\right)^2}$$

$$= \frac{9\times(1\bullet6)^2}{(5\bullet3)^2}\times10^{-7}\,N$$

$$= 8.2\times10^{-8}\text{ N}$$

From Newton's Law of Gravitation, the gravitation force

$$F_g = \frac{Gm_1m_2}{r^2}$$

$$= \frac{\left(6.7\times10^{-11}\,Nm^2kg^{-2}\right)\left(9.1\times10^{-31}kg\right)\left(1.7\times10^{-27kg}\right)}{\left(5.3\times10^{-11}m\right)^2}$$

$$= \frac{6.7\times9.1\times1.7}{(5.3)^2}\times10^{-67}\,N = 3.69\times10^{-47}\,N$$

$$\therefore \quad \frac{F_{es}}{F_g} = \frac{8.2\times10^{-8}}{3.69\times10^{-47}}\times2.2\times10^{39}$$

Thus the electrostatic force is about 1029 times greater than the gravitational force, so the gravitational force can be neglected in comparison to the former in nuclear and atomic problems.

Example 9. Two equal charge of +25μ coulomb are placed 1m apart. Compute (a) the force exerted by each charge on a+2μ coulomb charge placed midway between them, and (b) the resultant force on the test charge.

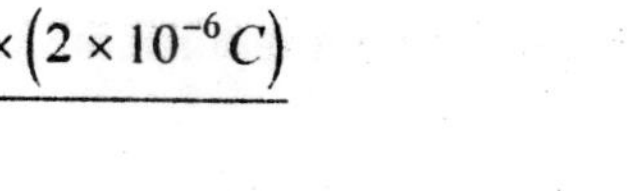

Solution. Let two charge q1 and q2 of +25 μC each, be placed at points A and B such that AB=1m. O be the mid point of the line joining A and B, where a charge of +2 μ C is placed (Fig. 2.13).

Fig. 2.13

By Coulomb's law force F1 on +2 μ C charge at O due to 25 μ C charge at A, is given by

$$F_1 = \frac{1}{4\pi\varepsilon_0}\frac{q_1q_2}{r^2} = \frac{\left(9\times10^9\,Nm^2C^{-2}\right)\times\left(25\times10^{-6}C\right)\times\left(2\times10^{-6}C\right)}{(.5m)^2}$$

$$= 1.8\text{ N along } OB$$

Force F_2 on +2 μC charge at O due to 25 μC charge at B will also be = 1.8 N along AO.

Thus froce F_1 is equal and opposite to force F_2.

Example 10. Calculate the electric intensity required to just support a water drop of mass 10 kg. and having a charge 4.9×10^{-10}C.

Solution. Weight of the body = mg = 10^{-7} kg × 9.8 ms^{-2}

= 9.8 × 10^{7} N

Fig. 2.14

∴ Force required to support the drop **F** = 9.8 × 10^{-7} N

If $\vec{E}$ be the electric field intensity in upward direction, then electrostatic force $\vec{F} = q\vec{E}$ (Fig. 2.14)

$$\therefore \qquad E = \frac{F}{q} = \frac{9.8\times10^{-7}\,N}{4.9\times10^{-10}\,C}$$

= 2×10^{3}NC^{-1} upward.

Example 11. Two pith balls, each weighting 100 mg are suspended from the same point by silk threads 20 cm long, are equally charged and repel each other to a distance of 100 mm. Calculate the charge on each ball.

Solution. Let A and B be two pith balls each having a charge 'q' are suspended from the same point P. Due to repulsion the two pith balls are at a distance r = 100 mm = 10^{-1}m (Fig. 2.9).

The various forces acting on each ball are the following:

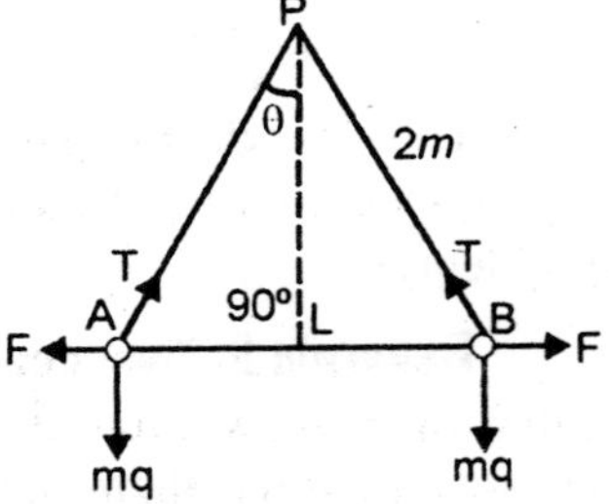

(*i*) The weight mg of the ball acting vertically downward i.e.

$$mg = 100\times10^{-6}\text{ kg}\times9.8\text{ ms}^{-2}$$

$$= 9.8\times10^{-4}\text{ N.}$$

(*ii*) Force of electrostatic repulsion ***F*** between the two balls i.e.

$$F = \frac{q^2}{4\pi\varepsilon_0 r^2} = \frac{9\times10^9\,Nm^2C^{-2}\times q^2}{(1m)^2}$$

$$= 9\times10^{11}\,q^2\text{N} \qquad \text{...(a)}$$

(iii) The tension T in the string acting along the string towards P. As the forces are in equilibrium, the sums of the moments of these forces about any point say P will be zero, hence

$$mg\left(\frac{r}{2}\right) - F(PL) = 0 \;(\text{where } PL \text{ is } \perp AB) \qquad \text{...(b)}$$

In rt. $\angle d\Delta ALP, \frac{PL}{AL} = cot\,\theta$

or $PL = AL cot\,\theta = \frac{r}{2} \bullet cot\,\theta$

Putting for PL in equation (2.17), we get

$$mg\left(\frac{r}{2}\right) - F\left(\frac{r}{2}\right)cot\,\theta = 0$$

or $mg - F\cot\theta = 0$

or $$F = \frac{mg}{cot\,\theta} = mg\,tan\,\theta \quad ...(c)$$

Comparing (a) and (c),

$$9\times10^{11}q^2 = \text{mg tan q} = 9.8\times10^{-4}\times\frac{AL}{PL}$$

$$= 9.8\times10^{-4}\times\frac{.05}{\sqrt{(.2)^2-(.05)^2}}$$

$$= \frac{9.8\times10^{-4}\times.05}{.1936} \quad \left[PL = \sqrt{AP^2 - OL^2}\right]$$

or $$q^2 = \frac{9.8\times10^{-4}\times.05}{.1936\times9\times10^{11}} = 2.8093\times10^{-16}$$

or $q = 1.676\times10^{-8}$ C.

Example 12. Two small charges +q and -q are placed at a distance 2a apart. Calculate electric field due to this charge at a point A, at a distace r along the perpendicular bisector of the line joining the charges.

Solution. Let $+q$ and $-q$ charge be situated at points B and C such that BC = 2a and AL be the right bisector of BC, so that $AL = r$ (Fig. 2.17), then

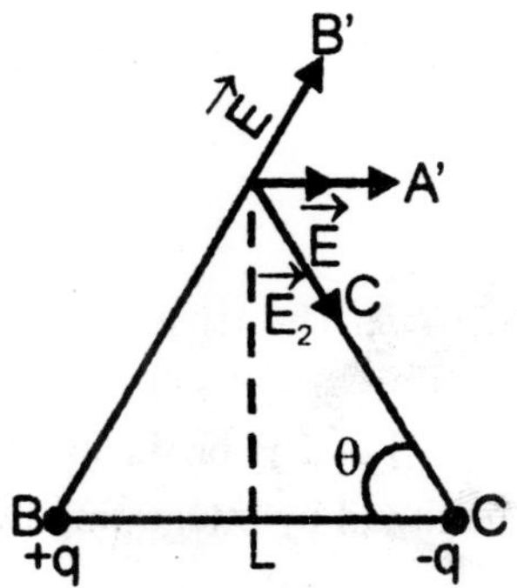

Electric field E_1 at A due to $+q$ charge at B *i.e.*

$$E_1 = \frac{1}{4\pi\varepsilon_0}\frac{q}{AB^2} = \frac{1}{4\pi\varepsilon_0}\frac{q}{\left(a^2+r^2\right)}$$

(along AB')

Electric field E_2 at A due to –q charge at C i.e.,

$$E_2 = \frac{1}{4\pi\varepsilon_0}\frac{q}{(a^2 + r^2)} \text{ along AC'}$$

Thus $E_1 = E_2$

The electric field at A= $\vec{E}_1 + \vec{E}_2 = \vec{E}$

The vector of= $\vec{E}_1$ and $\vec{E}_2$ points along AA' and has the magnitude

$$E = 2E \cos\theta \qquad \text{...(a)}$$

From Fig. (2.10), we see $\cos\theta = \frac{a}{\sqrt{a^2 + r^2}}$

Substituting the value of E_1 and $\cos\theta$ in equation (a), we have

$$E = \frac{2}{4\pi\varepsilon_0}\left(\frac{q}{r^2 + a^2}\right) \bullet \frac{a}{\sqrt{r^2 + a^2}} = \frac{1}{4\pi\varepsilon_0} \bullet \frac{2aq}{(r^2 + a^2)^{3/2}} \text{ along AA'}.$$

Example 13. Calculate the electric field at a point on the axis of a ring at a distance 'x' from its centre when the ring is uniformly charged and has a radius a and charge q.

Solution. Let the ring be divided into small elements. Let us consider one such element of length *dl* at the top of the ring as shown in Fig. 2·11. It contains an element of charge given by

$$dq = \frac{q}{2\pi a} dl$$

where $2\pi a$ is the circumference of the ring.

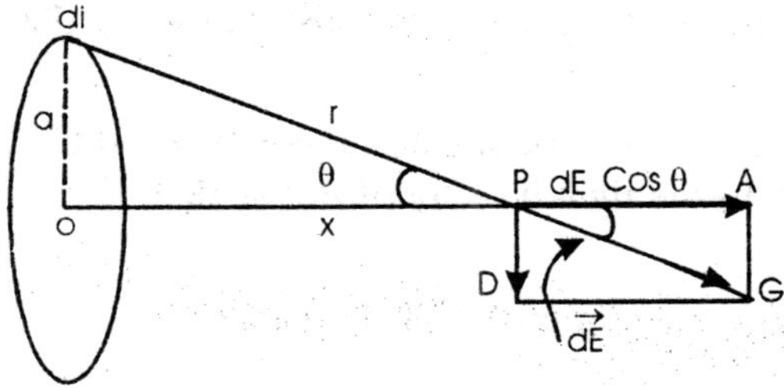

Fig. 2.17

Let d $\vec{E}$ be the electric field at *P* due to this small element. The resultant field $\vec{E}$ at P can be found by integrating the effects of all the elements that form the ring. From symmetry this resultant field must lie along the axis of the ring. Thus only the component of d $\vec{E}$ along PA contributes to the total field. The component perpendicular to axis is cancelled out by an equal but opposite component established by the charge element on the opposite side of the ring.

$$\therefore \qquad \vec{E} = \int d\vec{E}$$

or

$$E = \int dE \cos\theta$$

The quantity dE drom Coulomb's law is $= \dfrac{1}{4\pi\varepsilon_0} \bullet \dfrac{dq}{r^2}$

$$= \frac{1}{4\pi\varepsilon_0} \bullet \frac{dql}{2\pi a} \times \frac{1}{(a^2 + x^2)}$$

From Fig. 2.11, $\cos\theta = \dfrac{x}{r} = \dfrac{x}{\sqrt{a^2 + x^2}}$

$$\therefore \qquad E = \int dE \cos\theta = \int \frac{1}{4\pi\varepsilon_0} \bullet \frac{qdl}{(2\pi a)(a^2 + x^2)} \bullet \frac{x}{\sqrt{a^2 + x^2}}$$

As for the given point P, x has the same value for all charge elements and is not variable, so

$$E = \frac{1}{4\pi\varepsilon_0} \bullet \frac{qx}{2\pi a(a^2 + x^2)^{3/2}} \int dl \quad \text{along PA}$$

The integral $\int dl$ is simply the circumference of the ring, i.e., equal to $2\pi a$ so that

$$E = \frac{1}{4\pi\varepsilon_0} \bullet \frac{qx}{2\pi a(a^2 + x^2)^{3/2}} \times 2\pi a$$

$$= \frac{1}{4\pi\varepsilon_0} \bullet \frac{qx}{(a^2 + x^2)^{3/2}} \text{ along PA.}$$

Example 14. Calculate the electric field at a distance r from an infinite line charge having linear charge density γ.

Solution. Let AB be an infinite line charge having linear charge density γ (Fig. 2.18). Consider a small element dy of the line charge. Then the charge on the element is $dq = \gamma dy$.

The electric field at P (at a distance r from the line charge) due to the element is

$$d\vec{E} = \frac{\lambda dy}{4\pi\varepsilon_0 {}^{12}} \hat{i} \textit{ where } \hat{i} \text{ is unit vector along PN.}$$

Resolving the vector dE in two components along PL and PM, we have

(i) component along $PM = dE_x = -dE \sin \theta$

(ii) component along $PL = dE_x = dE \cos \theta$

The negative sign of dE_y indicates that dEy points in negative y–direction. The resultant field E_x and E_y cab be calculated by integration.

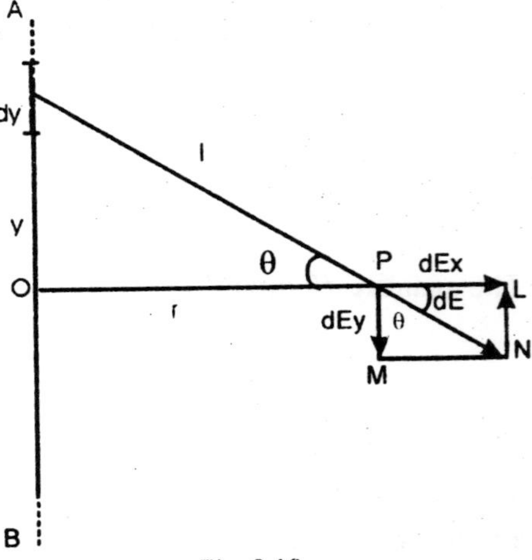

Fig. 2.18

$$E_y = \int dE_y = -\int_{-\infty}^{+\infty} \sin \theta dE$$

$$E_{\infty} = \int dE_{\infty} = -\int_{-\infty}^{+\infty} \cos \theta dE$$

E_y will be zero because for every charge element in the positive y-direction, there is corresponding element in the negative y-direction. Therefore, electric field contributions along $PM = 0$.

Hence the resultant field E will be normal to the line charge.

$\therefore$ Magnitude of $\vec{E} = E = E_{\infty} = \int_{-\infty}^{+\infty} \cos \theta dE$

$$= 2\int_0^{\infty} \cos \theta dE = 2\int_0^{\infty} \frac{\cos \theta \gamma dy}{4\pi\varepsilon_0 l^2}$$

$$= \frac{2\gamma}{4\pi\varepsilon_0}\int_0^{\infty} \frac{\cos \theta dy}{l^2} \qquad \text{...(a)}$$

Now from Fig. 2.12, $\tan \theta = \frac{y}{r}$ or $y = r \tan \theta$

or $$dy = r \sec^2\theta \, d\theta$$

When $y = 0$, $\theta = 0$; $y = \infty$, $q = \pi/2$

Putting the value of dy in equation (a), we have

$$E = \frac{\lambda}{2\pi\varepsilon_0}\int_0^{\pi/2} \frac{\cos \theta r \sec^2 d\theta}{l^2}$$

$$= \frac{\lambda}{2\pi\varepsilon_0}\int_0^{x/2} \frac{\cos \theta}{r}\left(\frac{r^2}{l^2}\right)\sec^2 \theta d\theta$$

$$= \frac{\lambda}{2\pi\varepsilon_0}\int_0^{x/2}\frac{\cos\theta}{r}\bullet\cos^2\theta\sec^2\theta d\theta$$

($\because$ r/l = cosθ from Fig. 2.18)

$$= \frac{\lambda}{2\pi\varepsilon_0 r}\int_0^{x/2}\cos\theta d\theta = \frac{\lambda}{2\pi\varepsilon_0 r}\sin\theta\Big|_0^{x/2}$$

$$= \frac{\lambda}{2\pi\varepsilon_0 r}\left[\sin\frac{\pi}{2} - \sin 0\right] = \frac{\lambda}{2\pi\varepsilon_0 r}$$

normal to the line charge.

Electric Flux

Let's assume an electric field $\vec{E}$ in space and in this space some arbitrary closed surface, like a ballon of any shape *S* be immersed in the electric field Fig. 2.19(a) shows such a surface, the field being suggested by few field lines. Let us suppose that the whole surface is divided into a large number of small patches or small elements [Fig 2.19(b)],

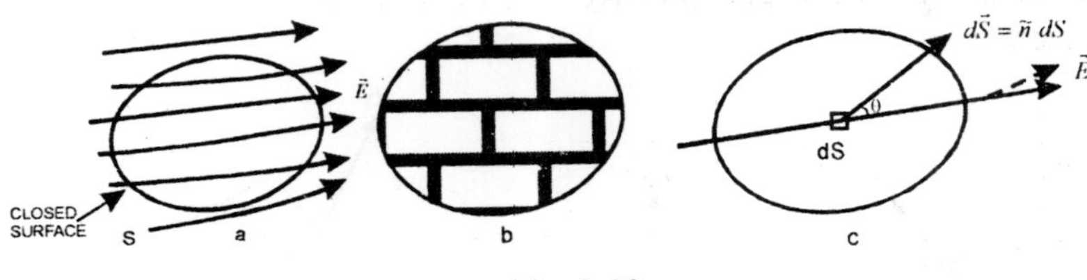

Fig. 2.19

which are so small that over any one patech the surface is practically flat and the field $\vec{E}$ does not change appreciably from one point of the patch to another. The area of each patch has a certain magnitude and defines an unique direction (art. 1.14b), which is outward normal to its surface (sine surface is closed, so normal is from inside to outside). If the magnitude and direction of area of one such patch be represented by a vector *d S* [Fig. 2.19 (c)], then

The electric flux over dS is defined as $d\phi = \vec{E}.d\vec{S}$

$$= \vec{E}.\hat{n}dS$$

The above product is a scalar number.

Now, $$\phi E = \sum_{alldS}\vec{E}.\hat{n}dS \quad ...(2.35)$$

Where ϕ_E is total flux through the entire surface which can be obtained by adding up the flux through all the patches.It is also a scalar quantity. Further let the patches become smaller and smaller in size i.e. *dS*—0, the sum given by equation (2.32) will tend to a surface integral taken over the entire closed surface. Thus

$$\phi E = \lim_{dS \to 0} \sum_{all\ dS} \vec{E}.\hat{n}dS = \oint_S \vec{E}\hat{n}dS$$

$= \oint_S E\cos\theta\, dS$, where q is the angle between vector

$= \vec{E}$ and vector $= \hat{n}$.

It should be noticed that the positivity, negativity or nullity of a flux through a particular patch depend on the relative orientations of $\vec{E}$ and d $\vec{S}$. Hence the flux, will be positive [Fig. 2.20(a)] if $\vec{E}$ and d $\vec{S}$ make an acute angle i.e., q<p/2 with each other, will be zero [Fig. 2.20 (b)] if $\vec{E}$ and d $\vec{S}$ are normal to each other i.e, q = $\pi/2$ and will be negative [Fig. 2.20 (c)] if $\vec{E}$ and d $\vec{S}$ make an obtuse angle with each other i.e., $\theta > \pi/2$.

Therefore, we must say that three conditions are posible, pertaining to the electric flux i.e. positive, negative or null.

The flux ϕ_E over the entire surface is positive If $\vec{E}$ is directed outward at every point on the surface because in this condition the angle θ between $\vec{E}$ and d $\vec{S}$ is acute every where on the surface. It means $\vec{E}$. d $\vec{S}$ is also positive. See Fig (2.21a)

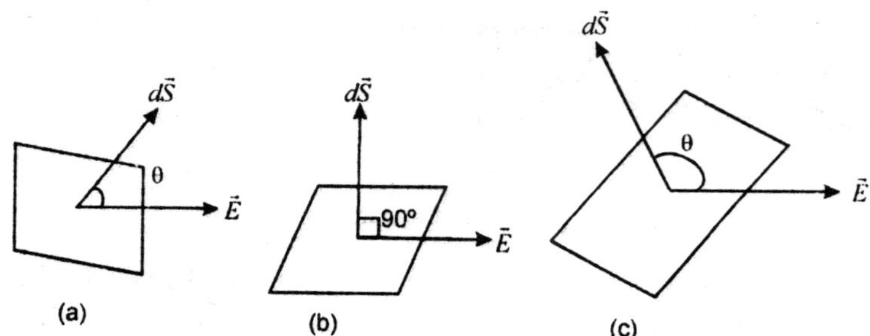

Fig. 2.20

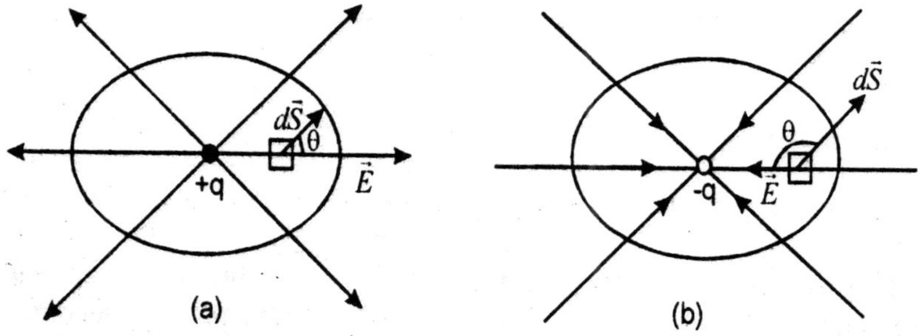

Fig. 2.21

(ii) if E is directed inward, then the angle q between E and d S is obtuse every where on the surface [Fig. 2.21 (b)] so E.dS is negative, hence the flux oE over the entire surface is negative.

Therefore, we must conclude; If E is directed in ward then the flux through closed surface is negative and if E is directed outward then the flux through surface is positive.

Gauss's Law

Gauss's. Law postulates the concepts about the relationship between electric flux and electric charges. It says that the total flux of the electric field E over any closed surface is equal to the total net charge enclosed by the surface divided by e if the charge is lying in vacuum.

Now, assume a single point charge +q enclosed by a hypothertical closed surface of any arbitrary shape at the point o. as shown in the Fig. 2.22.

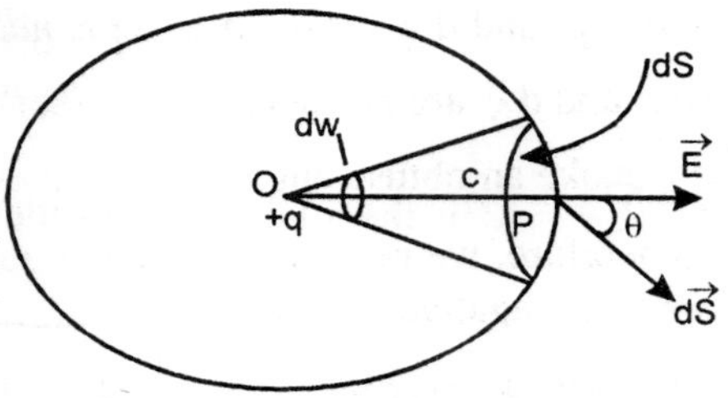

Let dS be a small area element at a distance 'r' from q and dw be a solid angle subtended by dS at 0.

Fig. 2.22

The electric flux over the closed surface is

$$\phi_E = \oint_S \vec{E}.d\vec{S} = \oint_S \vec{E}.\hat{n}dS$$

[Here, n is unit vector normal to dS.]

Now, intensity of Electric field at the point p on the surface is shown below:

$$\vec{E} = \frac{1}{4\pi\varepsilon_0}\frac{q\hat{r}}{r^2}$$

$$\therefore \quad \phi_E = \oint_S \frac{1}{4\pi\varepsilon_0}\frac{q}{r^2}(\hat{r}.\hat{n})dS$$

$$= \oint_S \frac{1}{4\pi\varepsilon_0}\frac{q}{r^2}\cos\theta dS\left(\because \hat{r}.\hat{n} = (1)(1)\cos\theta = \cos\theta\right)$$

$$= \oint \frac{q}{4\pi\varepsilon_0}dw$$

$$= \frac{q}{4\pi\varepsilon_0} \times 4\pi$$

($\because \oint_S dw = 4\pi$, as this integral represents the total solid angle subtencded by the closed surface at a point inside it)

Hence $$\phi_E = \frac{q}{\varepsilon_0}$$

There are a number of charges q1, q2 as shown in Fig. 2.23 enclosed by a surface *S*, then each charge will contribute to the flux. Let i^{th} charge's contribution to the flux.

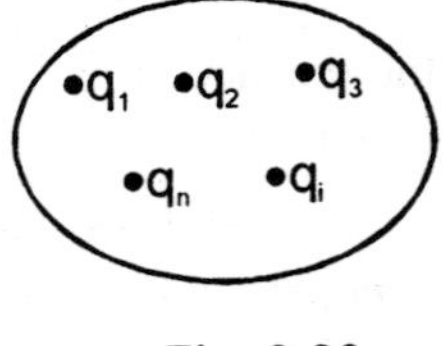

Fig. 2.23

$$\phi_{E6} = \frac{q_6}{\varepsilon_0}$$

Then the total flux

$$\phi_{E6} = \oint_S \vec{E}.\hat{n}dS = \sum_{i=1}^{n} \phi E6 = \sum_{i=1}^{n} \frac{q_6}{\varepsilon_0} = \frac{1}{\varepsilon_0}\sum_{i=1}^{n} q_1$$

As $\sum_{i=1}^{n} qi = Q$ is the total charge inside the surface, so

$$\boxed{\phi_E = \oint_S \vec{E} \bullet \hat{n}dS = \frac{Q}{\varepsilon_0}} \qquad \text{... (2.38)}$$

The total flux through a closed surface enclosing the net charge Q lying in vaccum is shown by the equation (2.38). Q is the total number of lines of force originating on Q and cutting a spherical surface surrounding it. It means that the total number of lines of force originating on a unit charge and cutting a sphere surrounding it is $\frac{1}{\varepsilon_0}$.

The closed surface taken in the electric field is called Gaussian surface.

Now, suppose that dielectric constant K is also in a medium in which the charge Q lies then the equation is as follows:

$$\phi_E = \oint_S \vec{E}.\hat{n}dS = \frac{Q}{K\varepsilon_0} \qquad \text{...(2.39)}$$

The modified form of equation (2.38) and (2.39) in cgs e.s.u. units is as follows.

$$\phi_E = \oint_S \vec{E}.\hat{n}dS = 4\pi Q$$

and

$$\phi_E = \oint_S \vec{E}.\hat{n}dS = \frac{4\pi Q}{K}$$

If the algebraic sum of the charges within the surface is zero i.e., $Q=0$, then $\phi E = 0$

Now, Let us examine : Do the charges outside the surface contribute to the value of ϕE. Assume a point charge $+q$ situated at '*O*' outside the closed surface and an elementary cone from O with small solid angle dw cut the closed surface at two elements of area dS_1 and dS_2 at the points A and B respectively (Fig. 2.24).

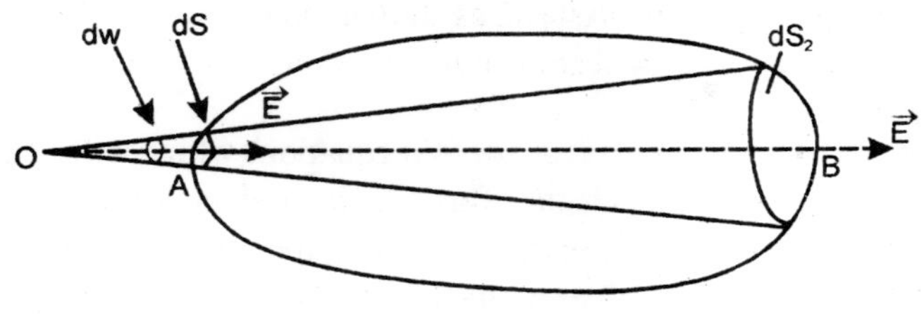

Fig. 2.24

As at A, the electric field E is directed inward, so the flux from the intercept $dS_1 = \dfrac{-qdw}{4\pi\varepsilon_0}$

$\therefore$ Total flux through these intercepts $= \dfrac{qdw}{4\pi\varepsilon_0} + \dfrac{qdw}{4\pi\varepsilon_0}$

$= 0.$

Thus the total flux over the whole closed surface is zero when the charge lies outside. This is quite evident because when the charge lies outside, the total number of lines of force entering the surface is the same as that leaving it, Hence, we must say that if some charges are outside the surface then they donot contribute to the value of ϕ E.

Gauss's Law–Maxwell's Equation

Karl Friedrich Gauss, a German mathematician who developed divergence theorem which we have discussed in previous chapter, popularly known by his name. Gauss's law constitutes one of the fundamental laws of electromagnetism.

According to Gauss Law that the total electric flux throught any closed surface is equal to the total charge enclosed by that surface.

Thus

$$\psi = Q_{enc} \qquad ...(2.40)$$

that is,

$$\Psi = \oint d\Psi = \oint_S D \bullet dS$$

$$= \text{Total charge enclosed } Q = \int \rho_v dv \qquad ...(2.41)$$

Where, D = $\vec{E}.\hat{h}$

or

$$Q = \oint_S D \bullet dS = \oint_v \rho_v dv \qquad ...(2.42)$$

By applying divergence theorem to the middle term in eqs. (2.42)

$$\oint_S D \bullet dS = \oint_v \nabla \bullet D dv \qquad '(2.43)$$

Comparing the two volume integrals in eqs. (2.42) and (2.43) results in

$$\rho_v = \nabla . D \qquad (2.44)$$

This is the first very equ. of maxwell out of four equations which will be derived. Accoridng to the equation, the divergence of the electric flux density and the volume charge density is the same. The definition of divergence given in previous chapter and the fact that ρv at a point is simply the charge per unit volume at that point should not be surprising for us.

Special Note :

(a) Here, In these above equations, Gauss law has been demostrated in different ways such as equation (2.44) is its differential form while equ (2.42) is its itegral form.

(b) Gauss's law is an alternative statement of Coulomb's law; proper application of the divergence theorem to Coulomb's law results in Gauss's law.

(c) For symmetrical charge distributions such as a point charge, an infinite line charge, an infinite cylinrical surface charge, and a spherical distribution of charge a simple method to find E or D is provided to us by Gauss's Law. A continuous charge distribution has rectangular symmetry if it depends only on x (or y or z), cylindrical symmetry if it depends only on p, or spherical symmetry if it depends only on r (independent of q and o). It must be stressed that awhether the charge distribution is symmetric or not, Gauss's law always holds. For example, consider the charge distribution in Figure 2.25 where v_1 and v_2 are closed surfaces (or volumes). The total flux leaving v_1 is Q–6 = 6nc because only 12nC and -6nC charges are enclosed by v_1. Although charges 30 nC and 15 nC outside v_1 do contribute to the flux crossing v_1, the net flux crossing v_1, according to Gauss's law, is irrespective of those charges outside v_1. Similarly, the total flux leaving v_2 is zero because no charge is enclosed by v_2. We may conclude here that even if the charge distribution is not symmetric, the Guass law, y = Q_{en} is used. We can not use the law to determine E or D. If the charge distribution is not symmetric then we use coulomb's Law to determine E or D.

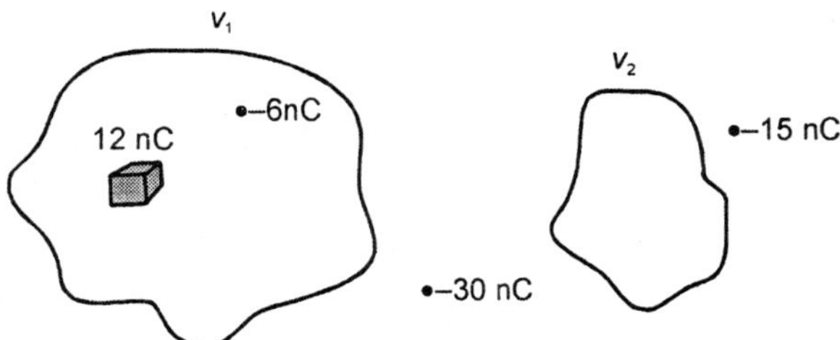

Fig. 2.25 Illustration of Gauss's law; flux leaving v_1 is 5 nC and that leaving v_2 is 0 C.

Applications of Gauss's Law

Before applying Gauss's Law, we first have to assess whether the symmetric charge distribution exists or not. If yes, then Gaussian surface is constructed. The surface is chosen such that D is normal ortangential to the Gaussian surface. When D is normal to the surface, D.dS = D dS because D is constant on the surface. When D is tangential to the surface, D.dS = 0. Thus we must choose a surface that has some of the symmetry exhibited by the charge distribution. We shall now apply these basic ideas to the following cases.

A. Point Charge

Assume a point charge Q is located at the origin. It is easy to see that choosing a spherical surface containing P will satisfy symmetry conditions to determine D at a point P. Hence, a spherical surface centered at the origin is the Gaussian surface in this case and is shown in Figure 2.26.

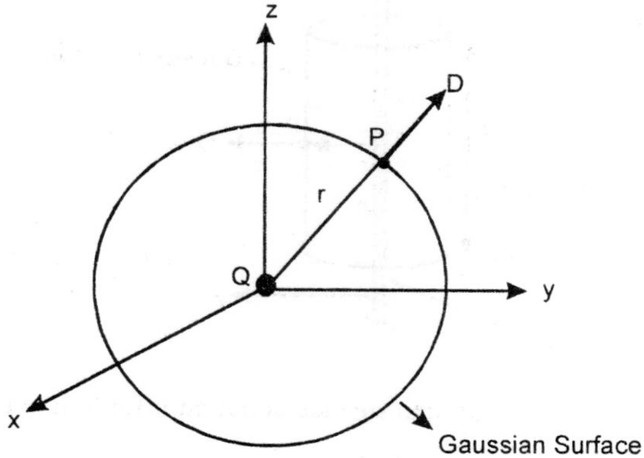

Fig. 2.26 Gaussian surface about a point charge

Since D is everywhere normal to the Gaussian surface, that is, D = D1ar, applying Gauss's law (Ψ= $Q_{enclosed}$) gives

$$Q = \oint D.dS = D_r \oint dS = D_r 4\pi r^2 \tag{2.45}$$

where is the surface area of the Gaussian surface is

$$\oint dS = \int_{\phi}^{2\pi} = 0 \int_{Q}^{\pi} = 0 \; r^2 \sin\theta d\phi = 4\pi r^2$$

Hence,

$$D = \frac{Q}{4\pi r^2} a_2 \tag{2.46}$$

B. Infinite Line Charge

We have to take a cylindrical surface containing P to satisfy symmetry condition as shown in the Fig (2.27) to determine D at the point P and also assume the infinite line of uniform charge PL C/m lies along the z-axis. D is constant on and normal to the cylindrical Gaussian surface; that is, D=$D_p a_p$. If we apply Gauss's law to an arbitrary length l of the line

$$\rho_L \ell = Q = \oint D.dS = D_\rho \oint dS = D_\rho 2\pi\rho\ell \qquad (2.47)$$

where $\oint dS = 2\pi\rho\ell$ is the surface area of the Gaussian surface. Note that D. dS evaluated on the top and bottom surfaces of the cylinder is zero since D has no z-component; that means that D is tangential to those surfaces. Thus

$$D = \frac{\rho_L}{2\pi\rho} a_\rho \qquad (2.48)$$

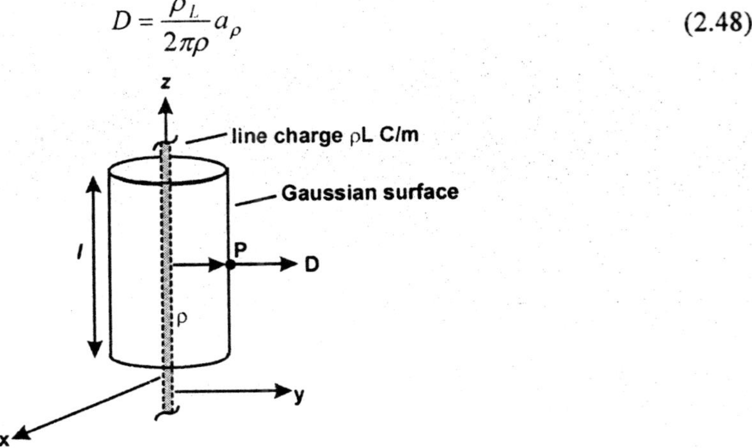

Figure (2.27) Gaussian surface about an infinite line charge.

C. Infinite Sheet of Charge

Now, we have to take a rectangular box which is cut symmetrically by the sheet of charge to determine D at a point p. The sheet of charge has two of its faces parallel to the sheet as shown in the Fig (2.28) and also assume that the infinite sheet of uniform charge ρs c/m² lying on the Z = 0 plane. As D is normal to the sheet, $D = D_z a_z$ and applying Gauss's law gives

$$\rho_S \int dS = Q = \oint D.dS = D_z \left[\int_{top} dS + \int_{bottom} dS \right] \qquad ...(2.49)$$

Note that **D**. *d*S evaluted on the sides of the box is zero because D has no components along a_x and a_y. If the top and bottom area of the box each has area A, eq. (2.49) becomes

$$\rho s A = D_z (A + A) \qquad (2.50)$$

and thus

$$D = \frac{\rho s}{2} a_z$$

or

$$E = \frac{D}{\varepsilon_0} = \frac{\rho s}{2\varepsilon_0} a_z \qquad (2.51)$$

D. Uniformly Charged Sphere

For two cases r≤a and r≥a, we have to prepare Gaussian surfaces separately to determine D every where. assume also a sphere of radius a with a uniform charge ρvC/M³. Since the charge has spherical symmetry, it is obvious that a spherical surface is an appropriate Gaussian surface.

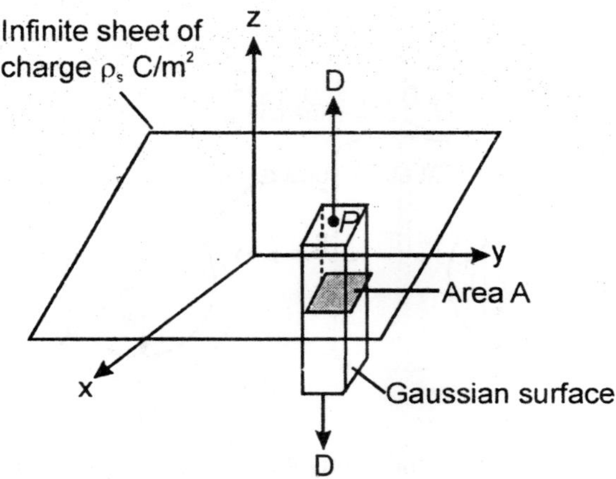

Figure (2.28) Gaussian surface about an infinite line sheet of charge.

For r ≤ a, the total charge enclosed by the spherical surface of radius r, as shown in Figure 2.29 (a), is

$$Q_{enc} = \int \rho_v dv = \rho_v \int_{\phi=0}^{2\pi} \int_{\theta=0}^{\pi} \int_{r=0}^{r} r^2 \sin\theta \, dr \, d\theta \, d\phi \qquad (2.52)$$

$$= \rho_v \frac{4}{3} \pi r^3$$

and

$$\psi = \oint D.dS = D_r \int_{\phi=0}^{2\pi} \int_{\theta=0}^{\pi} r^2 \sin\theta \, d\theta \, d\phi$$

$$= D_r 4\pi r^2 \qquad (2.53)$$

Hence, $\psi = Q_{enc}$ gives

$$D_r 4\pi r^2 = \frac{4\pi r^3}{3} \rho_v$$

or

$$D = \frac{r}{3} \rho_v a_r \qquad 0 < r \leq a \qquad (2.54)$$

For r ≥ a, the Gaussian surface is shown in Figure 2.29(b). The charge enclosed by the surface is the entire charge in this case, that is,

$$Q_{enc} = \int \rho_v dv = \rho_v \int_{\phi=0}^{2\pi} \int_{\theta=0}^{\pi} \int_{r=0}^{a} r^2 \sin\theta \, dr \, d\theta \, d\phi$$

$$= \rho_v \frac{4}{3} \pi a^3 \tag{2.55}$$

while

$$\psi = \oint D.dS = D_r 4\pi r^2 \tag{2.56}$$

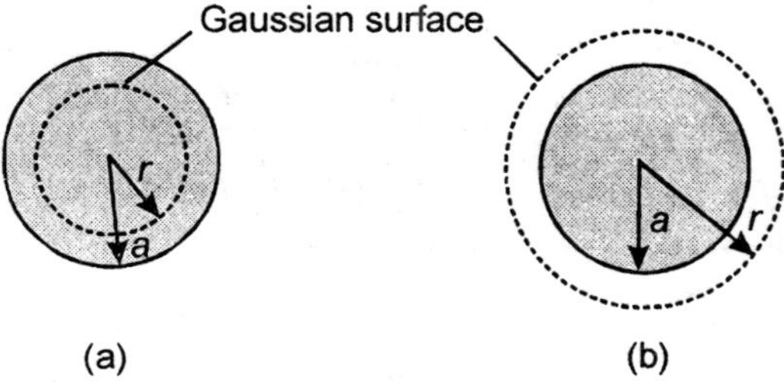

Figure 2.29 Gaussian surface for a uniformly charged sphere when: (a) r ≥ a and (b) r ≤ a.

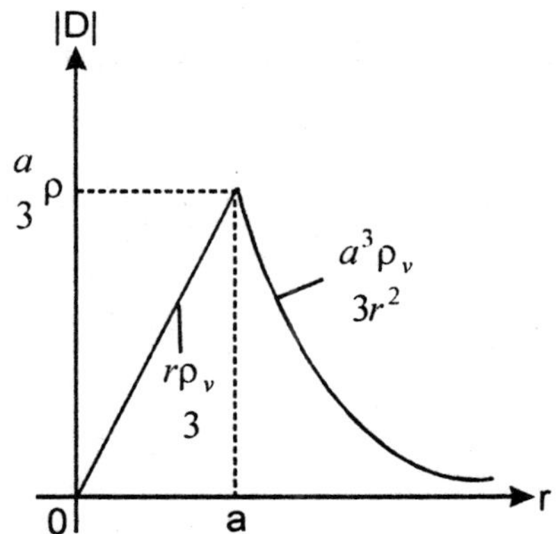

Figure 2.30 Sketch of |D| against r for a uniformly charged sphere.

just as in eq. (2.53) Hence:

$$D_r 4\pi r^2 = \frac{4}{3} \pi a^3 \rho_v$$

or

$$D = \frac{a^3}{3r^2} \rho_v a_r \qquad r \geq a \tag{2.57}$$

Thus from eqs. (4.53) and (4.56), D everywhere is given by

$$D = \begin{cases} \frac{r}{3}\rho_v a_r \\ \frac{a^3}{3r^2}\rho_v a_r \end{cases} \tag{2.58}$$

and |D| is as sketched in Figure 2.30

Notice from eqs. (2.45) (2.47), (2.49) and (2.53) that the ability to take D out of the integral sign is the key to finding D using Gauss's law. In other words, D must be constant on the Gaussian surface.

E. Field on the surface of a charged conductor : Assume acharged conductor of surface S with a surface charge density 6, placed in vacuum.

Now, assume an observation point p which is very close to surface the electric field $\vec{E}$ has to be found at this point P.

The direction of $\vec{E}$ for points close tot he surface is normal to the surface, pointing away from the surface if the charge is positive this is enidently proved by the symmetry. $\vec{E}$ would have a component lying in the surface If, it was not normal to the surface. Such a component would act on the charge carriers in the conductor and set up surface current. Since there are no such currents under the assumed electrostatic conditions, E must be normal to the surface.

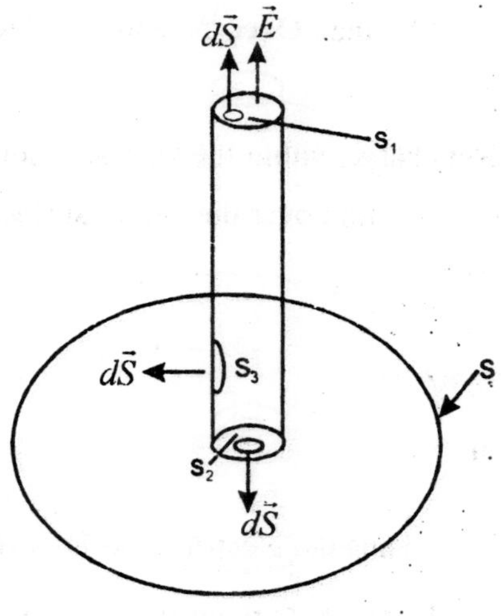

Fig. 2.31

By taking an imaginary cylinder (small in size), having area of cross-section Δ S The magniture of $\vec{E}$ anc be found, so that front plane surface, lies on point P and back plane surface inside the conductor, as Gaussian surface as shown in Fig. 2.26.

Our closed surface again consists of three surface S_1, S_2, and S_3. Therfore, total flux over the entire cylindrical surface.

$$\phi_E = \int \vec{E}.d\vec{S} = \int_{S_1} \vec{E}.d\vec{S} + \int_{S_2} \vec{E}.d\vec{S} + \int_{S_3} \vec{E}.d\vec{S}$$

The electric field is zero for S_2 because it lies within the conducting material

Hence, $\vec{E}.d\vec{S} = 0$

or $\int_{S_1} \vec{E}.d\vec{S} = 0$. Further $\vec{E}$ is parallel to curved surface and perpendicular to element area dS in surface S_3 so, on surface S_3, so, on surface s_3, $\vec{E}.d\vec{S} = 0$.

$$\int_{S_3} \vec{E}.d\vec{S} = 0$$

Therefore, we must say that S_1 is the only surface through which contribution to flux is possible.

$$\int_{S_1} \vec{E}.d\vec{S} = \int_{S_1} EdS \qquad (\because \vec{E} \text{ is directred parallel to element area } d\vec{S} \text{ on } S_1)$$

$$= E\int_{S_1} dS \qquad (\because \text{E is constant for all points in this small area})$$

$$= E\Delta S$$

where $\int_{S_1} dS$ gives the area of the surface S1 and is equal to ΔS.

So that, Over the entire cylindrical surface, the total flux is given by

$$\phi_E = E\Delta S \tag{2.59}$$

Net charge within the Gaussian surface = $\sigma\Delta S$

∴ flux over the closed surface as per the Gauss's Law :

$$= \frac{\sigma\Delta S}{\varepsilon_0} \tag{2.60}$$

or

$$\boxed{E = \frac{\sigma}{\varepsilon_0}} \tag{2.61}$$

Thus the electric field intensity at any point infinitely close to the surface of a charged conductor is $\frac{1}{\varepsilon_0}$ times the surface density of the charge on a surface. This is called Coulomb's Theorem. The direction of E is normal to the surface of the conductor.

In a media of dielectric constant *K*, the above equation becomes as

$$E = \frac{\sigma}{K\varepsilon_0} \tag{2.62}$$

In cgs e.s.u. units the above two equations (2.61) and (2.62) may be written as E = 4pq and $E = 4\pi\sigma \; and \; E = \frac{4\pi\sigma}{K}$

Mechanical force per unit area of a charged conducting surface **:** Assume that two points, lying in vacuum P_1 lies outside and P_2 lies inside the conductor as shown in the fig 2.32 which is infinitely close to the surface of charge density 6. Magnitude of Electric field at P1 is given by $E = \frac{\sigma}{\varepsilon_0}$ (according to Coulomb's theorem) and may be throught of as composed of the following two parts:

(i) Due to charge on a small part of the surface area say Δ S very near to the point P_1 the electric field at this point.

(ii) Due to charge on rest of the surface of the conductor the electric field at P_1.

As these two act in the same direction, so

$$E = E_1 + E_2 = \frac{\sigma}{\varepsilon_0}$$

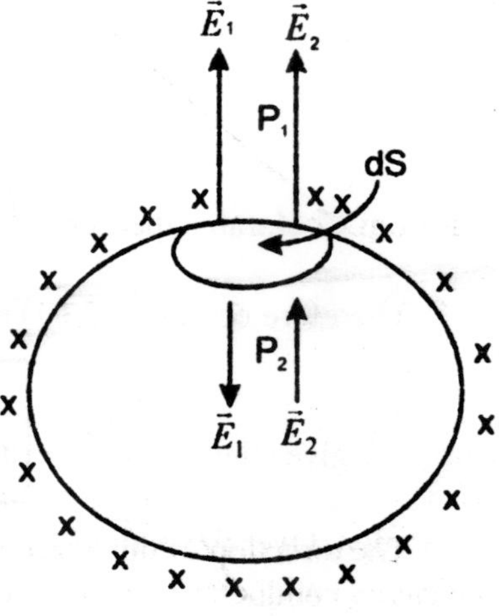

Fig. 2.32

Now consider the point P_2 lying just inside the conductor. The electric field at P_2. also consists of two parts.

(i) Due to the charge on the area element V S is $-E_1$ The electric field at point P_2. This part of the field is reversed in direction because P_2. lies on the opposite side of ΔS.

(ii) Due to rest of the surface is again E_2 and remains unaltered in direction the electric field at P_2.

Hence, the total field at P_2 is $=E_1+E_2$ awhich is equal to zero at P_2 lies inside the conductor i.e.

$$-E_1+E_2 = 0 \text{ or } E_1 = E_2$$

Putting this in equation (2.63), we get

$$E_1 + E_2 = 2E_1 = 2E_2 = \frac{\sigma}{\varepsilon_0}$$

or $$E_1 = E_2 = \frac{\sigma}{2\varepsilon_0}$$

Due to the charge present on the rest of the surface an upward force $\frac{\sigma}{2\varepsilon_0}$ is experienced by a unit positive charge on ΔS. This force acts along the outward drawn normal to the surface ΔS.

Total charge present on the area element $\Delta S = \sigma \Delta S$.

$\therefore$ force experienced by the area element ΔS of the charged surface =

$$F = \sigma \Delta S \times \frac{\sigma}{2\varepsilon_0} = \frac{\sigma^2 \Delta S}{2\varepsilon_0}$$

An area element ΔS of the charged conductor experiences the above mechanical force. Hence the mechanical force per unit area of the charged conductor is

$$P_E = \frac{\sigma^2}{2\varepsilon_0} \tag{2.64}$$

But from Coulomb's theorem, $E = \frac{\sigma^2}{\varepsilon_0}$ *or* $\sigma = \varepsilon_0 E$.

Therefore equation (2.64) may be written as

$$P_E = \frac{\varepsilon_0 E^2}{2} Nm^{-2} \tag{2.65}$$

The direction of this force will be along outward normal to the surface i.e. the surface of a charged conductor is thus alwyas under electrical pressure.

In a medium of dielectric constant K, the equation (2.65) becomes

$$P_E = \frac{\varepsilon_0 K E^2}{2} Nm^{-2} \tag{2.66}$$

In cgs esu units, the equations (2.65) and (2.66) become

$$P_E = \frac{E^2}{8\pi} dyne\ cm^{-2}$$

and $$P_E = \frac{KE^2}{8\pi} dyne\ cm^{-2}$$

Bulging of a Charged Soap Bubble

The gradual charging of a soap bubble when it will be found to increase in size demonstrate. The outward pressure experienced by the surface of a charged conductor theory of surface tension states that the pressure inside a soap bubble is more than the pressure outside it and this excess pressure (before being charged) inside a soap bubble of radius r due to surface tension T is given by

$$P = \frac{4T}{r}$$

If the bubble is now given a charge q, the surface density of charge is given by $\sigma = \frac{q}{4\pi r^2}$ and if the bubble is lying in vacuum the outward mechanical pressure experienced by the bubble is given by

$$\frac{\sigma^2}{2\varepsilon_0} = \frac{1}{2\varepsilon_0}\left(\frac{q}{4\pi r^2}\right)^2 = \frac{q^2}{32\varepsilon_0\pi^2 r^4}$$

∴ Net excess pressure inside the bubble over that outside, after being charged is

$$p = \frac{4T}{r} - \frac{q^2}{32\varepsilon_0\pi^2 r^4} = \frac{4}{r}\left[T - \frac{q^2}{128\varepsilon_0\pi^2 r^2}\right] \quad (2.67)$$

Thus, we may conclude here that the size of Bubble will increase because that the effective value of the surface tension of the bubble decreases on being charged.

Increase in radius of bubble can be calculated as follows. As the temperature of the bubble remains unaltered, Boyle's law must hold good during the expansion, so

$$p\alpha\frac{1}{V} \text{ or } p\alpha\frac{1}{r^3} = \frac{k}{r^3}$$

∴ $$\frac{dp}{dr} = \frac{-3k}{r^4} = \frac{-3p}{r}$$

or $$dr = -\left(\frac{r}{3p}\right)dp$$

Change in pressure due to charge 'q' (as calculated above)

$$-dp = \frac{q^2}{32\pi^2\varepsilon_0 r^4}$$

Hence $$dr = \frac{r}{3p}\frac{q^2}{32\pi^2\varepsilon_0 r^4} = \frac{q^2}{96\varepsilon_0 p r^3\pi^2} \quad (2.68)$$

$$P_E = \varepsilon_0\frac{E_2}{2}$$

Example 15. A charge distribution with spherical symmetry has density

$$\rho_v = \begin{cases} \frac{\rho_o r}{R} \\ 0, \end{cases} \quad 0 \le r \le R \; r > R$$

Determine E everywhere.

Solution : The charge distribution is similar to that in Figure 4.16. Since symmetry exists, we can apply Gauss's law to find E.

$$\varepsilon_0\oint E.dS = Q_{enc} = \int \rho_v dv$$

(a) For r < R

$$\varepsilon_0 E_r 4\pi r^2 = Q_{enc} = \int_0^r \int_0^\pi \int_0^{2\pi} \rho_v r^2 \sin\theta \, d\phi \, dr$$

$$= \int_0^r 4\pi r^2 \frac{\rho_0 r}{R} dr = \frac{\rho_0 \pi r^4}{R}$$

or

$$E = \frac{\rho_0 r}{4\varepsilon_0 R} a_r$$

(b) For r > R,

$$\varepsilon_0 E_r 4\pi r^2 = Q_{enc} = \int_0^r \int_0^\pi \int_0^{2\pi} \rho_v r^2 \sin\theta \, d\phi \, d\theta \, dr$$

$$= \int_0^R \frac{\rho_0 r}{R} 4\pi r^2 dr + \int_R^r 0.4\pi r^2 dr$$

$$= {}_{\pi\rho o}R^3$$

or

$$E = \frac{\rho_o R^3}{4\varepsilon_0 r^2} a_r$$

Exercise 5

A charge distribution in free space has ρv = 2r nC/m³ for $0 \leq r \leq 10$m and zero otherwise. Determine E at r = 2m and r = 12m.

Answer : 226ar V/m, 3.927ar kV/m/

Example 16. What is the charge density on the surface of a conducting sphere if the electric field near its surface is 1000 Vm-1.

Solution. Electric field near the surface is E = 1000 Vm-1.

Also electric field in air near the surface of a charged conductor with surface density σ is given by

$$E = \frac{\sigma}{\varepsilon_0} \quad \text{... (i)}$$

ε_0 is permittivity of free space = 8.85×10^{-12} C² N.

Putting for E and eo in (i), we have

$$1000 = \frac{\sigma}{8.85 \times 10^{-12}}$$

or $\sigma = 1000 \times 8.85 \times 10^{-12} = 8.85 \times 10^{-9}\ Cm^{-2}$

Example 17. A sphere of diameter 5 cm is charged to a potential of 1000 volts. Calculate the outward pull per unit area.

(K.U. 1986)

Solution. Potential of sphere, V = 1000 V

Radius of sphere $r = \frac{5}{2} = 2.5cm - .025m$

Now charge on the sphere

$$q = 4\pi\varepsilon_o Vr \qquad \left(\because Pot.\ V = \frac{q}{4\pi\varepsilon_0 r}\right)$$

$$= \frac{1000 \times .025}{9 \times 10^9} = \frac{25}{9 \times 10^9} C$$

∴ Surface charge density,

$$\sigma = \frac{q}{4\pi r^2} = \frac{25}{9 \times 10^9 \times 4\pi \times (.025)^2}$$

$$= \frac{1}{9\pi \times 10^5} cm^{-2}$$

Hence outward pull per unit area

$$= \frac{a^2}{2\varepsilon_0}$$

Putting for $\frac{1}{\varepsilon_0} = 36\pi \times 10^9$

$$= \left(\frac{1}{9\pi \times 10^5}\right)^2 \times \frac{36\pi \times 10^9}{2}$$

$$= \frac{1}{81\pi^2 \times (10^{10})} \times \frac{36\pi \times 10^9}{2}$$

$$= \frac{1}{9\pi \times 10^5} = .035 \times 10^{-5} Nm^{-2}$$

Example 18. A gold foil weighing 1 mg per square cm is placed on horizontal charged conductor. Calculate the surface density of the charge reduced by the conductor so that the gold foil may just rise?

Solution. Mass of gold foil per cm^2 = 1 mg = 10^{-6} kg

Mass of gold foil per m2 = $10^{-6} \times 10^4 = 10^{-2}$ kg

Let the surface charge density of the conductor be = σ

Mechanical force per unit area of a charged conductor

$$= \frac{\sigma^2}{2\varepsilon_0}$$

where e0 permittivity of free space

$$= 8.85 \times 10^{-12}\ C^2N^{-1}\ m^{-2}$$

As the weight of the gold foil is just balanced by the outward mechanical force, so

$$mg = \frac{\sigma}{2\varepsilon_0}$$

or $\sigma^2 = 2\varepsilon_0\ mg = 2 \times 8.85 \times 10^{-12} \times 10^{-2} \times 9.8$

or $\sigma = (2 \times 8.85 \times 9.8 \times 10^{-14})^{1/2}$

$\sigma = 1.317 \times 10^{-6}\ Cm^{-2}$.

Example 19. The air pressure is the same inside and outside a soap bubble of diameter 1 cm. Calculate the charge on the soap bubble if the surface tension is $3 \times 10^{-2}\ Nm^{-1}$.

Solution. Radius of the soap bubble

$$r = 1 \text{ cm} = .01 \text{ m}$$

Surface tension $T = 3 \times 10^{-2}$ Nm-1

As the pressure inside and outside is the same, so mechanical pressure of the charged bubble

$$P_E = \frac{4T}{r}$$

or
$$\frac{\sigma^2}{2\varepsilon_0} = \frac{4T}{r}$$

where s is the surface charge density.

If 'q' be the charge on the soap bubble, then

$$\sigma = \frac{q}{4\pi r^2}$$

$$\therefore \quad \left(\frac{q}{4\pi r^2}\right)^2 \times \frac{1}{2\varepsilon_0} = \frac{4T}{r}$$

or $$q^2 = \frac{16\pi^2 r^4 \times 2\varepsilon_0 \times 4T}{r} = \frac{128\pi^2 r^3 T}{36\pi \times 10^9}$$

$$\left(\because \varepsilon_0 = \frac{1}{36\pi \times 10^9}\right)$$

$$= \frac{32 \times \pi r^3 T}{9 \times 10^9}$$

or $$q = \left(\frac{32 \times \pi r^3 T}{9 \times 10^9}\right)^{1/2} = \frac{4}{3}\frac{r}{10^4}\left(\frac{2\pi r T}{10}\right)^{1/2}$$

$$= \frac{4}{3} \times \frac{.01}{10^4}\left(2 \times \frac{22}{7} \times \frac{.01 \times 3 \times 10^{-2}}{10}\right)^{1/2}$$

$$= \frac{4}{3} \times 10^{-6}\left(\frac{132 \times 10^{-4}}{70}\right)^{1/2} = \frac{4}{3} \times 10^{-8}\left(\frac{132}{70}\right)^{1/2}$$

$$= \frac{4}{3} \times 10^{-6} \times 1.373 = 1.830 \times 10^{-8} C$$

Example 20. An insulated soap bubble of 10 cm radius is charged with 7×10^{-9} C. Find the increase in radius due to the charge, neglecting surface tension effect. Given the atmospheric pressure = 10^5 Nm^{-2}.

Solution. If 'r' be the original radius of the bubble, then volume of the bubble.

$$V = \frac{4}{3}\pi r^3$$

and if P be the atmospheric pressure, then from Boyle's Law

PV=constant=k

or $$P = \frac{k}{V} = \frac{k}{\frac{4}{3}} = \frac{k}{r^3}$$

Putting $K = \dfrac{3k}{4\pi}$

If q be the charge given to the bubble, we can calculate increase in radius 'dr' of the bubble, proceeding in the same way as an article 2.14. Thus

$$dr = \frac{q^2}{96\pi^2 \varepsilon_0 P r^3}$$

Putting $q = 7\times10^{-9}$ C, $r = 10$ cm $= .1$m,

$P = 10^5$ Nm^{-2},

and $\frac{1}{\varepsilon_0} = 36\pi \times 10^9 \, Nm^2C^{-2}$, we get

$$dr = \frac{\left(7\times10^{-9}\right)^2 \times 36\pi \times 10^9}{96\pi^2 \times 10^5 \times (.1)^3} = \frac{49\times36}{96\times\frac{22}{7}} \times 10^{-11}$$

$= 5.84 \times 10^{-11}$ m.

Example 21. If the radius and surface tension of a spherical soap bubble be 'r' and 'T' respectively, prove that electric charge required to expand the bubble twice its linear dimensions would be $8\pi\sqrt{\varepsilon_0 r^3(7Pr + 12T)}$ where P is atmospheric pressure.

Solution. Given radius of bubble = r

Surface tension of bubble = T

Atmospheric pressure = P

If P1 be the internal pressure inside the bubble, then by the theory of surface tension.

$$P_1 = P + \frac{4T}{r}$$

Radius of the bubble after charging becomes 2r, so its volume will be 8 times the initial volume. Therefore, by Boyle's Law the pressure inside will become $\frac{P_1}{8}$. This decrease in pressure is because of the mechanical outward pressure due to the charge q given to the bubble.

Total outward pressure after charging the bubble

$$= \frac{P_1}{8} + \frac{\sigma^2}{2\varepsilon_0} = \frac{P_1}{8}\left(\frac{q}{4\pi(2r)^2}\right)^2 \frac{1}{2\varepsilon_0}$$

$$= \frac{1}{8}\left(P + \frac{4T}{r}\right) + \frac{q^2}{512\pi^2\varepsilon_0 r^4}$$

Total inward pressure

$$= P + \frac{4T}{2r}$$

But in equilibrium total outward pressure = total inward pressure

So $$\frac{1}{8}\left(P + \frac{4T}{r}\right) + \frac{q^2}{512\pi^2\varepsilon_0 r^4} = P + \frac{4T}{2r}$$

or $$q^2 = 512\pi^2\varepsilon_0 r^4\left(\frac{7}{8}P + \frac{12T}{8r}\right)$$

$$= 64\pi^2\varepsilon_0 r^3(7\,Pr + 12T)$$

or $$q = 8\pi\sqrt{\varepsilon_0 r^3(7\,Pr + 12T)}$$

Example 22. Find the flux throught the curved surface of a right circular cylinder of length 21 and radius 'b' produced by a single charge q at its geometrical centre.

Solution. Let us consider a closed cylinder of length 2/ and radius 'b' with) as the geometrical centre where charge q is placed (Fig. 2.33). For finding the flux throught the curved surface of the cylinder, let us first find the flux through the plane surfaces of the cylinder. Now to find the flux through, say, lower plane surfac of the cylinder, let us divide this plance surface into a large number of elementary rings concentric with the centre of lower plance surface (Fig. 2.39).

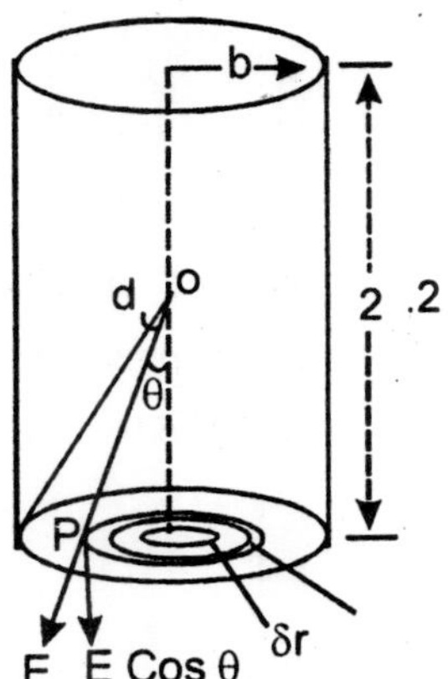

Fig. 2.33

This electric field at any point P of an annular ring between the radii r and r + 8r is

$$E = \frac{q}{4\pi\varepsilon_0(OP)^2}$$

$$= \frac{q}{4\pi\varepsilon_0\left(r^2 + l^2\right)}$$ directed along OP

Flux through a very small element of this ring is E cos q × area of element.

As E cos q is same ofor all such elements of this annular ring, so the flux through this annular ring is

$\delta\phi$ = E cos θ × area of ring

= E cos θ × 2 πr δr

$$= \frac{q \times 2\pi r \cos\theta \times \delta r}{4\pi\varepsilon_0 (r^2 + l^2)}$$

From Fig. 2.33, it is clear that

$$\cos\theta = \frac{l}{\sqrt{r^2 + l^2}}$$

$$\therefore \quad \delta\phi = \frac{qr(\delta r)}{2\varepsilon_0 (r^2 + l^2)} \frac{l}{\sqrt{r^2 + l^2}}$$

$$= \frac{qrl\delta r}{2\varepsilon_0 (r^2 + l^2)^{3/2}}$$

Total flux through the lower plane surface can be obtainied by integrating the above expression between the limits r = 0 to r = b. Hence

$$\int d\phi = \phi_{plane} = \int_0^b \frac{qrl\, dr}{2\varepsilon_0 (r^2 + l^2)^{3/2}}$$

$$= \frac{ql}{4\varepsilon_0} \frac{(r^2 + l^2)^{-1/2}}{\left(-\frac{1}{2}\right)}$$

$$= -\frac{ql}{2\varepsilon_0 \sqrt{r^2 + l^2}} \Bigg|_0^b$$

$$= \frac{ql}{2\varepsilon_0 l} - \frac{ql}{2\varepsilon_0 \sqrt{b^2 + l^2}}$$

$$= \frac{q}{2\varepsilon_0} \left(1 - \frac{1}{\sqrt{b^2 + l^2}}\right)$$

By symmetry consideration, we can say that flux throught the upper plane surface will also be the same. As the total flux throught the whole surface according to Gauss's Law is $\frac{q}{\varepsilon_0}$, therefore,

Flux through the curved surface

$$\phi_{euree} = \frac{q}{\varepsilon_0} - \frac{2q}{2\varepsilon_0}\left(1 - \frac{l}{\sqrt{b^2 + l^2}}\right)$$

$$= \frac{q}{\varepsilon_0}\left(1 - \frac{l}{\sqrt{b^2 + l^2}}\right)$$

Example 23. Find the solid angle in steradian, subtended at the centre of sphere of 2m diameter by a 100 cm2 area of its surface.

Solution. Radius of sphere = r = 1m

Area element, *ds* = 100 cm2 = 10^{-2} m^2

Solid angle subtended by ds at the centre of the sphere

$$= \frac{ds}{r^2} = \frac{10^{-2}}{1^2} = 10^{-2} \text{ steradian.}$$

Questions of Long Answer type

1. Write Coulomb's law in vector form. Give the meaning of each symbol. In what way is it different from Gauss's law? Explain.

(Pbi. U. 1986 A)

2. State and explain Coulomb's law of force between two point charges. Also explain the meaning of the term superposition principle.
3. Define electric charge and explain the principle of (i) conservation of charge and (ii) quantisation of charge.
4. Explain what do you understand by the term "electric field intensity"

(K.U. 1980, 1982)

5. (a) Explain the term solid angle. **(K.U. 1984 S)**

   (b) Define and explain the term electric flux. **(K.U. 1982)**
6. State and prove Gauss's law in electrostatics Hence derive an expression for mechanical force exerted o a unit area of a charged conductor.**(K.U. 1986 S; K.U. 1989 A, 93A)**
7. Using Gauss's law, prove that the magnitude of difference between the electric field injtensity vectors on either side of a very very large uniform plance sheet of charge (no conduction) is equal to $\frac{\sigma}{\varepsilon_0}$ where d is surface charge density.
8. Find the electrical intensity in front of a uniformly infinite plane. Indicate the effect on electrical intensity if the charge were spread on a plane conductor of infinite extent.

(K.U. 1987 A, 1995A)

9. Show that in the absence of any change in a given volume dive $\vec{E} = 0$

10. Derive an expression for electrical pressure acting normally outward on the surface of a charged conductor. **(K.U. 1995 A)**

11. Apply Gauss' law to calculate the electric field due to a charged infinite long straight wire carrying a charge y/metre at a point whose perpendicular distance from the wire is r metre.

(M.D.U 1986; K.U. 1993 A, 1995 S)

12. Prove Gauss's law of electrostatics. Write it in differential from. Prove that the electrical intensity at a point distant r from a charged wire of infinite length, with linear charge density y is given by $\vec{E} = \frac{\lambda}{2\pi\varepsilon_0 r}\hat{r}$ where r is unit vector along the distancer.

(M.D.U. 1980 : K.U. 1987 A, 92 S, 94A)

13. Gauss's law for gravitation is $\frac{1}{4\pi G}\oint g = \frac{1}{4\pi G}\int \vec{g}.\vec{ds} = m$, where m is the enclosed mass and G is the universal gravitational constant. Deduce Newton's law of gravitation from it.

14. Apply Gauss's law of find the electric field intensity inside and outside and uniformly charged spherical shell.

(K.U. 1993 A, 1994 S)

15. Show that mechanical force per unit area experienced by a charged surface is given by $\frac{\sigma^2}{2\varepsilon_0}$, where d is the surface charge density.

(K.U. 1994 S)

16. Obtain an expression for the electric field intensity just outside a charged conductor of any shape.

(M.D.U. 19784)

17. Prove that the energy associated with unit volume in an electric field is $E^2\varepsilon_0/2$ Jm^{-3}.

18. Find an expression for electrical energy stored in a charged conductor.

(K.U. 1984)

19. Prove that the electric charges outside a closed surface do not make any contribution to the electric flux.

(K.U. 1989 A)

20. Deduce an expression for the electrical pressure acting normally outward on the surface of a charged conductor.

(K.U. 1995 A)

21. State coulomb's law in electrostatics and deduce this law from Gauss law.

(K.U. 1992 S, 1993 S)

Questions of short Answer type.

1. A positively charged rod repels a suspended object. Can we conclude that the object is positively charged?
2. I the Coulomb force that one charge exert on another changes if a third charge is brought nearby?
3. A stone is placed in gravitational field of the earth. Can we say that the earth lies in the gravitational field of stone.
4. What is meant by point charge in physics? **(M.D.U. 1979)**
5. If the electric field E due to a point charge q is $\vec{E} = \frac{q}{4\pi\varepsilon_0}\frac{\hat{r}}{r^2}$ prove that curl $\vec{E}$ is zero.

   (M.D.U. 1982 S)
6. Will Gauss's law hold good, if we have an inverse cube law of forces?
7. If a charge q is moved out of a Gaussian surface, by how much will the electric flux increase, if at all
8. If the charge distribution within the Gaussian surface is altered in such a way that the total charge remains constant, will it change (a) the electic field distribution on the surface (b) the flux over the surface.
9. Is it possible to have a finite field due to an infinite charge distribution?
10. Give an example of a situation when the electric field is zero in a bounded region of space and different from zero in the rest of the region.
11. Why do not charge fly away from the charged surface, though the electric flux acts outward.

    (M.D.U. 1979)
12. A charged conductor has energy. Explain. ***(K.U. 1976)***
13. When charge is supplied to a hollow conductor the charge resides entirely ojn the outer surface. Explain.
14. A surface encloses an electric dipole. What you say about DE for this surface?
15. Is the superposition principle an independent principle or a part of Coulomb's law?
16. Two point charges are placed at a certain distance apart. The electric field strength is zero at one point between them. What can you conclude about the charges?
17. An electrostatic field is curl free. Why?

    (K.U. 1987 A)

    (Hint. E=-grad ϕ

    $\therefore$ curl E=-$\nabla\times(\nabla\phi)0$.
18. Show that there can be no electric field in the space within a hollow conductor if there is no charge within the surface.

    (M.D.U. 1982)

Problems

1. Point charges $Q_1 = 5$ μC and $Q_2 = -4$μC are placed at (3, 2,1) and (-4, 0, 6), respectively. Determine the force on Q_1.
2. Five identical 15-μC poit charge sare located at the center and corners of a square defined by -1 < x, y < 1, z = 0.
   (a) Find the force on the 10μC point charge at (0, 0, 2).
   (b) Calculate the electric field intensity at (0, 0, 2).
3. Point charges Q_1 and Q_2 are, respectively, located at (4, 0, -3) and (2, 0, 1). If $Q_2 = 4$ μC, find Q_1 such that
   (a) The E at (5, 0, 6) has no z-component
   (b) The force on a test charge at (5, 0, 6) has no x-component.
4. Charges + Q and +3Q are separated by a distance 2 m. A third charge is located such that the electrostatic systme is in equilibrium. find the location and the value of the third charge in terms of Q.
5. Determine the total charge
   (a) On line 0 < x < 5 m if $\rho_L = 12\ x^2$ C/m
   (b) On the cylinder p = 3, 0 < z < m if ps = ρz^2 nC/m²
   (c) Within the sphere r = 4 m if $\rho_v = \dfrac{10}{r \sin\theta} C/m^3$
6. Calculate the total charge due to the charge distributins labeled A, B, C in Fig. 2.34.
7. Find E at (5, 0, 0) due to charge distribution labeled *A* in Figure 2.34.
8. Due to the charge distribution labeled *B* in Figure 2.34
   (a) Find *E* at point (0, 0, 3) if $\rho s = 5$ mC/m²
   (b) find *E* at point (0, 0, 3) if $\rho s = 5 \sin f$ mC/m².
9. A circular disk of radius a carries charge $\rho_s = \dfrac{1}{\rho} C/m^2$. Calculate the potential at (0, 0, h).

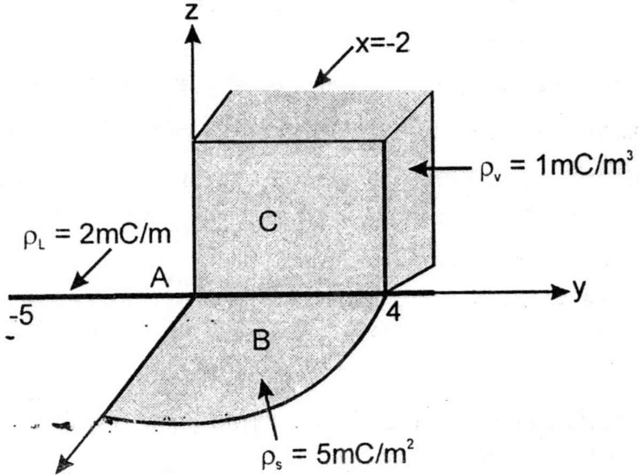

Figure 2.34 For Problem 6.

10. A ring placed along $Y^2 + z^2 = 4$, $x = 0$ carries a uniform charge of 5 μC/m.

    (a) Find **D** at $P(3, 0, 0)$.

    (b) If two identical point charges Q are placed at (0, -3, 0) adn (0, 3, 0) in addition to the ring, find the value of Q such that **D** = 0 at P.

11. (a) Show that the electric field at point (0, 0, h) due to the rectangle decribed by $-a < x < a$, $-b < y < b$, $z = 0$ carrying uniform charge ρ_s C/m^2 is

$$E = \frac{\rho_s}{\pi\varepsilon_0} tan^{-1}\left[\frac{ab}{h\left(a^2 + b^2 + h^2\right)^{1/2}}\right]a_z$$

    (b) If $a = 2$, $b = 5$, $\rho_s = 10^{-5}$, find the total charge on the plate and the electric field intensity at (0, 0, 10).

12. A point charge 100 pC is located at (4, 1, -3) while the x-axis carries charge 2 nC/m. If the plane z = 3 also carries charge 5 n C/m^2, find E at (1, 1, 1).

13. Line $x = 3$, $z = -1$ carries charge 20 nC/m while plane $x = -2$ carries charge 4 C/m2. Find the force on a point cdharge -5 mC located at the origin.

14. Point charges are placed at the corners of a square of size 4 m as shown in Figure 4.24. If Q = 15 μC, find **D** at (0, 0, 6).

15. Calculate the number of electrons flowing per second through a conductor of 16 when a potential difference of 19.2V is applied across the conductor.

    (Ans. 7.5×10^{18}**)**

16. How many electrons would have to be removed from a penny to leave it with a charge of + 10-7 C? **(Ans.** $625 \times 10^{+11}$**)**

17. In a neutral copper penny, the total positive and total negative charges be separated to a distance such that their force of attraction is 4.5 N. How far apart must they be?

    (Ans. 5.8×10^{10}m**)**

18. What is the electric field required to keep a liquid drop (density 900 kg m^{-8}) with one electron and of radius 10^{-6} cm just remain suspended?

19. Calculate the field intensity at the centre of a square of side r with four charges q, 2q, -4q and 2q situated in order at its four corners. [Ans. $\frac{10q}{4\pi\varepsilon_0 r^2}$ directed from q to -4q]

20. An electron is placed in an electric field and experiences an electric froce equal to its weight. Calculate the magnitude of the electric field intensity.

    [Ans. 5.6×10^{-11} **NC^{-1}]**

21. Two point charges 1 × 10-6 C and 2 × 10-6 C and are put at a distance of 10 cm from each other. Fing the position on the line joining the two charges where the electric field intensity may be zero. **[Ans.** 4142 cm from 1 × 10^{-6}C **charge]**

22. Three identical point charges of q coulomb are placed at the vertices of an equilateral triangle with 19 cm as side. Find the magnitude of force on each charge.

$$\left[Ans. \ \frac{\sqrt{3}}{4\pi\varepsilon_{a6}} \frac{q}{0.1} N \right]$$

23. What repulsive Coulomb force exists between two protons i a nucleus of iron. Assume a separation of 4.0×10^{-15} m.

[Ans. 14.4N]

24. PQRS is a square of 2 cm side, charges of 4×10^{-9}, -4×10^{-i} and 8×10^{-9} coulomb are put at the points P, R and S respectively. Find the electric field intensity due to these charges at the point Q.

$$\left[Ans. \ 9\sqrt{3} \times 10^4 \, NC^{-1}, \ at \ \angle\theta = 9^o - 44' \ with \ PQ \ produced \right]$$

25. Consider two identical particles of mass m and charge e. What must be the ratio of charge to mass if their gravitational attraction is to be equal in magnitude to the electrostatic repulsion? Calculate the ratio of electrostatic to gravitation force for an electron. The gravitation constant is 6.67×10^{-11} mks units, charge on electron is -1.6×10^{-19} C and its mass is 9.0×10^{-31} kg.

[Ans. (*i*) 8.608×10^{11}C kg-1 (*ii*) 4.27×10^{42}

26. Five thousand lines of force enter a certain volume of space and three thousand lines emerge from it. What is the total charge in coulombs within the volume ?

[Ans. 17.7 × 10+s C]

27. A-thin walled metal sphere has a radius of 25cm abd carries a charge of 2 × 10-7 C. Find electric field intensity for a point (a) inside the sphere, (b) just outside the sphere and (c) 3.00 m from the centre of sphere.

[Ans. (a)) (b) 2.88×10^4 NC^{-1} (c) 2×10^2 NC-1]

28. You are given two concentric spherical shells of radii a and b (b > a) and insulated from each other. The inner shell carries a charge +q and the outer shell carries a charge -q. Find the electric field at a point P lying at a distance r from the centre when (i) r > b (ii) a < r< b (iii) r < a. $\left[Ans. \ 0, \frac{q}{4\pi\varepsilon_0 r^2}, 0 \right]$

29. A point charge 0.1 × 10-6 is kept at the centre of a cube of side 10 cm. Calculate the electric flux through one of the cube faces.

[Ans. $6\pi \times 10^2$ S.I. units]

30. The electric field near the earth's surface is 300 Vm-1 directed downwards. What is the surface density of charge on the earth's surface.

[Ans. 26.55×10^{-70} C/m²]

31. 16 gm of copper has been beaten into a thin foil of area 1 m^2. A small piece is taken out of it and placed on the conductor. What must be the denisity of the charge so that the foil may just rise?

[**Ans.** 3.328×10^{-6} Cm^{-2}]

32. Calcuate the charge that must be put on a sphere of radius 20 cm so that the repulsion per m2 on the surface just balance the atmospheric pressure. Given atmospheric pressure=76 cm of Hg and density of Hg = 13.6 gm cm-3.

[**Ans.** 0.67×10^{-3} C]

33. An infinitely long wire has charge 2 × 10-6 Cm -1. Calculate the intensity of the electric field at a point distant 10 cm normal to the length of wire.

[**Ans.** 36×10^4 NC^{-1}]

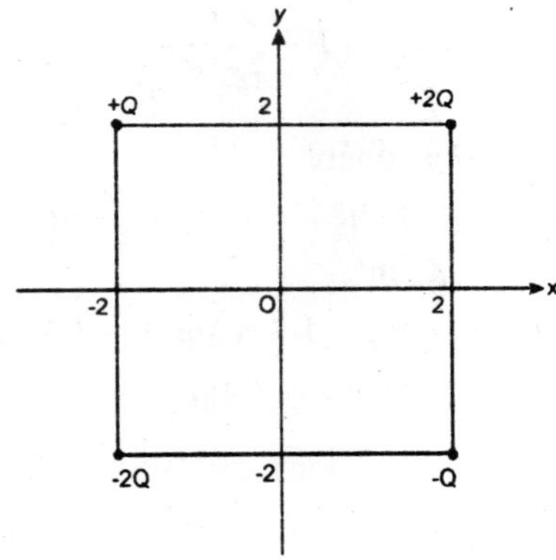

34. State Gauss's law. Deduce Coulomb's law from Gauss's law therby affirming that Gauss's law is an alternative statement of Coulomb's law and that coulomb's law is implicit in Maxwell's equation ∇. D = ρ_v.

35. Determine the charge density due to each of the following electric flux densities :

(a) D = 8xyax + 4x2ay C/m2

(b) D = r sin f ar + 2r cos Q a r + 2z2az C/m2

(c) D = $D = \frac{2\cos\theta}{r^3} a_r \frac{\sin\theta}{r^3} a_\theta \; C/m^2$

36. Let E = $xya_x + x^2a_y$, find

(a) Electric flux density **D**.

(b) The volume charge density ρ_v.

37. Plane $x + 2y = 5$ carries chargeρs =6 nC/m^2. Determining E at (-1, 0, 1).

38. In free space, D = $2y^2a_x + 4xy - a_z$ mC/m^2. Find the total charge stored in the region 1 < x < 2, 1 < y < 2, -1 <z <4.

39. In a certain region, the electri field is given by

$$D = 2\rho(z+1)\cos\phi a\rho - \rho(z+1)\sin\phi a_\phi + a_z \mu C / m^2$$

(a) Find the charge density.

(b) Calculate the total charge enclosed by the volume $0 < \rho < 2, 0 < \phi < \pi/2, 0 < z < 4$.

(c) Confirm Gauss's law by finding the net flux through the surface of the volume in (b).

40. The Thomson model of a hydrogen atom is a sphere of positive charge with an electron (a point charge) at its center. The total positive charge equals the electronic charge. Prove that when the electron is at a distance *r* from teh center of the sphere of positive charge, it is attracted with a force

$$F \frac{e^2 r}{4\pi\varepsilon_0 R^3}$$

where *R* is the radius of the sphere.

41. Three concentric spherical shells $r = 1$, $r = 2$, and $r = 3$ m, respectively, have charge distributions 2, -4, and 5 $\mu C/m^2$.

(a) Calculate the flux through r = 1.5m and r = 2.5 m.

(b) Find D at $r = 0.5$, $r = 2.5$, ad $r = 3.5$m.

42. Given that

$$\rho_v = \begin{matrix} 12\rho nC/nm^3 & 1<\rho<2 \\ 0, & otherwise \end{matrix}$$

Determine D everywhere.

43. Let

$$\rho_v = \begin{matrix} \frac{10}{r^2} mC/m^3 & 1 < r < 4 \\ 0, & \end{matrix}$$

r > 0

(a) Find the net flux crossing surface r = 2 m and r = 6 m.

(b) Determine *D* at r = 1 m and r = 5m.

44. Find the work done in carrying a 5-C charge from P(1, 2, -4) to R(3, -5, 6) in an electric field

$E = a_x + z^2 a_y + 2yza_z$ V/m

45. Given that the electric field in a certain region is

$E = (z + 1) \sin \phi a_\rho + (z + 1) \cos a_\phi + \rho \sin \phi a_z$ V/m determine the work done in moving a 4-nC charge from

(a) $A(1, 0\ 0)$ to $B(4, 0, 0)$

(b) $B(4, 0, 0)$ to $C(4, 30°, 0)$

(c) $C(4, 30, 0)$ to $D(4, 30°, -2)$

(d) A to D

46. In an electric field E = 20r sin q ar + 10r cos q aq V/m, calculate the energy expended in transferring a 10-nC charge

(a) From $A(5, 30°, 0°)$ to B(5, 90°, 0°)

(b) From A to C (10, 30°, 0°)

(c) From A to D (5, 30°, 60°)

(d) From A to $E(10, 90°, 60°)$

3

Electric Potential

Electric Potential

We have studied in previous classes that the change in potential energy of the body is equal to the work done while displacing a body from one point to another in the gravitational field. It can be shown in integral form in following way.

$$dw = \int_a^b \vec{F}.d\vec{l}$$

Here, the gravitational force acting on the body is $\vec{F}$ and displacement along the path a and b is dl. But we know that earth's gravitational field, a vector field, is a conservative field. Further for a conservative field the work done in moving a body from one point to another is independent of path taken between the two points, it depends only on the co-ordinates of the initial and final points.

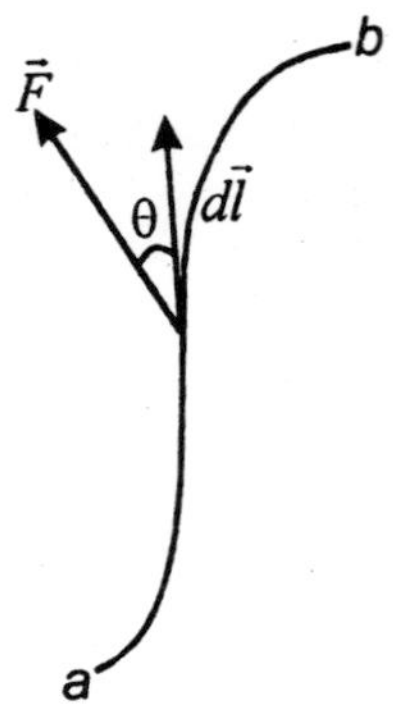

Fig. 3.1

Now, let's prove that the concept of work and energy is equally important for electrostates as for gravitational field. For this purpose we have to show that the electric field is conservative. Assume a test charge +q at the observation point P in the electrostatic field due to a single fixed point charge Q. Let $\vec{E}$ be the field intensity at P due to the point charge Q. The electric field exerts a force $q\,\vec{E}$ on the test charge as shown in Fig. 3.1.

It is well known concept that the body is accelerated when a force act on it. The velocity of the particle will increase in the process of acceleration which is produced by a text charge, left alone in the electricfield and the energy will be supplied by the electric field to generate this additional kinetic-energy. In order that the test charge may not be accelerated along the direction of field, an additional external force $\vec{F}$ exactly equal to $-Q\vec{E}$ must be applied on the test charge (Fig. 3.2). Then the total force on test charge will be zero and it will be in equilibrium.

Suppose test body is now moved through a displacement $d\vec{l}$, then the amount of work done by the external force in this displacement is $= -Q\vec{E}.d\vec{l}$

The negative sign indicdates that the work is being done by an external agent. Thus the total work done, or the potential energy required, in moving Q from A to B is

$$W = -Q\int_{A}^{B} E.dl \tag{3.1}$$

Dividing W by Q in eq. (4.59) gives the potential energy per unit charge. This quantity, denoted y V_{AB}, is known as the potential difference between points A and B. Thus

$$V_{AB} = \frac{W}{Q} = -\int_{A}^{B} E.dl \tag{3.2}$$

Specific Note

(a) A is the initial point while B is the final point in determining V_{AB}.

(b) there is a loss in potential energy in moving Q from A to B. If V_{AB} is negative this implies that the work is being done by the field. However, if V_{AB} is positive, there is a gain in potential energy in the movement; an external agent performs the work.

(c) V_{AB} is independent of the path taken (to be shown a little later).

(d) The measurement of VAB is done in volt which is also called joules/couloms.

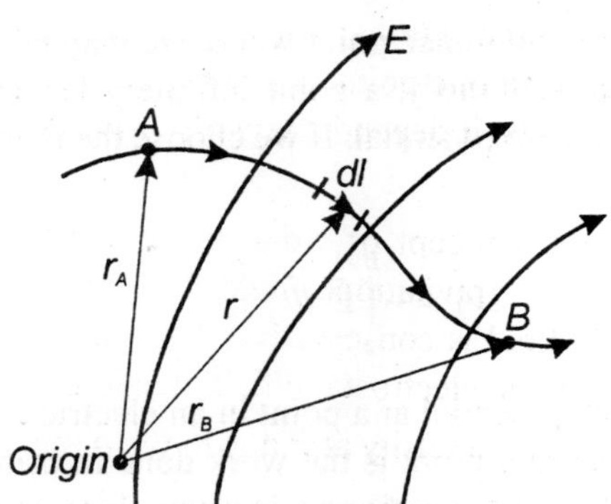

Figure 3.2 Displacement of point charge Q in an electrostatic field E.

As an example, if the E field in Figure 1.3 is due to a point charge Q located at the origin, then

$$E = \frac{Q}{4\pi\varepsilon_0 r^2} a_r \tag{3.3}$$

so eq. (4.60) becomes

$$V_{AB} = -\int_{r_A}^{r_B} \frac{Q}{4\pi\varepsilon_0 r^2} a_r . dra_r \tag{3.4 a}$$

$$= \frac{Q}{4\pi\varepsilon_0}\left[\frac{1}{r_B} - \frac{1}{r_A}\right]$$

or

$$V_{AB} = V_B - V_A \tag{3.4b}$$

where V_B and V_A are the potentials (or absolute potentials) at B and A, respectively. Thus the potential difference V_{AB} may be regarded as the potential at B with reference to A. In problems involving point charges, it is customary to choose infinity as reference; that is, we assume the potential at infinity is zero. Thus if $V_A = 0$ as $r_A \to \infty$ in eq. (3.4), the potential at any point ($r_B \to r$) due to a point charge Q located at the origin is

$$V = \frac{Q}{4\pi\varepsilon_0 r} \tag{3.5}$$

Note from eq. (3.4a) that because E points in the radial direction, any contribution from a displacement in the θ or ϕ direction is wiped out by the dot product E. $dl = E \cos q \, dl = E \, dr$. Hence the potential difference V_{AB} is independent of the path as aserted earlier.

Potential

Let's define potential, The potential at any point is the potential difference between that point and a point where the potential is zero.

We now choose a suitable reference point where we may take the potential to be zero. It is often convenient to choose arbitrarily, a point infinitely far away from any finite charge distribution as the reference of zero potential. If we choose the point 'a' at infinity, then Va = 0, so equation (3.6) reduces to

$$V_b = -\int_{\infty}^{B} \vec{E} d\vec{l} \quad ...(3.6)$$

This equation (3.6) defines the potential at a point in an electric field due to any finite charge distribution. Thus the potential at a point is the work done in bringing a unit positive charge from infinity to that point, without acceleration against electric force i.e. the physical significance of potential is that it is the potential energy of the unit charge brought from infinity to the specific point in space. Its unit in S.I. system is volt (V).

Bearing in mind the assumption made about the reference position, we find:

(i) The potential V near an isolated positive charge will be positive because positive work must be done by an outside agent on a unit positive charge to pust it in from infinity, against repulsive electrosatic force.

(ii) Similarly, the potential V near an isolated negative charge will be negative, as in this case the outside agent will have to work in pushing the unit charge away from the given negative charge against the attractive electrostatic force.

Potential due to a point charge

Assume that a point charge + Q is at point O and there is a point P at r distance from O at which we have to find the potential. According to equation (3.6) it means that we are to find out the potential at P which is by definition.

$$V = -\int_{\infty}^{P} \vec{E} d\vec{l}$$

Here the intensity of electric field at point P is $\vec{E}$. From Fig 3.3 it is clear that electric intensity vector $\vec{E}$ is directed towards right and $d\vec{l}$ which is always in the direction motion of the unit positive charge, points to the left. Thus the angle between $\vec{E}$ and $d\vec{l}$ is 180° so that

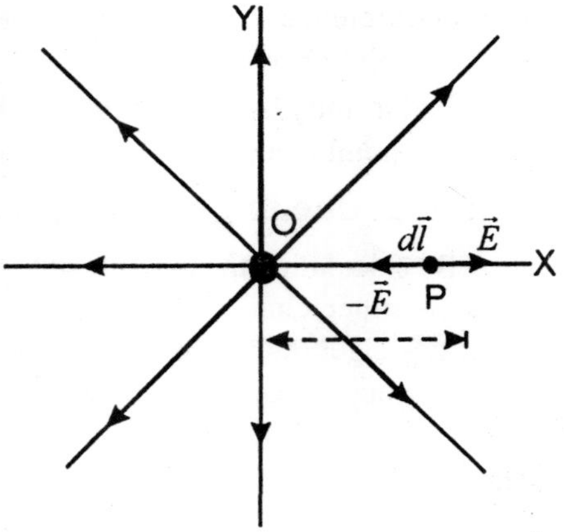

$$\vec{E}.d\vec{l} = E\, dl \cos 180^0 = -Edl$$

As we are considering r from Q, if we move a distance $d\vec{l}$ to the left, we are moving in the direction of decreasing r.

Hence, dl = – dr

so that, $\vec{E}.d\vec{l} = Edl = Edr$

Fig. 3.3

Therefore,
$$V = -\int_{\infty}^{r} Edr \qquad (3.7)$$

If the charge Q is situated in vacuum, then the field intensity at P is given by $E = \dfrac{Q}{4\pi\varepsilon_0 r^2} dr$ where ε_0 is the permittivity of free space.

Putting for E in (3.8), we have

$$\therefore \quad V = -\int_{\infty}^{r} \frac{Q}{4\pi\varepsilon r^2} dr$$

$$= -\frac{Q}{4\pi\varepsilon_0} \int_{\infty}^{r} \frac{dr}{r^2}$$

$$= -\frac{Q}{4\pi\varepsilon_0}\left[-\frac{1}{r}\right]_{\infty}^{r} = \frac{Q}{4\pi\varepsilon_0 r}$$

Therefore;
$$V = \frac{Q}{4\pi\varepsilon_0 r} \qquad ...(3.8)$$

If, there is medium of dielectric constant K in which Q is situated then the equation is as follows :

$$\boxed{V = \frac{Q}{4\pi\varepsilon_0 Kr}} \qquad ...(3.9)$$

The results (3.9) and (3.10) in cgs units are respectively written as

$$V = \frac{Q}{r} \text{ and } V = \frac{Q}{Kr}$$

Potential due to group of point charges

By calculating the potential due to each individual charge as if the other charges were absent and then adding these quantities algebraically we confind the potential at any point due to a group of point charges. Considering a group of charges $q_1, q_2 .... q_n$, we find the potential at P at a distance r_i from ith charge q_i (Fig. 3.4).

So potential at P due to ith charge q_i is

$$V_i = \frac{q_i}{4\pi\varepsilon_0 r_i}$$

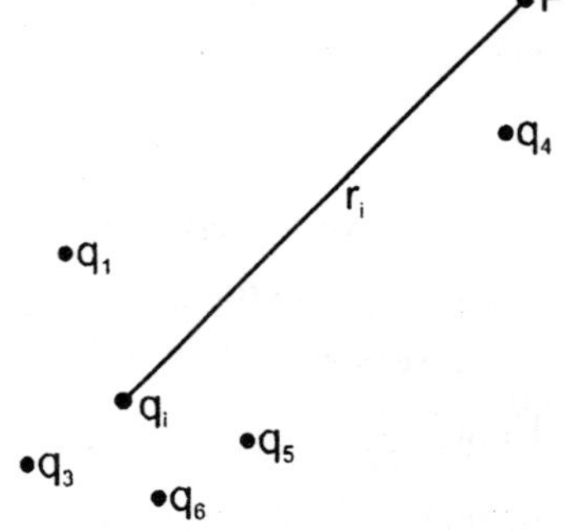

Fig. 3·4

Total potential at P due to group of charges like this is

$$V = \sum_{i=1}^{n} V_i = \sum_{i=1}^{n} \frac{q_i}{4\pi\varepsilon_0 r_i} \qquad ...(3.10)$$

This sum here is algebraic sum and not a vector sum like the one used to calculate $\vec{E}$ for a group of charges.

If, on the other hand, the charge distribution is continuous rather than being collection of points, the charge can be considered to be divided into small elements dq and the summation in equation (3.10) is replaced by an integral is

$$V = \int dV = \int \frac{dq}{4\pi\varepsilon_0 r} \qquad ...(3.11)$$

where r is the distance of the charge element dq from the point at which V is to be found and dV is the potential which charge element dq establishes at the point.

Relationship between E and V-Maxwell's Equation

We have already seen above that the value of potential is different at different points. It is a scalar quantity and also, the potential difference between points A and B is independent of the path taken. Therefore,

$$V_{BA} = -V_{AB}$$

that is,

$$V_{BA} + V_{AB} = \oint E.dl = 0$$

or

$$\boxed{\oint E.dl = 0} \qquad (3.12)$$

This shows that the line integral of E along a closed path as shown in Figure (3.5) must be zero. Physically, this implies that no net work is done in moving a charge along a closed path in an electrostatic field. Applying Stokes's theorem to eq. (3.12) gives

$$\oint E.dl = \int (\nabla \times E).dS = 0$$

or

$$\boxed{\nabla \times \mathrm{E} = 0} \qquad (3.13)$$

As we have discussed earlier, a field is said to be conservative if that vector field satisfy the equation (3.13) or equ (3.12). Thus an electrostatic field is a conservative field. Equation (3.12) or (3.13) is referred to as Maxwell's equation (the second Maxwell's equation to be derived) for static electric fields. Equation (3.12) is the integral form, and eq. (3.13) is the differential form; they both depict tha conservative nature of an electrostatic field.

From the way we defined potential, $V = -\int E.dl$, it follows that

$$dV = - E.dl = -E_x dx - E_y dy - E_z d_z$$

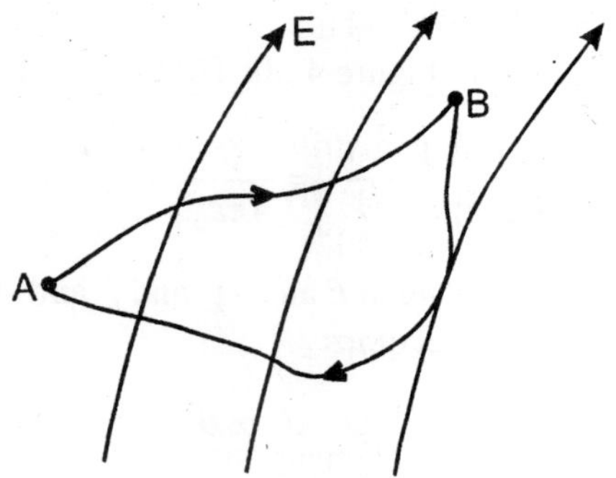

Fig. 3.5 Conservative nature of an electrostatic field.

But

$$dV = \frac{\partial V}{\partial x}dx + \frac{\partial V}{\partial y}dy + \frac{\partial V}{\partial z}dz$$

we obtain the following equations after comparing the two above equations of dV.

$$E_x = -\frac{\partial V}{\partial x}, \qquad E_y = -\frac{\partial V}{\partial y}, \qquad E_z = -\frac{\partial V}{\partial z} \tag{3.14}$$

Hence,

$$E = -\nabla V \tag{3.15}$$

Therefore, we must say that the gradient of V is the intensity of electric field and the negative sign demonstrates that E is directed from higher to lower levels of V. Since the curl of the gradient of a scalar function is is always zero ($\nabla \times \nabla V = 0$) eq. (3.13) obviously implies that E must be a gradient of some scalar function. Thus eq. (3.15) could have been obtained from eq. (3.13).

Equation (3.15) shows another way to obtain the E field apart from using Coulomb's or Gauss's law. That is, if the potential field V is known, the E can be found using eq. (3.15). One may wonder how one function V can possibly contain all the information that the three components of E carry. The three components of E are not independent of one another : They are explicitly interrelated by the condition $\nabla \times E = 0$.

Exercise 1

Given that $E = (3x^2 + y)\, a_x + x a_y$ k V/m, find the work done in moving a -2 μC charge form (0, 5, 0) to (2, -1, 0) by taking the path

(a) $(0, 5, 0) \rightarrow (2, 5, 0) \rightarrow (2, -1, 0)$

(b) $y = 5 - 3x$

Answer: (a) 12mJ, (b) 12 mJ.

An Electric Dipole and Flux Lines

If, two point charges of equal magnitude but opposite sign are kept at a small distance then they tunction as an electric dipole.

Consider the dipole shown in Figure 4.20. The potential at point $P(r, \theta, \phi)$ is given by

$$V = \frac{Q}{4\pi\varepsilon_0}\left[\frac{1}{r_1} - \frac{1}{r_2}\right] = \frac{Q}{4\pi\varepsilon_0}\left[\frac{r_2 - r_1}{r_1 r_2}\right] \tag{3.16}$$

where r_1 and r_2 are the distances between P and $+Q$ and P and $-Q$, respectively. If $r >> d$, $r_2 - r_1 = d \cos \theta$, $r_2 r_1 = r_2$, and eq. (3.16) becomes

$$V = \frac{Q}{4\pi\varepsilon_0}\frac{d\cos\theta}{r^2} \tag{3.17}$$

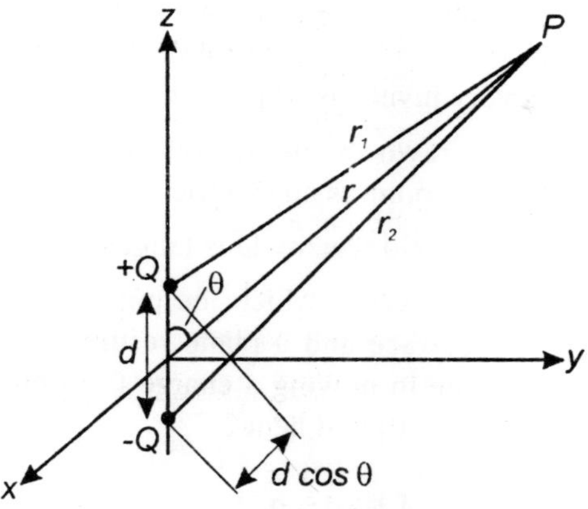

Figure 3.6 An electric dipole.

Since $d \cos \theta = d.\, a_r$, where $d = da_z$, if we define

$$p = Qd \tag{3.18}$$

as the dipole moment, eq. (3.17) may be written as

$$V = \frac{p.a_r}{4\pi\varepsilon_0 r^2} \quad ...(3.19)$$

The equation (3.19) converts as follows if dipole centre is at r' rather than at origin. Point should be noted that the dipole moment P is directed from -Q to +Q.

$$V(r) = \frac{p.(r - r')}{4\pi\varepsilon_0 |r - r'|^3} \quad ...(3.20)$$

Due to the dipole with center at the origin, shown in Figure 3.60 We be obtained readily from eqs. (3.16) and (3.17) as

$$E = -\nabla V = -\left[\frac{\partial V}{\partial r} a_r + \frac{1}{r}\frac{\partial V}{\partial \theta} a_\theta\right]$$

$$= \frac{Qd \cos\theta}{2\pi\varepsilon_0 r^3} a_r + \frac{Qd \sin\theta}{4\pi\varepsilon_0 r^3} a_\theta$$

or

$$\boxed{E = \frac{p}{4\pi\varepsilon_0 r^3}(2\cos\theta\, a_r + \sin\theta\, a_\theta)} \tag{3.21}$$

where $p = |p| = Qd$.

We must conclude here that the electricfield due to dipole varies inversely as r^3 whereas electric potential due to dipole varies ar r^2. as it is evident from equ. (3.19) and (3.21). The electric fields due to successive higher-order multipoles (such as a quadrupole consisting of two dipoles or an octupole consisting of two quadrupoles) very inversely as r^4, r^5, r^6, . . . while their corresponding potentials very inversely as r^3, r^4, r^5, ...

An imaginary path or line drawn in such a way that its direction at any point is the direction of the electric field at that point is an electric flux line.

We must say that the electric field density D is tangential at every point of these lines.

Equipotential surface is a surface on which the potential is the same throughout. The intersection of an equipotential surface and a plane results in a path or line known as an equipotential line. No work is done in moving a charge from one point to another along an equipotential line or surface (V_A-V_B = 0) and hence

$$\int E.dl = 0 \tag{3.22}$$

on the line or surface. From eq. (3.22), we may conclude that the lines of force or flux lines (or the direction of E) are always normal to equipotential surfaces. Examples of equipotential surfaces for point charge and a dipole are shown in Figure 3.7. Note from these examples that the direction of E is everywhere normal of the equipotential

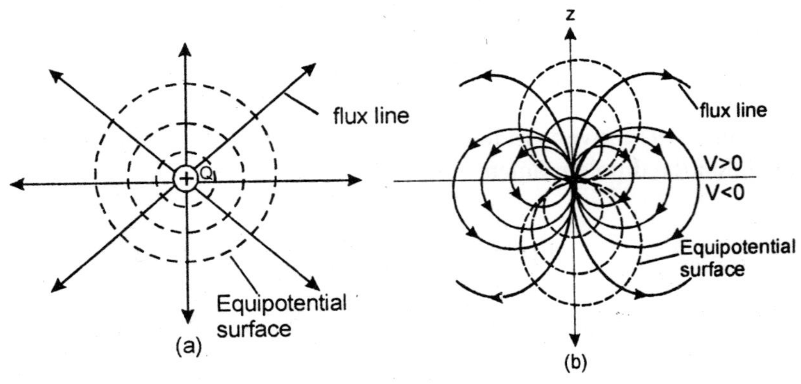

Figure 3.7 Equipotential surfaces for (a) a point charge and (b) an electric dipole

Figure 3.7 Equipotential surfaces for (a) a point charge and (b) an electric dipole lines. We shall see the importance of equipotential surfaces when we discuss conducting bodies in electric fields; it will suffice to say at this point that such bodies are equipotential volumes.

Special Note : Michael Faraday in his experiment used the idea of electric flux lines. Hence, he was the first who gave its concept.

Energy Density in Electrostatic Fields

Assume that in an initially empty space we want to place three point charges Q_1, Q_2 and Q_3, Since the space is initially empty, no work is done to place Q_1 at P_1 as shown in Fig. 3.8. The work done in transferring Q_2 from infinity to P_2 is equal to the product of Q_2 and the potential V_{21} at P_2 due to Q_1. Similarly, the work done in positioning Q_3 at P_3 is equal to Q_3 (V_{32}

$+ V_{31}$), where V_{32} and V_{31} are the potentials at P_3 due to Q_2 and Q_1, respectively. Hence the total work done in positioning the three charges is

$$W_E = W_1 + W_2 + W_3$$
$$= 0 + Q_2 V_{21} + Q_3 (V_{31} + V_{32}) \quad (3.23)$$

If the charges were positioned in reverse order,

$$W_E = W_3 + W_2 + W_1$$
$$= 0 + Q_2 V_{23} + Q_1 (V_{12} + V_{13}) \quad (3.24)$$

where V_{23} is the potential at P_2 due to Q_3, V_{12} and V_{13} are, respectively, the potentials at P_1 due to Q_2 and Q_3. By the addition of equation (3.23) and (3.24) we get,

$$2W_E = Q_1 (V_{12} + V_{13}) + Q_2(V_{21} + V_{23}) + Q_3(V_{31} + V_{32})$$
$$= Q_1V_1 + Q_2V_2 + Q_3V_3$$

or

$$W_E = \frac{1}{2}(Q_1V_1 + Q_2V_2 + Q_3V_3) \quad (3.25)$$

where V_1, V_2 and V_3 are total potentials at P_1, P_2 and P_3 respectively. In general, if there are n point charges, eq. (3.25) becomes

$$W_E = \frac{1}{2}\sum_{k=1}^{n} Q_k V_k \quad \text{(in joules)} \quad (3.26)$$

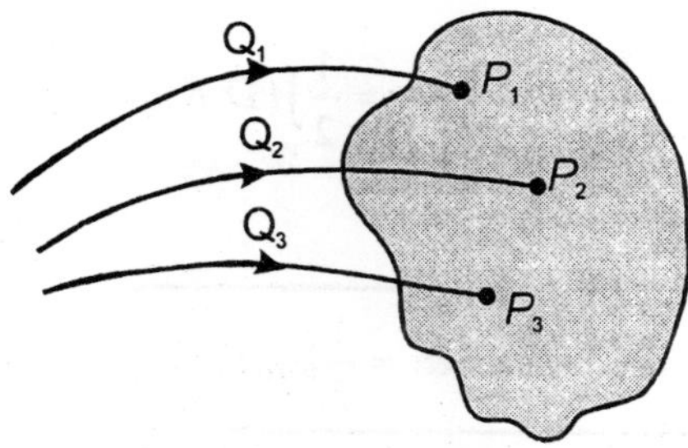

Figure 3.8 Assembling of charges.

If, the region has a continuous charge distribution rather than point charges then addition of equation (3.26) converts in integral form as given below

$$W_E = \frac{1}{2}\int \rho_L V\, dl \quad \text{(line charge)} \quad (3.27)$$

$$W_E = \frac{1}{2}\int \rho_L V\, dS \quad \text{(surface charge)} \quad (3.28)$$

$$W_E = \frac{1}{2}\int \rho_V V\, dV \quad \text{(volume charge)} \quad (3.29)$$

Since rv = V . D, eq. (3.29) can be further developed to yield

$$W_E = \frac{1}{2}\int_v (\nabla . D)V\, dv \qquad (3.30)$$

But for any vector A and scalar V, the identity

$$\nabla .\, VA = A . \nabla V + V(\nabla . A)$$

or

$$(\nabla .A)\, V = \nabla .\, VA\text{-}A.\nabla V \qquad (3.31)$$

holds. Applying the identity in eqs. (3.30) to (3.31), we get

$$W_E = \frac{1}{2}\int_v (\nabla .VD)dv - \frac{1}{2}\int_v (D.\nabla V)dv \qquad (3.32)$$

By applying divergence theorem to the first term on the right-hand side of this equation, we have

$$W_E = \frac{1}{2}\oint_v (\nabla .VD)dv - \frac{1}{2}\int_v (D.\nabla V)dv \qquad (3.33)$$

We recall that V varies as $1/r$ and D as $1/r^2$ for point charges; V varies as $1/r^2$ and D as $1/r^3$ for dipoles; and so on. Hence, V**D** in the first term on the right-hand side of eq (3.33) must vary at least as $1/r^3$ while dS varies as r^2. Consequently, the first integral in eq. (3.33) must tend to zero as the surface S becomes large. Hence, eq. (3.33) reduces to

$$W_E = -\frac{1}{2}\int_v (D.\nabla V)dv = \frac{1}{2}\int_v (D.E)dv \qquad (3.34)$$

and since E = - ∇V and D = ε_oE

$$\boxed{W_E = -\frac{1}{2}\int D.E\, dv = \frac{1}{2}\int \varepsilon_0 E^2 dv} \qquad (3.35)$$

From this, we can define electrostatic energy density w_E (in J/m^3) as

$$W_E = \frac{dW_E}{dv} = \frac{1}{2}D.E = \frac{1}{2}\varepsilon_0 E^2 = \frac{D^2}{2\varepsilon_0} \qquad (3.36)$$

so eq. (3.34) may be written as

$$W_E = \int w_E dv \qquad (3.37)$$

Mislleneous Examples

Example 1. Three point charges - 1 nC, 4 nC, and 3 nC are located at (0, 0, 0), (0, 0, 1), and (1, 0, 0), respectively. Find the energy in the system.

Solution.

$$W = W_1 + W_2 + W_3$$

$$= 0 + Q_2 V_{21} + Q_3(V_{31} + V_{32})$$

$$= Q_2 \cdot \frac{Q_1}{4\pi\varepsilon_0 |(0,0,1) - (0,0,0)|}$$

$$+ \frac{Q_3}{4\pi\varepsilon_0}\left[\frac{Q_1}{|(1,0,0) - (0,0,0)|} + \frac{Q_2}{|(1,0,0) - (0,0,1)|}\right] + \frac{1}{4\pi\varepsilon_0}\left(Q_1Q_2 + Q_1Q_3 + \frac{Q_2Q_3}{\sqrt{2}}\right)$$

$$= \frac{1}{4\pi.\dfrac{10^{-9}}{36\pi}}\left(-4 - 3 + \frac{12}{\sqrt{2}}\right).10^{-18}$$

$$= 9\left(\frac{12}{\sqrt{2}} - 7\right) \text{ nJ} = 13.37 \text{ nJ}$$

Alternatively,

$$W = \frac{1}{2}\sum_{k=1}^{3} Q_k Q_k = \frac{1}{2}(Q_1V_1 + Q_2V_2 + Q_3V_3)$$

$$= \frac{Q_1}{2}\left[\frac{Q_2}{4\pi\varepsilon_0(1)} + \frac{Q_3}{4\pi\varepsilon_0(1)}\right] + \frac{Q_2}{2}\left[\frac{Q_1}{4\pi\varepsilon_0(1)} + \frac{Q_3}{4\pi\varepsilon_0\left(\sqrt{2}\right)}\right] + \frac{Q_3}{2}\left(\frac{Q_1}{4\pi\varepsilon_0(1)} + \frac{Q_2}{4\pi\varepsilon_0\left(\sqrt{2}\right)}\right)$$

$$= \frac{1}{4\pi\varepsilon_0}\left(Q_1Q_2 + Q_1Q_3 + \frac{Q_2Q_3}{\sqrt{2}}\right)$$

$$= 9\left(\frac{12}{\sqrt{2}} - 7\right) \text{ nJ} = 13.37 \text{ nJ}$$

as obtained previously.

Example 2. What is the potential at a point situated at a distance of 3 m from a positive charge of 1 uC in vacuum.

Solution. Here charge $q = 1\ \mu\text{C} = 10^{-6}$ C

Distance $r = 3$ m

Permittivity of free space $\varepsilon_0 = \frac{1}{36} \times 10^{-9}$ MKS units

$$\therefore \quad \text{Potential, } V = \frac{q}{4\pi\varepsilon_0 r} = \frac{10^{-6}}{4\pi \times \frac{1}{36\pi} \times 10^{-9} \times 3} V$$

$$= 3 \times 10^8 \text{ V}$$

This example emphasizes that a large potential arises in the field of a very small static charge.

Example 3. Eight charges having values as shown in Fig. 3.7 are arranged symmetrically on a circle of radius 10 cm in air. What is the potential at the centre O ?

Solution. Potential at O contributed by each charge is given by equation (3.5) with $r = 10$ cm $= 1.1$ *m, i.e.,*

$$V_n = \frac{q_n}{4\pi\varepsilon_9 \times .1}$$

$$= \frac{q_n \times 9 \times 10^9}{.1}$$

$$= q_n \times 9 \times 10^{10} \text{ volt}$$

Total potential at O is

$$V = \sum_{n=1}^{8} q_n 9 \times 10^{10}$$

$$= 9 \times 10^{10} (+3 - 1 - 7 + 5 + 7 + 15 + 10^{-6} \text{ V}$$

$$= 27 \times 9 \times 10^{10} \times 10^{-6}$$

$$= 243 \times 10^4 \text{ V.}$$

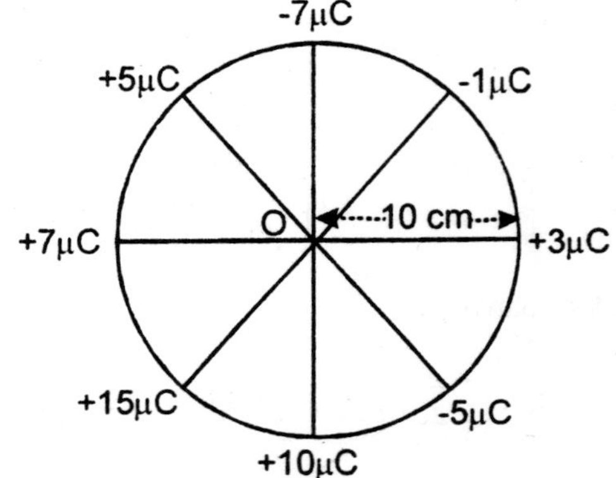

Figure 3.9

Example 4. Following Vector Function represents a possible electrostatic field

$$E^x = 6xy,\ E_y = 3x^2 - 3y2\ Es = 0$$

Calculate the line integral of E from the point $A(0, 0, 0)$ to the point $C(x_1, y_1, 0)$ along the path that runs straight from $A(0, 0, 0)$ to $B(x, 0, 0)$ and then to $C(x_1, y_1, 0)$

Solution. The electrostatic field $E = 6xy, \hat{i} + (3x^2 - 3y)\hat{j}$. The vector representing the path element $d\vec{l} = \hat{i}dx + \hat{j}dy$ *where* $\hat{i}$ and $\hat{j}$ are the unit vectors along x and y directions respectively.

$$\therefore \quad \vec{E}.d\vec{l} = \left(6xy\,\hat{i} + (3x^2 - 3y^2)\hat{j}\right).\left(\hat{i}dx + \hat{j}dy\right)$$

$$= 6xydx + (3x^2 - 3y^2)dy$$

Along the path A(0, 0, 0) to B(x_1, 0,0), y=0, *i.e.*, $dy = 0$

$$\therefore \quad \int_A^B \vec{E}.d\vec{l} = \int_{(0,0,0)}^{(x_1 0,0)} 6xydx = \left.\frac{6x^2y}{2}\right|_{(0,0,0)}^{(x_1,0,0)} = 0$$

Along the path from B(x_1, 0, 0) to C (x_1, y_1, 0) ; $x = x_1$, $dx = 0$

$$\therefore \quad \int_B^C \vec{E}.d\vec{l} = \int_{(x_1,0,0)}^{(x_1 y_1,0)} \left(3x^2 - 3y^2\right)dy = 3x^2y - \left.\frac{3y^3}{3}\right|_{(x_1,0,0)}^{(x_1,y_1,0)}$$

Hence total line integral along the path is $y_1\ (3x_1^2 - y_1^2)$

Example 5. Prove that the field given by $\vec{F} = \left[a\sqrt{x}\hat{i} + \sqrt{y}\hat{j}\right]$ represents an electrostatic field.

Solution. An electrostatic field must be conservative for which line integral along two suitable paths between two points must be the same.

Let us consider two paths (i) one from (0, 0) to (x, 0) and then from (x, 0) to (x, y) (ii) other from (0, 0) to (x, y)

(i) Line integral for the first path where

$$\vec{F}.d\vec{l} = a\left[\sqrt{x}\hat{i} + \sqrt{y}\hat{j}.\right]\left(\hat{i}dx + \hat{j}dy\right)$$

$$= a\left(\sqrt{x}dx + \sqrt{y}dy\right) \text{ is given by}$$

$$\int_{first\ path} \vec{F}.d\vec{l} = a\int_{0,0}^{x,0}\left(\sqrt{x}dy + \sqrt{y}dy\right) + a\int_{(x,0)}^{(x,y)}\left(\sqrt{x}dy + \sqrt{y}dy\right)$$

$$= a\left(\frac{x^{3/2}}{\frac{3}{2}} + \frac{y^{3/2}}{3/2}\right) + a\left.\left(\frac{x^{3/2}}{3/2} + \frac{y^{3/2}}{3/2}\right)\right|$$

$$= a\frac{2}{3}x^{3/2} + a\frac{2}{3}y^{3/2} = \frac{2}{3}a\left(x^{3/2} + y^{3/2}\right)$$

Line integral for the second path

$$\int \vec{F}.d\vec{l} = a\int_{0,0}^{(x,y)}\left(\sqrt{x}dx + \sqrt{y}dy\right) = a\left.\left(\frac{x^{3/2}}{3/2} + \frac{y^{3/2}}{3/2}\right)\right|_{0,0}^{x,y}$$

Second path

$$= \frac{2a}{3}\left(x^{3/2} + y^{3/2}\right)$$

Thus the line integral for both the paths between the points A and B is the same. Hence $F = a\left(\sqrt{x}\hat{i} + \sqrt{y}\hat{j}\right)$ represents an electrostatic field.

Example 6. A point charge 5×10^{-9} C is situated at the origin of co-ordionates. Fing the potential difference between the points A (1, 2, 2) and $B\left(\sqrt{5}, 2, 4\right)$

Solution. The position vector of the point A is

$$\vec{r}_A = \hat{i} + 2\hat{j} + 2\hat{k}$$

$$\therefore \qquad r_A = \sqrt{1^2 + 2^2 + 2^2} = \sqrt{9} = 3.$$

The position vector of the point B is

$$\hat{r}_B = \sqrt{5}\hat{i} + 2\hat{j} + 4\hat{k}$$

$$\therefore \qquad = \sqrt{(5)^2 + 2^2 + 4^2} = \sqrt{25} = 5$$

potential difference between the points A and B is

$$V_B - V_A = \frac{q}{4\pi\varepsilon_O r_B} - \frac{q}{4\pi\varepsilon_O r_A}$$

$$= \frac{q}{4\pi\varepsilon_O}\left(\frac{1}{r_B} - \frac{1}{r_A}\right)$$

Substituting for q, ε_o r_A and r_B, we have

$$V_B - V_A = 5 \times 10^{-9} \times 9 \times 10^9\left(\frac{1}{5} - \frac{1}{3}\right)$$

$$= 45 \times \left(-\frac{2}{15}\right) = -6V$$

Example 7. What must the magnitude of an isolated positive point charge before the electric potential at 10 cm from the charge to be + 100 V?

Solution. Potential due to a point charge q at a distance

r is given by $\qquad V = \dfrac{q}{4\pi\varepsilon_0 r}$

Here V= 100 V

r = 10 cm = .1m

$$\therefore \qquad q = V \times 4\pi\varepsilon_0 r = \frac{100 \times .1}{9 \times 10^9}\left(\because \frac{1}{4\pi\varepsilon_0} = 9 \times 10^9\, Nm^2C^{-2}\right)$$

$$= 1 \cdot 1 \times 10^{-9}\ C.$$

Example 8. A circular disc of radius 'R' is uniformly charged with a charge a per unit area. Determine the potential at a point P on the axis of the disc at a distance r from the centre.

Solution. Let the flat circular disc of radius 'R' be divided into a large number of annular rings concentric with the disc. Let P be the observation point on the axis of the disc at a distance r from its centre. Consider one such annular ring of radius y and width dy (shown by dotted line in Fig. 3·8). Now amount of charge contained in the ring=dq=a($2\pi y$) dy where ($2\pi y$) dy is the area of the ring.

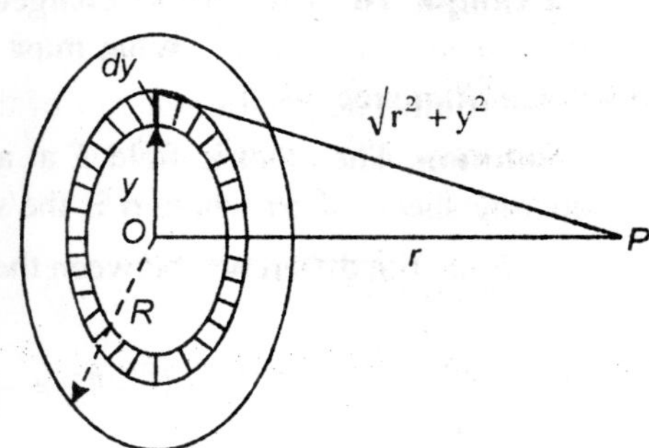

Fig. 3.10

All the points lying on the ring which we have considered as charge element are at the same distance $r' = \sqrt{r^2 + y^2}$ from the axial point P, so their contribution to the electric potential at P be

$$dV = \frac{1}{4\pi\varepsilon_0}\frac{dq}{r'} = \frac{1}{4\pi\varepsilon_0}\frac{2\sigma\pi y dy}{\sqrt{y^2 + r^2}} = \frac{\sigma}{2\varepsilon_0}\frac{ydy}{\sqrt{y^2 + r^2}}$$

Potential V due to whole of the disc is found by integration.

$$\therefore \qquad V = \int dV = \frac{\sigma}{2\varepsilon_0}\int_0^R \frac{ydy}{\sqrt{y^2 + r^2}} = \frac{\sigma}{2\varepsilon_0}\left(y^2 + r^2\right)^{1/2}\Bigg|_0^R$$

$$= \frac{\sigma}{2\varepsilon_0}\left(\sqrt{R^2 + r^2} - r\right)$$

Example 9. Potential due to a point charge q at a distance r is given by $V = \frac{1}{4\pi\varepsilon_0}\frac{q}{r}$, calculate $\vec{E}(r)$

Solution. From symmetry, $\vec{E}$ must be directed radially outward for a positive point charge. Consider a point P in the field at a distance r from the charge. It is clear that $\frac{dV}{dr}$ at P

has the greatest value. So $\left(-\frac{dV}{dr}\right)$ will give the magnitude of $\vec{E}(r)$ at the point P.

$$\therefore \quad E = -\frac{dV}{dr} = -\frac{d}{dr}\left(\frac{1}{4\pi\varepsilon_0}\frac{q}{r}\right) = -\frac{q}{4\pi\varepsilon_0}\frac{d}{dr}\left(\frac{1}{r}\right)$$

$$= \frac{1}{4\pi\varepsilon_0}\frac{q}{r^2}$$

Example 10. A uniformly charged infinite non-conducting sheet of charge has surface charge density = e.s.u./cm^2. What must be the separation between sucessive planes having a potential difference 300 V.

Solution. The electric field E at a point near a uniformly charged infinite plance non-conducting sheet = $2\pi\sigma$ where σ is the surface charge density.

$\therefore$ Potential difference between the points A and B at distances r_1 and r_2 from the sheet is

$$V_b - V_a = -\int_A^B \vec{E}.d\vec{l} = -\int_{r_1}^{r_2} 2\pi\sigma(r_2 - r_1)$$

Here $V_b - V_a = 300\ V = 1$ e.s.u.

$\sigma = 1$ esu/cm2

$\therefore \quad 1 = -2\text{p} \times 1\ (r_2 - r_1)$

$$\therefore \quad r_1 - r_2 = \frac{1}{2\pi} cm.$$

Example 11. The potential function at a point is given by $V = x(3y^2 - x^2 + z)$. Find the components of the electrostatic field at that point.

Solution. Electrostatic field $\vec{E} = -$ grad V.

$$= \left[\hat{i}\frac{\partial V}{\partial x} + \hat{j}\frac{\partial V}{\partial y} + \hat{k}\frac{\partial V}{\partial z}\right]$$

$$= -\left[\hat{i}\frac{\partial}{\partial x}\left(-x^3 + 3y^2x + xz\right) + \hat{j}\frac{\partial}{\partial y}\left(-x^3 + 3y^2 + xz\right) + \hat{k}\frac{\partial}{\partial z}\left(-x^3 + 3y^2x + xz\right)\right]$$

$$= -\left[\hat{i}\left(-3x^2 + 3y^2 + z\right) + \hat{j}(6xy) + \hat{k}x\right]$$

$\therefore \quad Ex = 3x^2 - 3y^2 - z$

$Ey = -6xy$

$Ez = -x$

Example 12. Two protons in a nucleus of U^{238} are 6.0×10^{-15} m apart. What is their mutual electric potential energy?

Solution. P.E. is given by

$$U = \frac{1}{4\pi\varepsilon_0}\frac{q_1 q_2}{r}$$

for charges q_1 and q_2, r distance apart

Here $q_1 = q_2 = 1{\cdot}6 \times 10^{-19}$ C

$r = 6{\cdot}0 \times 10^{-15}$m

$$U = \frac{\left(9\times 10^{9}\, Nm^2/C^2\right)\left(1\bullet 6\times 10^{19}C\right)^2}{6\bullet 0\times 10^{-15}m}$$

$= 3.8 \times 10^{-14}$

$= 2{\cdot}4\times 10^{5}\, eV\, (\because\ 1eV = 1\bullet 6\times 10^{-19})$

Example 13. Show that the potential function

$V = a(x^2 + y^2 + z^2)^{1/2}$

does not satisfy the Laplace's equation

Solution. The Laplace's equation is $\nabla^2 V = 0$

or
$$\frac{\partial^2 V}{\partial x^2} + \frac{\partial^2 V}{\partial y^2} + \frac{\partial^2 V}{\partial z^2} = 0$$

$$\text{L.H.S.} = \frac{\partial^2}{\partial x^2}\left[b\left(x^2+y^2+z^2\right)^{1/2}\right] + \frac{\partial^2}{\partial y^2}\left[b\left(x^2+y^2+z^2\right)^{1/2}\right]$$

$$+\frac{\partial^2}{\partial z^2}\left[b\left(x^2+y^2+z^2\right)^{1/3}\right]$$

Considering first term

$$\frac{\partial^2}{\partial x^2}\left[b\left(x^2+y^2+z^2\right)^{1/2}\right] = \frac{\partial}{\partial x}\left[\frac{\partial}{\partial x}\left\{x^2+y^2+z^2\right\}^{1/2}\right]$$

$$= \frac{\partial}{\partial x}\left[\frac{1}{2}b\left(x^2+y^2+z^2\right)^{1/2} 2x\right]$$

$$= \frac{\partial}{\partial x}\left[bx\left(x^2+y^2+z^2\right)^{-1/2}\right]$$

$$= bx\left(x^2+y^2+z^2\right)^{-3/2}\left(-\frac{1}{2}\right)2x + b\left(x^2+y^2+z^2\right)^{-1/2}$$

$$= \frac{-bx^2}{\left(x^2+y^2+z^2\right)^{3/2}} + \frac{b}{\left(x^2+y^2+z^2\right)^{3/2}}$$

$$= \frac{-bx^2 + b\left(x^2+y^2+z^2\right)}{\left(x^2+y^2+z^2\right)^{3/2}} = \frac{b\left(y^2+z^2\right)}{\left(x^2+y^2+z^2\right)^{3/2}} \quad \text{...(i)}$$

Similarly

$$\frac{\partial^2}{\partial y^2}\left[b\left(x^2+y^2+z^2\right)^{1/2}\right] = \frac{b\left(x^2+z^2\right)}{\left(x^2+y^2+z^2\right)^{3/2}} \quad \text{...(ii)}$$

and

$$\frac{\partial^2}{\partial z^2}\left[b\left(x^2+y^2+z^2\right)^{1/2}\right] = \frac{b\left(x^2+y^2\right)}{\left(x^2+y^2+z^2\right)^{3/2}} \quad \text{...(iii)}$$

Adding (i), (ii) and (iii), we get

$$\nabla^2 V = \frac{2b\left(x^2+y^2+z^2\right)}{\left(x^2+y^2+z^2\right)^{3/2}} = \frac{2b}{\left(x^2+y^2+z^2\right)}$$

$$\neq 0$$

$\nabla^2 V \neq 0$, so potential function $V = b\left(x^2+y^2+z^2\right)^{1/2}$ does not satisfy Laplace's equation.

Example 14. The electric potential at any point in x–y place is given by $V = 5(x^2 + y^2)1/$ 2+y $(x^2+y^2)^{-1/2}$. Find the cartesian components of the electric field intensity at that point.

Solution.

Here $V = 5x\left(x^2+y^2\right)^{1/2} + y\left(x^2+y^2\right)^{-1/2}$

and $E = -\text{grad } V$

$$= -\left[\hat{i}\frac{\partial V}{\partial x} + \hat{j}\frac{\partial V}{\partial y} + \hat{k}\frac{\partial V}{\partial z}\right]$$

$$\therefore \quad E_a = -\frac{\partial V}{\partial x}, E_y = -\frac{\partial V}{\partial y}, E_z = -\frac{\partial V}{\partial z}$$

$$\therefore \quad E_a = -\frac{\partial V}{\partial x} = -\frac{\partial}{\partial y}\left[5x\left(x^2+y^2\right)^{-1/2}\right]$$

$$= -\left[5x \times \frac{1}{2}\left(x^2+y^2\right)^{-1/2} 2x + 5\left(x^2+y^2\right)^{1/2} + y\left(-\frac{1}{2}\right)\left(x^2+y^2\right)^{-3/2} 2x\right]$$

$$= \frac{-1}{\left(x^2+y^2\right)^{3/2}}\left[5x^2\left(x^2+y^2\right) + 5\left(x^2+y^2\right)^2 - xy\right]$$

$$E_y = -\frac{\partial V}{\partial y} = -\frac{\partial}{\partial y}\left[5x\left(x^2+y^2\right)^{-1/2}\right]$$

$$= -\left[5x\left(x^2+y^2\right)^{-1/2} \frac{1}{2} \times 2y + y\left(-\frac{1}{2}\right)\left(x^2+y^2\right)^{-3/2} 2y + \left(x^2+y^2\right)^{-1/2}\right]$$

$$= -\frac{1}{\left(x^2+y^2\right)^{3/2}}\left[5xy\left(x^2+y^2\right) - y^2 + \left(x^2+y^2\right)\right]$$

$$= -\frac{1}{\left(x^2+y^2\right)^{3/2}}\left[5x^3 y + 5xy^3 + x^2\right]$$

$$E_z = -\frac{\partial}{\partial z}\left[5x\left(x^2+y^2\right)^{+1/2} + y\left(x^2+y^2\right)^{-1/2}\right] = 0$$

Exercise 2

Point charges Q1 = 1 nC, Q2 = –2 nC, Q3 = 3 nC, and Q4 = –4 nC are positioned one at a time and in that order at (0, 0, 0), (0, 0, -1), and (0, 0, 1) respectively. Calculate the energy in the system after each charge is positioned.

Answer : 0, –18 nJ, –68.27 nJ.

Example 15. A charge distribution with spherical symmetry has density

$$\rho_v = \begin{cases} \rho_o & 0 \le r \le R \\ 0, & r > R \end{cases}$$

Determine V everywhere and the energy stored in region $r < R$.

Solution. The D field has already been found in Section 4.6D using Gauss's law.

(a) For $r \geq R$, $E = \dfrac{\rho_o R^3}{3\varepsilon_0 r^2} a_r$

Once E is known, V is determined as

$$V = -\int E \bullet dl = -\frac{\rho_o R^3}{3\varepsilon_o} \int \frac{1}{r^2} dr$$

$$= \frac{\rho_o R^3}{3\varepsilon_o r} + C_1, \qquad r \geq R$$

Since $V(r = \infty) = 0$, $C_1 = 0$.

(b) For $r \geq R$, $E = \dfrac{\partial_o r}{3\varepsilon_o} a_r$

Hence,

$$V = -\int E \bullet dl = -\frac{\rho_o}{3\varepsilon_o} \int r dr$$

$$= -\frac{\rho_o r^2}{6\varepsilon_o} + C_2$$

From part (a) $V(r = R) = \dfrac{\rho_o R^2}{3\varepsilon_0}$. Hence,

$$\frac{R^2 \rho_o}{3\varepsilon_o} = \frac{\rho_o R^2}{6\varepsilon_o} + C_2 \rightarrow C_2 = \frac{R^2 \rho_o}{2\varepsilon_o}$$

and

$$V = \frac{\rho_o}{6\varepsilon_0}\left(3R^2 - r^2\right)$$

Thus from parts (a) and (b)

$$V = \begin{cases} \dfrac{\rho_o R^3}{3\varepsilon_o r} & r \geq R \\ \dfrac{r_o}{6e_o}\left(3R^2 - r^2\right), & r \leq R \end{cases}$$

(c) The energy stored is given by

$$W = \frac{1}{2}\int D.E\,dv = \frac{1}{2}\varepsilon_o \int E^2 dv$$

For r ≤ R,

$$E = \frac{\rho_0 r}{3\varepsilon_0} a_r$$

Hence,

$$W = \frac{1}{2}\varepsilon_o \frac{\rho_o^2}{9\varepsilon_o^2}\int_{r=0}^{R}\int_{\theta=0}^{\pi}\int_{\phi=0}^{2\pi} r^2 . r^2 \sin\,\theta\, d\phi\, d\theta\, dr$$

$$= \frac{\rho_o^2}{18\varepsilon_0} 4\pi . \frac{r^5}{5}\bigg|_0^R = \frac{2\pi\rho_0^2 R^5}{45\varepsilon_0} J$$

Exercise 3

If V = x − y + xy + 2z V, find E at (1, 2, 3) and the electrostatic energy stored in a cube of side 2 m centered at the origin.

Answer : $-3a_x - 2a_z$ V/m, 0.2358 nJ.

Example 16. Two point charges −4μC and 5μC are located at (2, −1, 3) and (0, 4, −2), respectively. Find the potential at (1, 0, 1) assuming zero potential at infinity.

Solution:

Let

$$Q_1 = -4\mu C,\ Q_2 = 5\mu C$$

$$V(r) = \frac{Q_1}{4\pi\varepsilon_o |r - r_1|} + \frac{Q_2}{4\pi\varepsilon_o |r - r_2|} + C_o$$

If $V(\infty) = 0$, $C_o = 0$,

$$|r - r_1| = |(1,0,1) - (2,-1,3)| = |(-1,1,-2)| = \sqrt{6}$$

$$|r - r_2| = |(1,0,1) - (0,4,-2)| = |(1,-4,3)| = \sqrt{26}$$

Hence

$$V(1,0,1) = \frac{10^{-6}}{4\pi \times \frac{10^{-9}}{36\pi}}\left[\frac{-4}{\sqrt{6}} + \frac{5}{\sqrt{26}}\right]$$

$$= 9 \times 10^3(-1.633 + 0.9806)$$

$$= -5.872 \text{ kV}$$

Practice Exericse 4

If point charge $3\mu C$ is located at the origin in addition to the two charges of example 4.10, find the potential at (–1, 5,2) assumig $V(\infty)=0$.

Answer : 10.23 kV.

Example 17. A point charge 5 nC is located at (–3, 4, 0) while line $y=1$, $z=1$ carries uniform charge 2 nC/m

(a) If V = 0 V at O (0, 0, 0), find V at A (5, 0, 1).

(b) If V = 100 V at B(1, 2, 1), find V at C(–2, 5, 3).

(c) If V = –5 V at *O*, find V_{BC}.

Solution :

Let the potential at any point be

$$V = V_Q + V_L$$

where V_Q and V_L are the contributions to V at that point due to the point charge and the line charge, respectively. For the point charge.

$$V_Q = -\int E - dl = -\int \frac{Q}{4\pi\varepsilon_o r^2} a_r.dr\, a_r$$

$$= \frac{Q}{4\pi\varepsilon_o r} + C_1$$

For the infinite line charge,

$$V_L = -\int E.dl = -\int \frac{\rho_L}{2\pi\varepsilon_o \rho} a_p.d\rho\, a_\rho$$

$$= -\frac{\rho_L}{2\pi\varepsilon_o}\; In\, \rho + C_2$$

Hence,

$$V = -\frac{\rho_L}{2\pi\varepsilon_O}\; In\, \rho + \frac{Q}{4\pi\varepsilon_O r} + C$$

where $C = C_1 + C_2 =$ constant, p is the perpendicular distance from the line $y = 1$, $z = 1$ to the field point, and r is the distance from the point charge to the field point.

(a) If V = 0 at *O* (0, 0, 0) and *V* at A (5, 0, 1) is to be determined, we must first determine the values of ρ and r at O and A. Finding r is easy; we use eq. (2.31). to find ρ for any point (*x*, *y*, *z*), we utilize the fact that *p* is the perpendicular distance from (x, y, z) to line $y = 1$, $z = 1$, which is parallel to the *x*-axis. Hence ρ is the distance between (*x*, *y*, *z*) and (*x*, 1, 1) because the distance vector between the two points is perpendicular to a_x, Thus

$$\rho = |(x, y, z) - (x, 1, 1)| = \sqrt{(y-1)^2 + (z-1)^2}$$

Applying this for p and eq. (2.31) for r tat points O and A, we obtain

$$\rho_O = |(0,0,0) - (0,1,1)| = \sqrt{2}$$

$$r_O = |(0,0,0) - (-3,4,0)| = 5$$

$$\rho_A = |(5,0,1) - (-5,1,1)| = 1$$

$$r_A = |(5,0,1) - (-3,4,0)| = 9$$

Hence,

$$V_O - V_A = -\frac{\rho_L}{2\pi\varepsilon_0} \, In \, \frac{\rho_O}{\rho_A} + \frac{Q}{4\pi\varepsilon_O}\left[\frac{1}{r_O} - \frac{1}{r_A}\right]$$

$$= \frac{-2 \bullet 10^{-9}}{2\pi . \frac{10^{-9}}{36\pi}} \, In \, \frac{\sqrt{2}}{1} + \frac{5 \bullet 10^{-9}}{4\pi . \frac{10^{-9}}{36\pi}}\left[\frac{1}{5} - \frac{1}{9}\right]$$

$$0 - V_A = -36 \, In \, \sqrt{2} + 45\left(\frac{1}{5} - \frac{1}{9}\right)$$

or

$$V_A = 36 \, In \, \sqrt{2} - 4 = 8.477 \, V$$

Notice that we have avoid calculating the constant C by subtracting one potential from another and that it does not matter which one is subtracted from which.

(b) If V = 100 at B(1, 2, 1) and V at C (-2, 5, 3) is to be determined, we find

$$\rho_B = |(1,2,1) - (1,1,1)| = 1$$

$$r_B = |(1,2,1) - (-3,4,0)| = \sqrt{21}$$

$$\rho_C = |(-2,5,3) - (-2,1,1)| = \sqrt{20}$$

$$r_C = |(-2,5,3) - (-3,4,0)| = \sqrt{11}$$

$$V_C - V_B = -\frac{\rho_L}{2\pi\varepsilon_O} \, In \, \frac{\rho_O}{\rho_B} + \frac{Q}{4\pi\varepsilon_O}\left[\frac{1}{r_C} - \frac{1}{r_B}\right]$$

$$V_C - 100 = -36\ In\ \frac{\sqrt{20}}{1} + 45.\left[\frac{1}{\sqrt{11}} - \frac{1}{\sqrt{21}}\right]$$

$$= -50.175\ \text{V}$$

or

$$V_C = 49.825\ \text{V}$$

(c) To find the potential difference between two points, we do not need a potential reference if a common reference is assumed.

$$V_{BC} = V_C - V_B = 49.825 - 100$$

$$= -50.15\ \text{V}$$

as obtained in part (b)

Practice Exercise (5)

A point charge of 5 nC is located at the origin. If V = 2 V at (0, 6, –8), find

(a) The potential at A (–3, 2, 6)

(b) The potential at B (1, 5, 7)

(c) The potential difference VAB

Answer : (a) 3.929 V, (b) 2.696 V, (c) –1.233 V

Check Yourself

1. Point charges Q_1 = 1 nC and Q_2 = 2 nC are at a distance apart. Which of the folloiwng statements are incorrect?

(a) The force on Q_1 is repulsive.

(b) The force on Q_2 is the same in magnitude as that on Q_1.

(c) As the distance betweenm them decreases, the force on Q_1 increases linearly.

(d) the force of Q_2 is along the line joining them.

(e) A point charge Q_3 = –3 nC located at the midpoint between Q_1 and Q_2 experiences no net force.

2. Plane z = 10 m carries charge 20 nC/m2. The electric field intensity at the origin is

(a) – 10az V/m (b) – 18 p qz V/m

(c) – 72 p az V/m (d) –360 p az V/m

3 Point charges 30 nC, –20 nV, and 10 nC are located at (–1, 0, 2), (0, 0, 0), and (1, 5, –1), respectively. The total flux leaving a cube of side 6 m centered at the origin is :

(a) – 20 nC (b) 10 nC

(c) 20 nC (d) 30 nC

(e) 60 nC

4. The electric flux density on a spherical surface r = b is the same for a point charge Q located at the origin and for charge Q uniformly distributed on surface r = a(a < b).

   (a) Yes (b) No (c) Not necessarily

5. The work done by the force $F = 4a_x - 3a_y + 2a_z$ N in giving a 1 nC charge a displacement of $10a_x + 2a_y - 7a_z$ m is

   (a) 103 nJ (b) 60 nJ

   (c) 64 nJ (d) 20 nJ

6. By saying that the electrostatic field is conservative, we do not mean that

   (a) Is is the gradient of a scalar potential.

   (b) Its ciorculation is identically zero.

   (c) Its curl is identically zero.

   (d) The work done in a closed path inside the field is zero.

   (e) The potential difference between any two points is zero.

7. Suppose a uniform electric field exists in the room in which you are working, such that the lines of force are horizontal and at right angles to one wall. As you walk toward the wall from which the lines of force emerge into the room, are you walking toward

   (a) Points of higher potential ?

   (b) Points of lower potential?

   (c) Points of the same potential (equipotential line)?

8. A charge Q is uniformly distributed throughout a sphere of radius a. Taking the potential at infinity as zero, the potential at r = b < a is

   (a) $-\int_{\infty}^{b} \frac{Qr}{4\pi\varepsilon_O a^3} dr$ (b) $-\int_{\infty}^{b} \frac{Qr}{4\pi\varepsilon_O r^2} dr$

   (c) $-\int_{\infty}^{a} \frac{Qr}{4\pi\varepsilon_O r^2} dr - \int_{a}^{b} \frac{Qr}{4\pi\varepsilon_O a^3} dr$ (d) $-\int_{\infty}^{a} \frac{Q}{4\pi\varepsilon_O r^3} dr$

9. A potential field is given by $V = 3x^2y - yz$. Which of the following is not true?

   (a) At point (1, 0, –1) V and E vanish.

   (b) $x^2y = 1$ is an equipotential line on the xy-plane.

   (c) The equipotential surface $V = -8$ passes through point P (2, –1, 4).

   (d) The electric field at P is 12 ax – 8ay – az V/m.

   (e) A unit normal to the equipotential surface $V = -8$ at P is $-0.83a_x + 0.55a_y + 0.07a_z$.

Long answer Questions

1. What do you understand by the conservative nature of the electric field? Explain how the conservative nature of electric field leads to the concept of electric potential.
2. Explain the physical interpretation of the line integral of the electric field $\vec{E}$.

(P.U. 1979)

3. What is a conservative field? Show that for a conservative field the line integral for a closed path is zero.

(K. U. 1981, 1984)

4. Define electric potential. Derive an expression for the relation between electric field and electric potential.

(K.U. 1993 A)

5. Show that the electric field at a point in an electriostatic field is equal to the negative gradient of potential at that point. Hence prove that $E = \left(\frac{dV}{dl}\right)_{max}$

(M.D.U. 1979; K.U. 1980, 92 S)

6. Explain potential difference as line integral of the electric field.

(K.U. 1984, 88 S)

7. Deduce the relation = – grad V between electric field and electric potential. *(M.D.U.* **1986; K.U. 1984 A, 88 S)**
8. Derive Poisson's and Laplace's equation, starting from the differential form of Gauss's Law.

(K.U. 1979, 82, 84, 92 S)

9. Derive Poisson's and Laplace's equation. What is their importance is physics?

(M.D.U. 1986; K.U. 1986 A, 93 A)

10. What is an electric dipole? Calculate potential and field due to it at any point in space.
11. Define the electric potential at a point in an electric field. Show that the electric potential at a point is equal to the line integral of the intensity of electric field.

(K.U. 1989 A)

Short Answer Questions

1. What type of field is defined by electrostatic potential?
2. What is more basic, potential difference or potential?
3. An electrostatic field is curl-free. Explain why?

(P.U. 1980: K.U. 1986 A, 87 A)

4. If there is no charge distribution anywhere in space, what can be said about potential in space.

5. What are the circumstances in which zero level potential at infinity is justified?
6. If potential is known at a given point in space, how can you find the electric field at that point.
7. What is the significance of negative sign in equation

$$\vec{E} = -\nabla V$$

8. What is the amount of work done in moving a unit positive test charge form infinity to origin where a stationary positve point charge is lying?
9. What can you say about potential in a given region of space where electric field is uniform?
10. What can you say about $\vec{E}$ (i) in a region where potential V is constant, (ii) at a point where potential V= 0, (iii) at a point where V is increasing uniformly ?

(M.D.U. 1984; K.U. 1980)

11. Show that no work is done in moving a test charge from point to point on the surface of the metal.

(M.D.U. 1984)

Problems

1. Let $V = xy^2z$, calculate the energy expended in transfering a 2–μC point charge from (1, –1, 2) to 2, 1, –3).
2. Determine the electric field due to the following potentials :
   (a) $V = x^2 + 2y^2 + 4z^2$ (b) $V = \sin(x^2 + y^2 + z^2)^{1/2}$
   (c) $V = \rho^2 (z + 1) \sin \phi$ (d) $V = e^{-r} \sin \theta \cos 2\phi$
3. Three point charges Q1 = 1 mC, Q2 = -2 mC, and Q3 = 3 mC are, respectively, located at (0, 0, 4), (–2, 5, 1) and (3, –4, 6).
   (a) Find the potential VP at P (–1, 1, 2).
   (b) Calculate the potential difference VPQ if Q is (1, 2, 3).
4. In free space, V = x2y(z +3) V. Find
   (a) E at (3, 4, –6) (b) the charge within the cube 0 < x, y, z < 1.
5. A spherical charge distribution is given by

$$\rho_v = \begin{cases} \rho_O \dfrac{r}{a}, & r < a \\ 0, & r > a \end{cases}$$

   Find V everywhere.
6. To verify that E = yzax + xzay + xyaz V/m is truly an electric field, show that
   (a) $\Delta \times E = 0$
   (b) $\oint_L E.dl = 0$, where L is the edge of the square defined by 0 < x, y < 2, z = 1.

7. (a) A total charge Q = 60 mC is split into two equal charges located at 180° intervals around a circular loop of radius 4 m. Find the potential at the center of the loop.

   (b) If Q is split into three equal charges spaced at 120° intervals around the loop, find the potential at the center.

   (c) If in the limit $\rho_L = \dfrac{Q}{8\pi}$, find the potential at the center.

8. For a spherical charge distribution

$$\rho_V = \begin{cases} \rho_O(a^2 - r^2), & r < a \\ 0, & r > a \end{cases}$$

   (a) Find E and V for r ≥ a

   (b) Find E and V for r ≤ a

   (c) Find the total charge

   (d) Show that E is maximum when r = 0. 145a.

9. (a) Prove that when a particle of constant mass and charge is accelerated from rest in an electric field, its final velocity is proportional to the square root of the potential difference through which it is accelerated.

   (b) Find the magnitude of the proportionality constant if the particle is an electron.

   (c) Through what volage must an electron be accelerated, assuming no change in its mass, to require a velocity one-tenth that of light? (At such velocities, the mass of a body becomes apreciably larger than its "rest mass" and cannot be considered constant.)

10. An electron is projected with an initial velocity uo = 10-7 m/s into the uniform field between the parallel plates of Figure 3.11. It enters the field at the midway between the plates. If the electron just misses the upper plate as it emerges from the field.

    (a) Find the electric field intensity.

    (b) Calculate its velocity as it emerges from the field. Neglact edge effects.

11. An electric dipole with p = paz C·m is placed at $(x, z) = (0, 0)$. If the potential at (0, 1) nm is 9 V, find the potential at (1, 1) nm.

12. Point charges Q and -Q are located at (0, d/2, 0) and (0, -d/2, 0). Show that at point (r, θ, ϕ), where r >> d,

$$V = \frac{Qd \sin\theta \sin\phi}{4\pi\varepsilon_O r^2}$$

    Find the corresponding E field.

13. Determine the work necessary to transfer charges Q_1 = 1mC and Q_1 = –2 mC from infinity to points (–2, 6, 1) and (3, –4, 0), respectively.

14. A point charge Q is placed at the origin. Calculate the energy stored in region r > a.

15. Find the energy stored in the hemispherical region r < 2m, $0 < \theta < \pi$, where

$$\mathbf{E} = 2r \sin \theta \cos \phi \, \mathbf{a}_r + r \cos \theta \cos \phi \, \mathbf{a}_\theta - r \sin o \, \mathbf{a}_\phi \text{ V/m} \quad \text{exists.}$$

16. If V = p2z sin f, calculate the energy within the region defined by 1 < p < 4, -2 < z < 2, 0 < f < π/3.

17. Calculate the electric potential at the surface of the nucleus of gold atom whose radius is 6.6 × 10-15m. Given atomic number of gold 79 and charge on proton = $1{\cdot}6 \times 10^{-19}$ C.

[Ans. 17.2×10^6 V]

18. Find the potential 1 cm distant from an electron in vacuum. The charge of an electron is negative and = 1·6 × 10-19 C.

[Ans. -1.44×10^{-7} V]

19. Three positive charges 1, 2, 3×10-9 C are arranges at the corners of an equilateral triangle of side 1 m. Find the potential at the centre of triangle.

[Ans. $3\sqrt{3} \times 10^{-9} V$]

20. Prove that field $\vec{F} = a\left(\sqrt{y}\hat{i} + \sqrt{x}\hat{j}\right)$ represents an electrostatic field.

21. What is the potential at the centre of a square of side 1 m when charge

$q_1 = +1.0 \times 10^{-8} C, q_2 = -2.0 \times 10^{8} C, q_3 = +3.0 \times 10^{-8} C$

and $q_4 = +2.0 \times 10^{-8}$ C are placed at its corners? **[Ans.** 507.V]

22. Consider a point charge with q = 1·5 × 10-8 C. What is the radius of equipotential surface having potential of 30 V?

[Ans. 4.5 m]

23. Potential function at a point is given by V = 4x + 3y–z. Obtain the expression for potential gradient and intensity.

[Ans. $4\hat{i} + 3\hat{j} - \hat{k}, -\left(4\hat{j} + 3\hat{j} - \hat{k}\right)$]

24. The electric potential in space is given by jV = x(y2 –4x2). Calculate the cartesian components of the electric field at that point

[Ans. $E_\infty = \left(12x^2 - y^2\right), E_y = -2xy, E_z = 0$]

25. If (x, y, z) = 3x2y–y3 x2 be any scalar function of x, y, z, Find grad f at (1, 2, 2).

[Ans. $-4\hat{i} - 9\hat{j}$]

26. Calculate the electric potential energy of an electron proton system of an atom. The radius of an orbit of electron is 2.116×10^{-10} m. Charge on electron is 1.6 × 10^{-10} C.

[Ans. -1.09×10^{-18} J]

27. Select the electrostatic field from the followong, keeping in mind the electrostatic field must essentially be conservative.

(i) $\vec{F} = a(3x\,\hat{i} - 3y\hat{j})$ (ii) $\vec{F} = a(\sqrt{y}\hat{i} + \sqrt{x}\hat{j})$

(iii) $\vec{F} = a\,xy^2(y\hat{i} + x\hat{j})$

[**Ans.** (i), (ii)]

28. A charge –2e is placed at the origin and two charges +e at points (b, o, o) respectively. Prove that the potential V at a distance r large compared to 'b' is approximately given by

$$V = \frac{eb^2}{4\pi\varepsilon_0}\left(\frac{3\cos^2\theta - 1}{r^3}\right)$$

where q is the angle which r makes with x-axis.

29. What is the charge density function in a region of space where the electrostatic potential

V = a – b (x2 + y2)– $c \log\sqrt{x^2 - y^2}$? [**Ans.** $\frac{b}{\pi} esu\ cm^{-3}$]

30. Two positive point charges, each of magnitude θ are fixed on the x-axis at poits $x = \pm$ a, prove that the potential at any point on y-axis is

$$V = \frac{1}{4\pi\varepsilon_0}.\frac{2q}{\sqrt{a^2 + y^2}}$$

31. Six equal positive charges 'q' are placed at six corners of a regular hexagon of side 'a'. If a charge –6q is placed at the centre, calculate the potential at a distance r.

[**Ans.** $\frac{3qa^2}{8\pi\varepsilon_0 r^3}$]

4

Electric Fields in Material Space

Till now, we have discussed electric field in free space or vacuum that is generally called vacuum theory of electrostates. But now, we are going to space. There are not going to be much more changes from older one as most of the theory of electric phenomena in material space is based on the older one. Here the free space is filled with materials. It means the electric field is going to exist in material space rather than free space.

Materials are broadly classified in terms of their electrical properties as conductors and nonconductors. Nonconducting materials are usually referred to as insulators or dielectrics. A brief discussion of the electrial properties of materials in general will be given to provide a basis for understanding the concepts of conduction, electric current, and polarization. Further discussion will be on some properties of dielectric materials such as susceptibility, permittivity, linearity, homogeneity, dielectric strength, and relaxation time.

We will also discuss how the two different mediums change the boundary conditions for electric field.

Properties of Materials

At such an advance level of the discussion of the concepts of physics, reopening the chapter of discussion on electrical properties of materials is not suitable but there are some queries such as why an electron does not leave a conductor surface, why a current-carrying wire remains uncharged, why materials behave differently in an electric field, and why waves travel with less speed in conductors than in dielectrics can broadly be solved .

In a broad sense, materials may be classified in terms of their conductivity σ, siemens per meter (S/m), as conductors and nonconductors, or technically as metals and insulators (or dielectrices). The conductivity of a material usually depends on temperature and frequency. A material with high conductivity (σ>>1) is referred to as a metal whereas one with low conductivity (σ<<1) is referred to as an insulator: A material whose conductivity lies somewhere between those of metals and insulatator. A material whose conductivity lies somewhere between those of metals and insulators is called a semiconductor. Materials such as copper and aluminum are metals, silicon and germanium are semiconductors, and glass and rubber are insulators.

The conductivity of metals generally increases with decrease in temperature. At temperatures near absolute zero (T=0°k), some conductors exhibit infinite condectivity and are called superconductors. Lead and aluminum are typical examples of such metals. The conductivity of lead at 4°k is of the order of 10^{20} mhos/m. The interested reader is referred to the literature on superconductivity.[1]

Here the matter of concern for us is the difference between metals and insutators. The mojor difference between a metal and an insulator lies in the amount of electrons available for conduction of current. Dielectric materials have few electrons available for conduction of current in contrast to metals, which have an abundance of free electrons. We will have extensive discussions on the presence of conductors and dielectrics in the electric field in coming topics.

Convection and Conduction Currents

Before moving on to discuss how the electric field behave in material space let's study electric current.

Electric Current

The phenomenon of moving charges i.e., electric current can be understood broadly in following ways. Some electrons in the outermost orbit lose their association with their parent atom and become free in on isolated metallic conductor such as length of a copper wire because of electrostatic interaction between interaction between various atoms of copper. These electrons do not remain attached to a particular atom of copper but move freely within the boundary of the copper wire. These free electrons move like molecules of a gas, collide among themselves and also with the positive ions with great frequency and changing direction at each collision. The velocities of these free electrons are of the order of 10^6 ms^{-1} even at normal temperatures and are distributed at random in all directions, so average velocity of free electrons is zero. Number of electrons crossing the palne per second in one direction will be the same as in the opposite direction; It means the net rate which electrons move through a hypothetical plane passing through the wire will be zero.

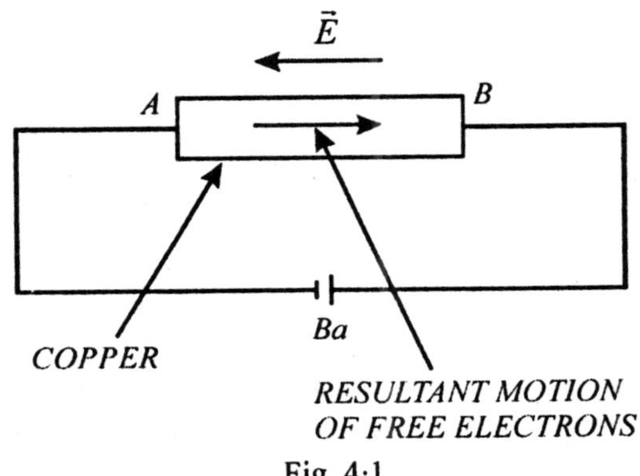

Fig. 4·1

If the two ends of the wire are connected to battery B*a* as shown in Fig.4·1, a steady electric field $\vec{E}$ is established along the length of the wire from the end B to the end A. This field will act on the free electrons in the copper wire and will give them a resulatant motion in a direction opposite to that of the field $\vec{E}$, i.e., in the direction A to B. This flow of electrons through the wire constitutes an electric current and is defined as the rate of flaw of electric charge through any section of the wire. Thus if the net charge q passes through any cross-section of the wire in a time t, the average current i is given by

$$t = \frac{q}{t} \tag{4·1}$$

If equal quantities of charge q pass through a specified section of the conductor in succesive equal intervals of time t, then the current i is said to be constant magnitude or steady.

If the rate of flow of charge with time is not constant i.e., the current varies with time, at an y time t, the current is given by the differential limit of Eq. (4.1), i.e.,

$$t = \underset{\delta t \to 0}{Limit} \frac{\delta q}{\delta t} = \frac{dq}{dt} \tag{4·2}$$

As there is no accumulation of charge at any point along the conductor, the current i is the same for all cross-sections of a conductor of non-uniform cross-sections, in as much the same way as the rate of flow of water (assumed to be incompressible) remains the same, through any cross-section of a pipe even if cross-section of the it should be noticed that if the conductor is smaller in cross-section then the charge flows faster and if charge in cross-section then charge flows shower.

Direction of Current

The charge carriers are negatively charged electorns and positively charged holes in the case of semiconductors while the charge the carriers consist of positive and negative ions in case of eletrolyte and gaseous conductors. Therefore; we must say that the charge carriers electrons in metals.

Under the action of an externally applied field, the positive and negative ions in case of electrolytes and gaseous condusctors (and electrons and holes in case of semiconductors), move in opposite directions. Whichever direction is assigned to the current, we find the charges moving the opposite direction also. So some convention must be adopted for labelling the direction of current. It has been agreed for the sake of consistency to take the direction of any curreat as if carriers were all positive charges i.e., positive direction of the current is taken conventionally as the direction of flow of positive charges. If are charge carriers are negative, the direction of current is opposite to the flow of negative charges. Hence in a metallic conductor the eleectrons move in the oppsite direction to the conventional the electrons move in the oppsite direction to the conventional current. The sign convention of a current is demonstrated in Fig 4.2. point should be noted that the current is a scalar quatity and the arrow demonstrated in the Fig (4.2) is just the disply of direction of the charge flow. the arrow does not signify current a vector quantity.

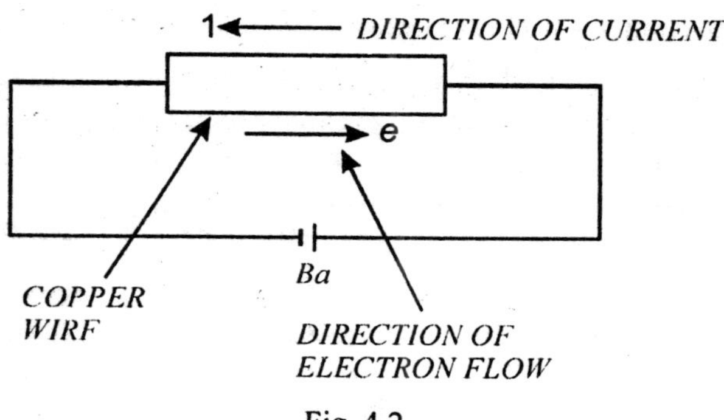

Fig. 4.2

Unit of Electric Current

Ampere is the S.I unit of current. A charge of one coulomb (3×10^9 e.s.u.) flows through any section of a conductor in one second is called one ampere it is also the practical unit of current

Hence 1 Ampere (A)=1 coulomb (C)/1 second (s).

An e.m.u. of current is the c.g.s. unit of current. A charge of 1 e.m.u. (3×10^{10} e.s.u.) flows through any section of a conductor per second is called an e.m.u. An e.m.u. of current is also called *ab*-ampere.

$$1 \text{ e.m.u. unit of current} = \frac{1 \text{ e.m.u. of charge}}{1 \text{ second}}.$$

Relation between Ampere and e.m.u

$$1 \text{ ampere} = \frac{1 \text{ coulomb}}{1 \text{ second}}.$$

$$= \frac{3\times10^9 \text{ e.s.u}}{1 \text{ s}}$$

$$= \frac{3\times10^9}{3\times10^{10}\times1} \text{ e.m.u. of current}$$

$$\therefore \quad 1\text{A} = \frac{1}{10}$$

Description of Current Flow in Terms of Drift Velocity

From the above exploration of concepts, we have examined that once the conductor's ends are connected a battery, a steady electric field say $\vec{E}$ is set up at every point within the conductor (here a copper wire). Each free electron experiences force$-e\vec{E}$ directed opposite to the electric field, where –e is the charge on the electron. Each electron will, therefore, be accelerated. This acceleration will be in addition to the random velocity possessed by the electron at that instant under the influence of the electric field, the electrons flow continuously from negative terminal through the wire to the positive terminal and then to battery. As they move, they suffer frequent collisions among themselves and with the positive ions. The average time between two successive collision is 10^{-14} s and even less. During this short time between the collision, an electron gets accelerated and gains an extra velocity towards positive terminal of the battery as shown in Fig. 4.3. but during each collision, this extra velocity is destroyed. This means that whatever momentum is built up in the direction of electrical force in between collisions, is destroyed during collision. This loss of momentum during collisions can be considered to be equivalent to a backward force. This force may be called collision drag. A steady state is reached when the collision drag completely cancels the average electric force on the free electron. In this situation, the electrons drift slowly towards the positive end with a velocity of the order of 10^{-3}m/s, called drift velocity, and give rise to organised transport of charge which constitute a current we may thus think an electron moving at random with high velocity and often colliding with positive ions and suffering change in direction, but at the same time drifting slowly towards positive terminal.

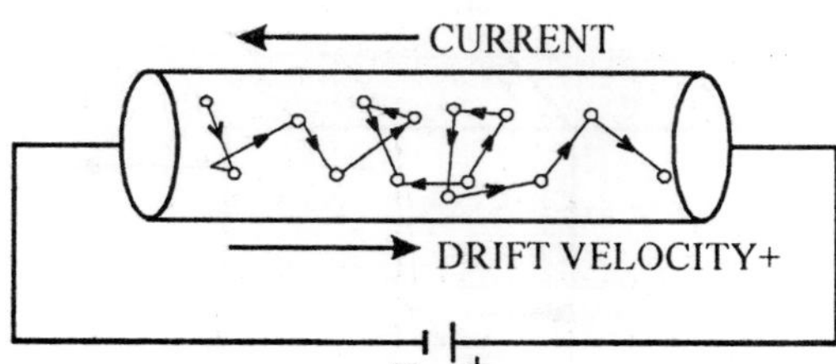

Fig 4.3 Electrons drift in a direction opposite to the electric field in a conductor.

Current Density

Now, the concept of current density is going to be introduced. It is denoted by j. If' through a surface ΔS the current ΔI flows then the current density can be written as follows

$$j_n = \frac{\Delta I}{\Delta S}$$

or

$$\Delta I = j_n \Delta S \tag{4·3}$$

Here the current density is perpendicular to the surface.

Now, take the current density normal to the surface then the equation:

$$\Delta I = j.\Delta S \tag{4·4}$$

Thus, the total current flowing through a surface S is

$$I = \int_S J.ds \tag{4·5}$$

there are different kinds of current densities: convection current density, conduction current density, and displacement current density all this is depending on how I is produced we need to keep in mind that eq. 4·5) applies to any kind of current density. If we compare the general definition of flux eq. (4·5) then we find that the current I through S is merly the flux of the current density J.

Convection current, as distinct from conduction current, does not involve conductors and consequently does not satisfy Ohm's law. It occurs when current flows through an insulating medium such as liquid, rarefied gas, or a vacuum. A beam of electrons in a vacuum tube, for example, is a convection current.

Consider a filament of Figure 4·4. If there is a flow of charge, of density ρ_v., at velocity u = a_y, a_y, from eq (4.1), the current the filament is

$$\Delta I = \frac{\Delta Q}{\Delta t} = \rho_v \; \Delta S \frac{\Delta \partial}{\Delta t} = \rho_v \; \Delta S \; u_y \tag{4·6}$$

The current through a point of a unit normal area is called current density at that point.

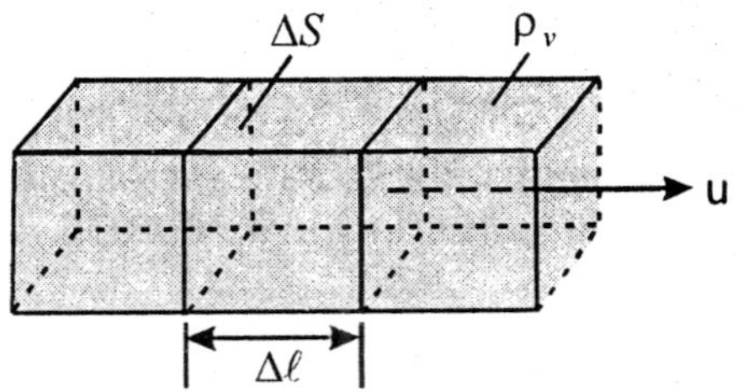

Fig. 4·4 Current in a Filament

The y-directed current density J_y is given by

$$j_y \frac{\Delta I}{\Delta S} = \rho_v u_y \tag{4·7}$$

Hence, in general

$$\boxed{J = \rho_v u} \tag{4·8}$$

The current I is convection current and J is the convection current density in amperes/ square meter (A/m^2).

A conductor is required for conduction current. Conductor electrons that provide conduction current a powerful electric field. Now the force on an electron with charge -e when an electric field E is applied.

$$F = -eE \tag{4·9}$$

It suffers constant collision with the atomic lattice and drifts from one atom to another rather than the acceleration under the influence of the electric field because the electron is not in free space. If the electron with mass m is moving in an electric field E with an average drift velocity u, according to Newton's law, the average change in momentum of the free electron must the applied force thus,

$$\frac{mu}{\tau} = -eE$$

or

$$u = -\frac{e\tau}{m}E \qquad (4{\cdot}10)$$

where τ is the average time interval between collision. This indicates that the drift velocity of the electron is directly proportional to the applied field. If there are n electrons per unit volume, the electronic charge is given by

$$\rho_v = -ne \qquad (4{\cdot}11)$$

Thus the conduction current density is

$$J = \rho_v u = \frac{ne^2\tau}{m}E = \sigma E$$

or

$$\boxed{J = \sigma E} \qquad (4{\cdot}12)$$

Here, the conductivity of the conductor, $\sigma = \frac{ne^2\tau}{m}$ and this equ (4·12) is known as the point form of ohm's law.

where $\sigma = ne^2\tau / m$ is the conductivity of the conductor. As mentioned earlier, the values of for common materials are provided in Table B. 1 in Appendix B. The relationship in eq. (5·11) is known as the point form of *Ohm's law*.

Conductors

In the Fig 5·2 (a) we have considered an isolated conductor which has abundance of charge that is free to move. The positive free charge are pushed along the same direction of the field where as the negative free charges move in the opposite dirction when an electric field Ee is applied externally.

This charge migration takes palce very quickly. The free charges do two things. First, they accumulate on the surface of the conductor and form an induced surface charge. Second, the induced charges set up an internal induced field Ei, which cancels the eternally applied fieldee. The result is illustrated in Figure (4·5.b). This leads to an important property of a conductor:

An electorstatic field within a perfact conductor can not be contained the fact that E = – ΔV =0. Signifies that the potental is the same evey that the conductor. It means an the conductor is an equipental body.

Another way of looking at this is to consider Ohm's law, $J = \sigma E$. To maintain a finite current density J, in a perfect conductor ($\sigma \rightarrow \infty$) , requires that the electric field inside the conductor must vanish. In other words, $E \rightarrow 0$ because $\sigma \rightarrow \infty$ in a perfect conductor. If some charge are introduced in the interior of such a conductor, the charges will move to the surface and redistribute themselves quickly in such a manner that the field inside the conductor vanishes. According to Gauss's law, if $E = 0$, the charge density ρ_v must be the follow static conditions , a perfact conductor can contain an electorstatic field with it.

$$\boxed{E = 0,\ \rho_v = 0,\ V_{ab} = 0 \text{ inside a conductor}} \tag{4·13}$$

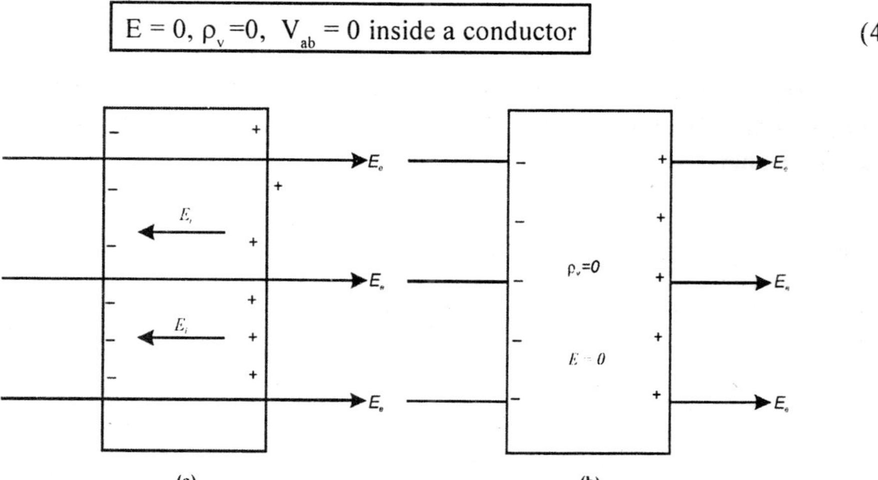

Figure (4·5) (a) An isolated conductor under the influence of an applied field; (b) a conductor has zero electric field under static conditions.

Now, we take a new conductor as shown in the Fig 4·6. The ends of this conductor is maintained at a potential difference V and here electric field inside the conductor is not zero. But the question arises how we can differentiate between then there are few following points.There is no static equilibrium in Figure (4.6) since the conductor is not isolated but wired to a source of electromotive force, which compels the free charges to move and prevents the evenmtual establishment of electrostatic equilibrium. Thus in the case of Figure (4·6), an electric field must exsit inside the conductor to sustain the flow of current. As the electrons move, they encounter some damping forces called resistance. Based on OPhm's law in eq. (4·12) , we will derive the resistance of the conducting material. Suppose the conductor has a uniform cross section S and is of length *l*. The direction of the electric field E produced is the same as the direction of the flow of positive charges or current I. This direction and its magnitude is given by

$$E = \frac{V}{l} \tag{4·14}$$

Since the conductor has a uniform cross section,

$$J = \frac{I}{S} \tag{4·15}$$

Substituting eqs. (5.11) and (5·13) into eq. (5·14) gives

$$\frac{I}{S} = \sigma E = \frac{\sigma V}{l} \tag{4·16}$$

Hence

$$R = \frac{V}{I} = \frac{l}{\sigma S}$$

or

$$R = \frac{\rho_c \partial}{S} \tag{4·17}$$

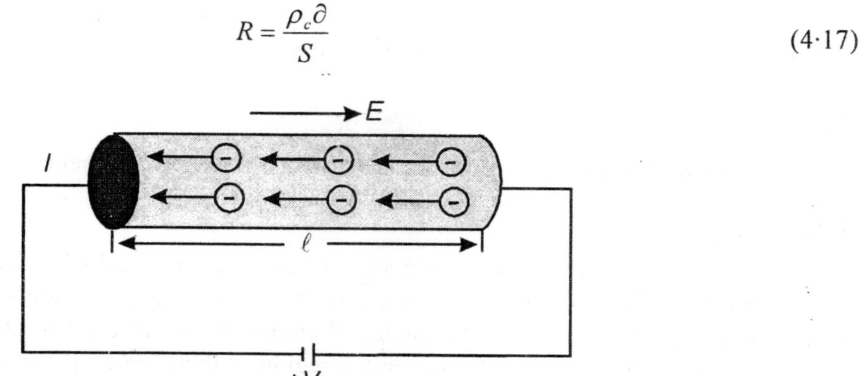

Figure (4.6). A conductor of uniform cross section under an applied E field.

Here the resistiveity of the material $\rho_c = \frac{1}{\sigma}$. To determine the resiance of any conductor of uniform cross section equ 4.17 is very useful.

If the cross section of the conductor is not uniform, eq. (4·17) is not applicable. However, the basic definition of resistance *R* as the ratio of the potential difference *V* between the two ends of the conductor to the current I through the conductor still applies. Therefore. Therefore, the resistance of a conductor of nonuniform cross section; is as follows:

$$R = \frac{V}{I} = \frac{\int E.dI}{\int \sigma E.dS}$$

Note that the negative sign before V= - $\int E.dI$ is dropped in eq. (4·18) because $\int E.dI$ <0 if I >0.

Power P (in watts) is defined as the rate of change of energy W (in joules) or force times velocity. Hence.

$$\int \rho_v d_v E.u = \int E.\rho_v u\, d_v$$

or

$$\boxed{p = \int E.Jdv} \tag{4·19}$$

Which is known as Joule's law: The power density wp (In watts/m³) is given by the intgreand in eq. (4·19); that is,

$$wp = \frac{dp}{dv} = E.J = \sigma|E|^2 \tag{4·19}$$

For a conductor with uniform cross section, dv=ds dl, so eq. (4·19) becomes

$$p = \int_L E\,dl \int_S J\,ds = VI \tag{4·20}$$

or

$$P = I^2R \tag{4·21}$$

Which is the more common form of Jaule's law in electric circuit theory.

Dielectric

A material that contains no free electrons to carry current is called dielectric. In a dielectric the conduction of electric current is not possible. It is a kind of material in which all the electrons are tightly bonded to the nuclei of atoms. When an electric charge is given to a dielectric, it remains localised of fixed in the region of their initial distribution or application in contrast to conductors where the charge immediately spreads over the surface. The properties of a dielectric may not be same in all directions i.e. it may not be isotropic or the properties of a dielectric may very from point to point i.e, it may not be homogeneeous. However, in the following discussion, we shall assume that we are dealing with a perfectly *isotropic* and *homeogeneous* dielectric. Is sotropic dielectric is one which has the same properties in all directions the dielectric which has the same properties at all points, It means same kind all through the region is cauld homogeneous dietectric.

Classification of Dielectrics

There are two types of the atoms of dielectric ,and the fundamental conceptral support to understand these two polar and non-polarcomes from their electrical structure. Now a substance which is normally electrically neutral as a whole, contains positive and negative charges in equal amounts. The positive charge exists in the nucleus in the form of protons and the negative charge may be considered (according to modern quantum mecharics) to reside in the electron cloud of

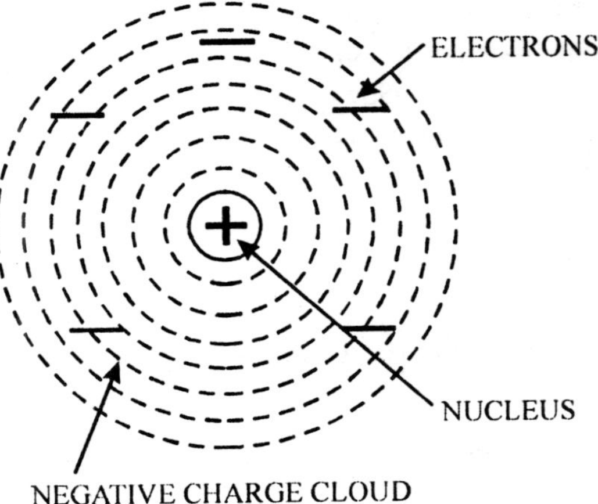

Fig. 4·6 Effective centres of positive and negative Charge coincide.

smoothly verying charge density round the nucleus as shown in Fig. 4·6. The shape of the cloud and variation of electron density is different for different molecules. Further as far as electric effects of whole molecule are concened, the positive charge may be regarded as acting from a specific point within the molecule. This point is called effective centre or centre of effective centre or centre of gravity of positive charge. Similarly there is another specific point in the molecule, where whole of the negative charge is regarded to be concentrated and its electrial effects are equivalant to the effects due to distributed electrons in the molecule. This point is called effective centre or dentre of gravity of negative charge distibution. I the absence of any external elctric field, if the centre of gravity of positive chages coincides with the centre of gravity of gravity charges, the molecule is called non-polar and if the two centre of gravity do not coincuide i.e.., they are separated from each other by a finite distance the molecule is called polar. The polar molecule, because of the separation of two centres of gravity, possess dipole*8888 moment which is the product of amount of either charge and distance between effective centre of positive and negative charges.

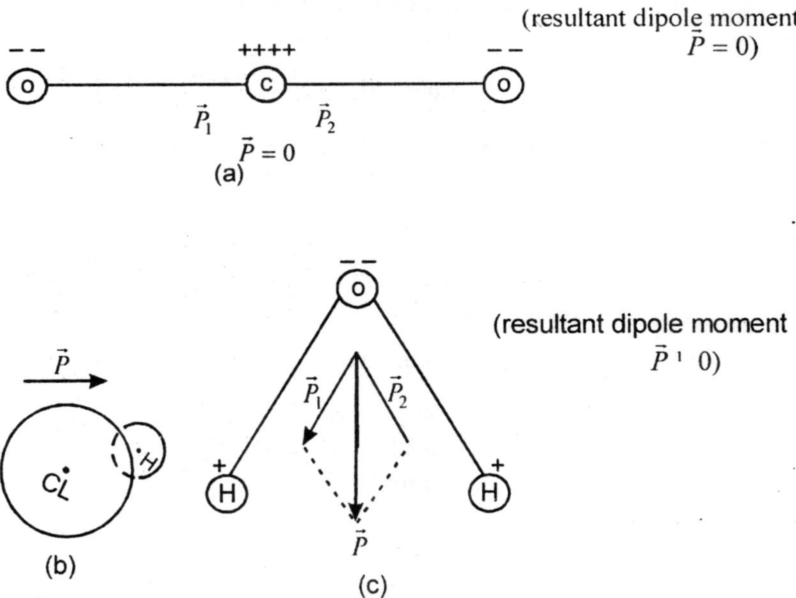

Fig. 4·7

Molecules having symmetrical structure are non-polar. The example for which are H_2, O_2, N_2, CO_2, CH_4 (Methane) CCl_4, (Carbontetrachloride) molecules etc. For such molecules inherent dipole moment is zero (Fig 4·7a). If an atom of hydrogen molecule is replaced by chlorine atom, it becomes polar, becasue chlorine being elector-negative, the electron or negative cloud of hydrogen atom shifts slightly uncoverd. So a dipole moment is produced in a direction from chlorine atom towards the hydrogen atom (Fig 4·7b). Other examples ofr polar molecules are NH_3 (ammonia), NaCl (Sodium chloride), H_2O (water), NO_2 (Nitrite) and NaOH (Sodium hydroxide) etc. Such molecules possess an inherent dipole moment (Fig. 4·2c).

The materials with non-polar moleculs are calling non-polar dielectrics while the materials with polar molecules are called polar dielectries.

Polarization in Dielectrics

We have already discussed that the conduction of current is not possible in dielects because electorons are not free, placement if extrernal force is working on them. Now, Let's consider a Atom of the dielectric as consisting of a negative charge-Q (Electron cloud) and a positive charge +Q (nucleus) as in Figure 4·8 (a). A similar picture can be adopted for a dielectric molecule; we can treat the nuclei in molecules as point charge and the electronic and negative charge, the whole atom or omolecule is electrially neutral. When an electric field E is applied , the positive charge is displaced from is displaced from its equilibrium positions in the direction E by thye force F + = QE while the negative charge is displaced in the opposite direction by the force F_=QE. A dipole results from theb displacement of the displacement of the charge and the dielectric is said to be polarized. In the polarized state, the electron cloud is distorted by the applied electric field E. This distorted charge distribution is equivalent, by the principle of superposition, to the original distribution plus a dipole whose moment is

$$\mathbf{p} = Q\mathbf{d} \tag{4·86}$$

where **d** is the distance vector from-Q to + Q of dipole as in Figure (4·86). If there are N dipoles in a volume Δv of the dielectric, the total dipole moment due to the electric field is

$$Q^1d^1 + Q^2d^2 + \ldots\ldots + Q_Nd_N = \sum_{k=1}^{N} Q_k d_k \tag{4·23}$$

As a measure of intensity of the polarization, we define polarization **P** (in coulomds/meter square) as the dipole moment per unit volume of the dielectric; that is,

$$P = \frac{\lim\limits_{\Delta v \to 0} \sum_{k=1}^{N} Q_k d_k}{\Delta v} \tag{4·24}$$

Thus we conclude that the major effect of the electric field E on a dielectric is the creation of dipole moment that align themselves in the direction of E. This type of dielectric is said to be nonpolar.

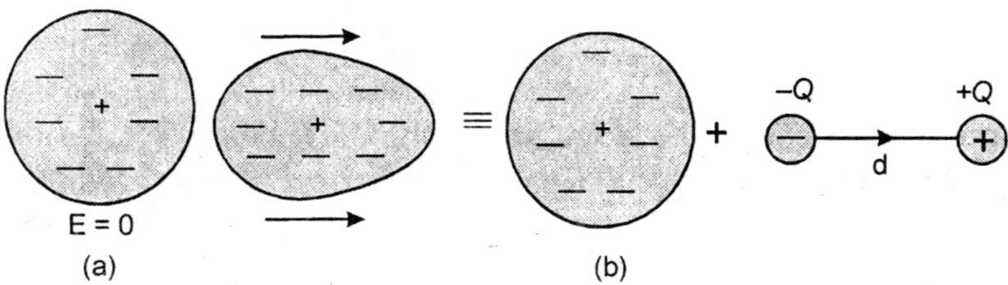

Fig. 4·8 Polarization of a nonpolar atom or molecule.

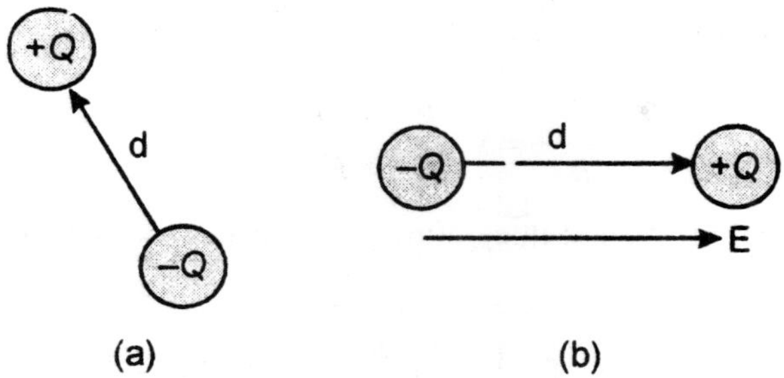

Fig. 4·9 Polarization of a polar molecule (a) permanent dipole (E = 0), (b) alignment of permanent dipole (E ≠ 0)

Let us now calculate the field due to a polarized : Consider the dielectric material shown in Figure 4·9 as consisting of dipoles with dipole moment P per unit volume. We know from previous chapter that the potential dV at an exterior point O due to the dipole moment o dv' is

$$dV = \frac{p.a_R dv'}{4\pi\varepsilon_o R^2} \tag{4·25}$$

where $R^2 = (x-x')^2 + (y-y')^2+(z-z')^2$ and R is the distance between the volume element dv' at (x', y', z') and the field point O (x,y,z). We can transform eq. (4·25) into a form that facilitates physical interpretation. The gradient of 1/R with respect to the primed coordinates is

$$\nabla' = \frac{1}{R} = \frac{a_R}{R^2}$$

Thus,

$$\frac{P.a_R}{R^2} = P.\nabla'\left(\frac{1}{R}\right)$$

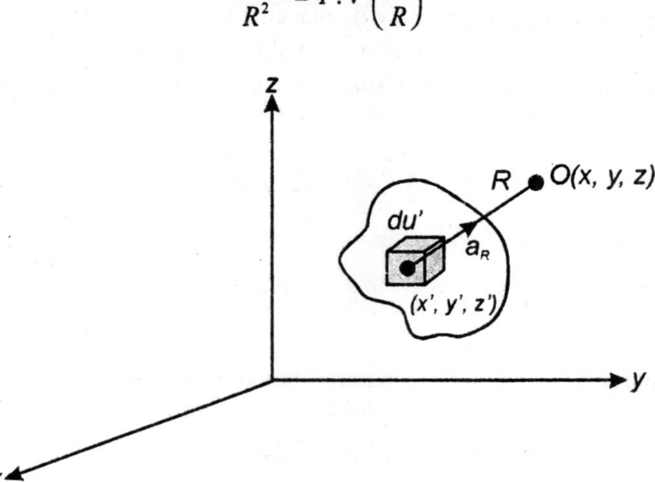

Figure 4·9. A block of dielectric material with dipole moment P per unit volume.

Applying the vector identity $\nabla'.f\nabla'.A + A.\nabla' f$

$$\frac{P.a'_R}{R^2} = \nabla'.\frac{P}{R} - \frac{\nabla'.P}{R} \tag{4.26}$$

Substituting this into eq. (4·25) and integrating over theentrr volume' v' of the dielectric, we obtain

$$V = \int_{v'} \frac{1}{4\pi\varepsilon_o}\left[\nabla'\frac{P}{R} - \frac{1}{R}\nabla'.P\right]dv'$$

Applying divergence theorem to the first term leads finally to

$$V = \int_{S'} \frac{P.a'_n}{4\pi\varepsilon_o R}ds' + \int_{v'} \frac{-\nabla'.P}{4\pi\varepsilon_o R}dv' \tag{4·27}$$

where a'n is the outward unit normal to surface ds' of the dielectric. Comparing the two terms on the right side of eq. (4·27) with the potential due to surface and volume charge disributions with densities (upon dropping primes)

$$\rho_p s = p.a_n \tag{4·28a}$$

$$\rho_q v = -\nabla.P \tag{4·28b}$$

In other words, eq. (4·27) reveals that where polarization occurs, an equivalent volume charge density $\rho_p v$ is formed through out the dielectric while an equivalant surface charge dendity $\rho_p s$ is formed over the surface of the dielectric. We refer to $\rho_p s$ and $\rho_p v$ as bound (or polarization) surface and Volume charge densities, respectively, as distinct from free surface and volume charge densities ρs and ρv. Bound charge are those that are not free to move within the dielectric material; they are caused by the displacement that occurs on a move within the dielectric material; they are caused by the displacement that occurs on a molecular scale during polarization. Free charges are those that are capable of moving over macroscopic distance as electrons in a conductor; they are the stuff we control. The total positive bound charge on surface S bounding the dielectric is

$$Q_b = \oint P.dS = \int \rho_{ps} dS \tag{4·29a}$$

while the charge that remains inside S is

$$-Q_b = \int_v \rho_{pv} dv = -\int_v \nabla\, P dv \tag{4.29b}$$

Thus the total charge of the dielectric material remains zero, that is,

$$\text{Total charge } = \oint_s \rho_{ps} dS + \int_V \rho_{pv} dv = Q_b - Q_b = 0$$

This is expected because the dielectric was electrically neutral before polarization.

We now consider the case in which the dielectric region contains free charge. If ρ_v is the free charge volume density, the total volume charge densisty ρ_t is given by

$$\rho_t = \rho_v + \rho_{pv} = \nabla.\varepsilon_\circ E \quad (4{\cdot}30)$$

Hence

$$\rho_v = \nabla.\varepsilon_\circ E - \rho_{pv} \quad (4{\cdot}31)$$

$$= \nabla.(\varepsilon_\circ E + P)$$

$$= \nabla.D$$

where

$$\boxed{D = \varepsilon_\circ E + P} \quad (4{\cdot}32)$$

We conclude that the net effect of the dielectric on the electric field E is to increase D inside it by amount P. In other wiords, due to the application of E to the dielectric material, the flux density is greater than it would be in free space.

We would expect that the polarization P would very directly as the applied electric field E. For some dielectrics, this is usually the case and we have

$$\boxed{P = X_e \varepsilon_\circ E} \quad (4{\cdot}33)$$

where X_e, known as the electric susceptibility of the material, is more or less a measure of how susceptible (or sensistive) a given dielectric is to electric fields.

Reduced Value of Field Inside Dielectric and Dielectric Constant.

We now have come to know from above explanations that the external field $\vec{E}_\circ$ is opposed by the electric field E_p*s et* up by induced surface charges which appear at the ends of the dielectric.

The vector sum of $\vec{E}_\circ$ and $\vec{E}_p$ (i.e., $\vec{E}_\circ - \vec{E}_p$) gives the resultant field $\vec{E}$ in the dielectric. Clearly the resultant field of $\vec{E}$, the external field, but will be smaller in magnitude. Thus when a dielectric is put in an electric field, the induced surface charges weaken the original field inside the dielectric.

In the Fig 4·11, lines of force can depicte the reduced value the electric field . We can observe that a very few number of lines are passing through the dielectric because some of the lines of force starting from positive free charge q terminate on the -ve polarisation charges q'. The lines of force starting from positive polarisation charges +q' charges on the conducting plate. Hence the resultant field $\vec{E}$ within the dielectric is less than $\vec{E}_0$, the field in free space.

The extent to which the field inside a dielectric can be reduced depends on the degree of polarisation of the dielectric material which in turn depends on the dielectric constant, K (which takes care of the effect of nature of the media). Thus K can be definded as

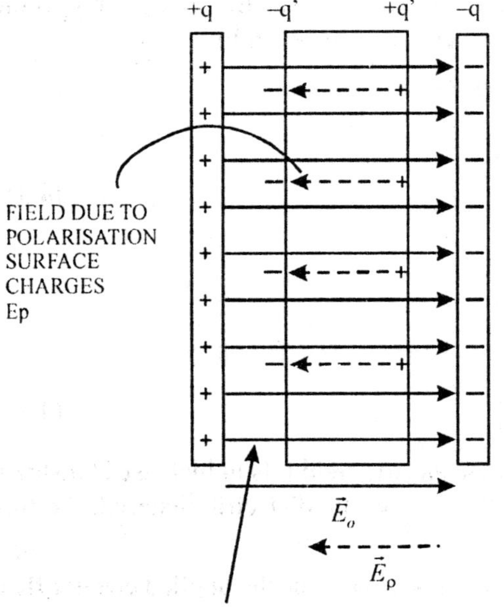

Fig. 4·10

Fig. 4·11 Depicts that number of lines of force within the dielectric reduces.

$$K = \frac{E_0}{E_0 - E_p} = \frac{E_0}{E} \tag{4·34}$$

K is a dimensionless quantity. It is clearly demonstrated by the above equation (4·34).

In the absence of dielectric constant, Vo. The potential difference between the plates of the capacitor is given below

$$V_0 = E_0 t \tag{4·35}$$

and potential difference V with dielectrics is

$$V = E\,t \quad ...(4·36)$$

where 't' is the thickness of the dielectric slab as also the separation between the capacitor plates as shown in fig. (4·12).

Dividing (4·35) and (4·36), we get

$$\frac{V_0}{V} = \frac{E_0 t}{Et} = \frac{E_0}{E} = k \quad ...(4·37)$$

But $E_0 > E$, (proved above), so the dielectric constant $k>1$ and hence $V_0 > V$.

Once the dielectric is placed between the plates of capacitor, the electric field is reduction which causes the reduction of potential difference.

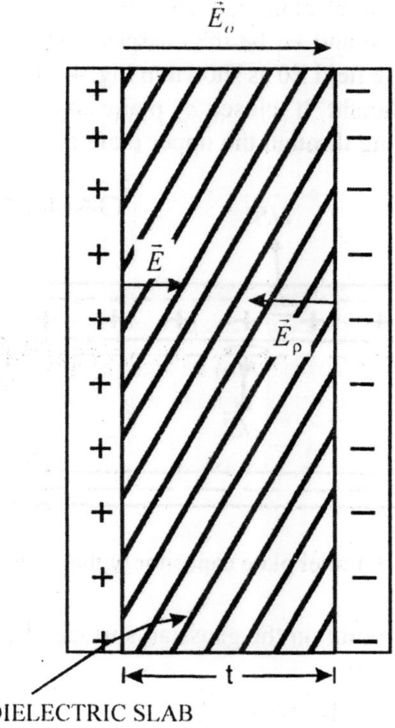

Fig. 4·12

Further if C_0 be the capacitance of the capacitor in the absence of dielectric and C be its capacicance in the presence of dielectric, then

$$C_0 = \frac{q}{V_0} \text{ and } C = \frac{q}{V} \tag{4·38}$$

Therefore
$$\frac{C}{C_0} = \frac{q}{V} \bigg| \frac{q}{V_0} = \frac{V_0}{V} = K. \tag{4·39}$$

The ratio of the capacitance of a capacitor with dielectric and its capacitance without dielectric is also an appropriate definition of the dielectric constant of a dielectric material.

From equation (4·7), it is clear that $C > C_0$ because K>.

Thus the capacitance of a capacitor increases, when a dielectric material is put in between its plates.

Gauss's Law in the Presence of Dielectric

Till now, we have discuss gauss law for the free space where no dielectric is applied but now we will discuss the law for paralled plate capacitor with a dielectric.

Let us first consider a parallel plate capacitor without the presence of any dielectric media between the plates. Let +q and –q be the charges on the plates of a capacitor and let these charges establish an electric field E0 as shown in Fig. 4·19. Further let a Gaussian surface (shown dotted in Fig. 4·19) of height 'h' closed by plane caps having same size and shape as that of capacitor plate and passing through the upper plate of the capacitor.

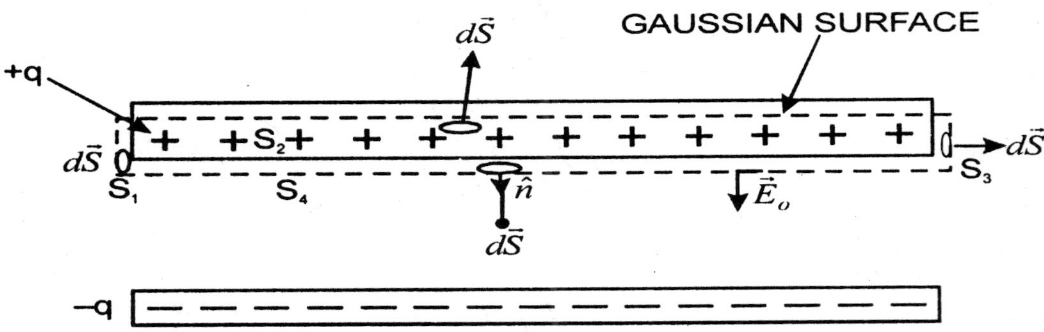

Fig. 4·13 Parallel plate capacitor without dielectric

If E_0 be the electric field at any point on the gaussian surface, then

$$\oint_S \vec{E}_0 \hat{n}\, ds = \frac{q}{\varepsilon_0} \quad (4.40)$$

where $\hat{n}$ is unit vector normal to surface element of area dS.

$$\oint_S \vec{E}_e \hat{n}\, dS = \oint_{S_1} \vec{E}_0 \hat{n}\, dS + \oint_{S_2} \vec{E}_0 . \hat{n}\, ds + \int_{S_3} E_0 \hat{n}\; dS + \int_{S_4} \vec{E}_0 \hat{n}\, dS \quad (4\cdot41)$$

Each of integrals $\oint_{S_1} \vec{E}_0 \hat{n}\, dS, \int_{S_2} \vec{E}_0 \hat{n}\, dS$ and $\int_{S_3} \vec{E}_0 \hat{n}\, dS$ is equal to zero because no electric field is through S_1, S_2, and S_3.

∴ From (4·40) and (4.41),

$$\oint_{S_4} E_0 \hat{n}\, dS = -\frac{q}{\varepsilon_0}$$

$$\oint_{S_4} E_0 dS = -\frac{q}{\varepsilon_0} \left(\because \vec{E}_0 \text{ is parallelto '}\hat{n}\text{'}; so\ \vec{E}_0 . \hat{n} = E_0(1)\cos 0^\circ = E_0 \right)$$

or $$E_0 \int_{S_4} dS = \frac{q}{\varepsilon_0} \quad \text{(as } E_0 \text{ is constant)}$$

or
$$E_0 A = \frac{q}{\varepsilon_0}$$

$$\left(\because \int_{S_4} dS = A \text{ where } A \text{ is area of capacitor plates} \right)$$

or
$$E_0 = \frac{q}{A\varepsilon_0} \quad (4{\cdot}42)$$

In terms of free charge on the plates of the capacitor, the magnitude of electric field $\vec{E}_0$ in free space is demonstrate by the above equ (4·42).

The dielectric slab gets polarised with the appearance of induced surface a material of dielectric constant K is introduced between its two plats. This is the second very case: The resulting electric field will reduce to $\vec{E}$. Applying the Gauss's Law to the same Gaussian surface in this case, we have

$$\oint_S \vec{E}.n\, dS = \frac{q - q'}{\varepsilon_0} \quad (4{\cdot}43)$$

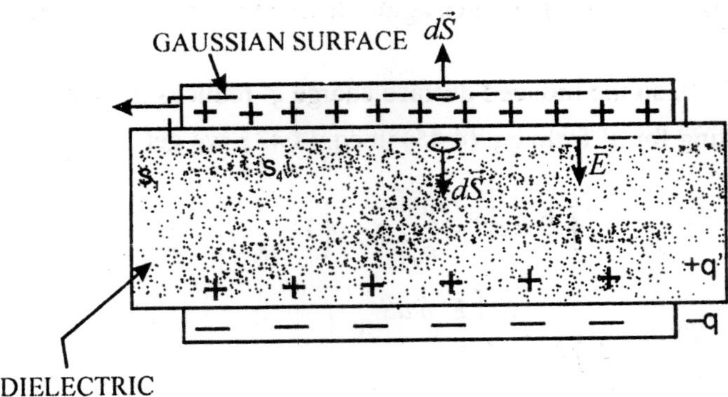

Fig. 4·14 Parallel plate capacitor with dielectric slab in the intervening space between the plates.

where q-q' is the net charge within the Gaussian surface because the two charges, q the original free charge and q, the induced surface charge, both of which lie within the Gaussian surface are opposite in sign.

Proceeding as in the first case, we get

$$EA = \frac{q - q'}{\varepsilon_0}$$

or
$$EA = \frac{q}{\varepsilon_0 A} - \frac{q'}{\varepsilon_0 A} \quad (4{\cdot}44)$$

which gives the magnitude of the electric field intensity $\vec{E}$ within the dielectric in terms of free charge q and induced polarisation surface charge q'.

But $$E = \frac{E_0}{K}$$ (From equation (4·37)

Therefore from equation 4·42, we have

$$E = \frac{q}{A\varepsilon_0 K} \tag{4·44a}$$

Substituting this value of E in (4·12), we get

$$\frac{q}{A\varepsilon_0 K} = \frac{q}{\varepsilon_0 A} - \frac{q'}{\varepsilon_0 A}$$

or $$\frac{q}{K} = q - q'$$

or $$q' = q - \frac{q}{K} = q\left(1 - \frac{1}{K}\right) \tag{4·45}$$

which gives the magnotude of the induced ssurface charge q' in terms of free charge q and the dielectric constant K of the dielectric slab. The above equation (4·45) shows that the induced surface charge q' is always less than (in magnitude) the free charge q. In the absence of dielectric i.e., for K=1.

$q = q$ (1–1) =0, so the induced durface charge q' reduces to zero.

Substituting for q' from eq. (4·45) in eq. (4·43), we get

$$\oint_S \vec{E}.\hat{n}\, dS = \frac{1}{\varepsilon_0}\left[q - q\left(1 - \frac{1}{K}\right)\right] = \frac{q}{K\varepsilon_0}$$

or $$K\oint_S \vec{E}.\hat{n}\, dS = \frac{q}{\varepsilon_0}$$

The above relation (4·46), though derived for a parallel plate capacitor, is true in general. Therefore it represents general form of Gauss's law when dielcteric is present. Here we see that

(a) the flux integral is now multiplied by a factor K;

(b) the charge q enclosed by the Gaussian surface is taken to be the free charge only. Induced surface charge q' is ignored on the right hand side of this equation, having been taken into account by multiplying left hand side of this equation by K; and

(c) on the basis of Ganuss's law equation (4·46) every positive charge may be regarded to be associated with $1/\varepsilon_0$ lines emerging from it regardless of the medium surrounding it.

Relation Between Three Electric Vector:

$(\vec{E}, \vec{P}\ and\ \vec{D})$

$$D = \frac{q}{A} = K\varepsilon_0\left(\frac{q}{K\varepsilon_0 A}\right)$$

or

$$D = K\varepsilon_0 E \tag{4·47}$$

$\left(\because\ E = \frac{q}{K\varepsilon_0 A}\right.$ from eqn, (4·44a).

or

$$D = \varepsilon E \tag{4·48}$$

where $\varepsilon = \varepsilon_0 K$ is the absolulte permitivity of medium.

In vector (4·49) is written as

$$\boxed{\vec{D} = \varepsilon\vec{E}} \tag{4·49}$$

E quation (4·49) shows that displacement vector $\vec{D}$ is directly proportional to electric field $\vec{E}$ in the dielectrics.

Now using relations $p = \frac{q'}{A}$ and $q' = q\left(1 - \frac{1}{K}\right)$, we can write

$$p = \frac{q'}{A} = \frac{q}{A}\left(1 - \frac{1}{K}\right)$$

$$= \frac{q}{A}\left(\frac{K-1}{K}\right) = \frac{q}{KA\varepsilon_0}(K-1)\varepsilon_0$$

or

$$P = (K-1)\,\varepsilon_0 E \quad \left(\because \frac{q}{KA\varepsilon_0} = E\right) \tag{4·50}$$

In cgs Gaussian units above result (4·51) can be written as

$\vec{P} = (K-1)\varepsilon_0\vec{E}$ (4·52)

This shows cleraly that in free space K=1, the polarisation vector $\vec{p} = 0$.

linear, Isotropic, and Homogeneous Dielectrics

There are three kinds of dielectrics which can be identified on relationship of three electric vectors, A material is said to be *linear* if **D** varies linearly with **E** and nonlinear otherwise. Materials for which ε (or σ) does not vary in the region being considered and is therefore the same at all points (i.e., independent of x, y, z) are said to be homogeneous. They are said to be *inhomogenous* (or nonhomogeneous) when ε is dependent of the space coordinates. The atmosphere is a typical example of an inhomogeneous medium; its permittivity varies with altitude. Materials for which **D** and **E** are in the same direction are said to be isotropic. That is, isotropic dielectrics are those which the same properties in all directions. For *anisotropic* (or nonisotropic) materials, **D**, **E**, and **P** are not parallel; ε or X_e has nine components that the collectively referred to as a tensor. For example, insteand of eq. (4·48), we have

$$\begin{bmatrix} D_x \\ D_y \\ D_z \end{bmatrix} = \begin{bmatrix} \varepsilon_{xx} & \varepsilon_{xy} & \varepsilon_{xz} \\ \varepsilon_{yz} & \varepsilon_{yy} & \varepsilon_{yz} \\ \varepsilon_{zx} & \varepsilon_{zy} & \varepsilon_{zz} \end{bmatrix} \begin{bmatrix} E_x \\ E_y \\ E_z \end{bmatrix} \qquad (4{\cdot}53)$$

for anisotropic materials, Crystalline materials and magnetized plasma are anisotropic.

A **dielectric material** (in which **D**=ε **E** applies) is linear if ε does not chage with the applied E field, homogeneous if ε does not change from point to point, and isotropic if ε does not change with direction.

The same idea holds for a conducting materical in which **J** = σ**E** applies. The material is linear if σ does not very with **E**, homogeneous if σ is the same at all points, and isotropic if σ does nnot very with direction.

Relation between X_e and K :

Now; compare equ (4·33) with equ (4·51) then, we get:

$$X_e = K-1 \qquad (4{\cdot}53)$$

Permitivity and Dielectric Constant

Permitivity is another most important quantity which is used for the characterisation of the dielectoric behavious of matter. This is contrasted with the permitvity of free space ε_0. The relationship is defined as

$$\varepsilon=\varepsilon_0 K$$

Therefore equation (4·53) can be written as

$$X_e=K-1=\frac{\varepsilon}{\varepsilon_0}-1 \qquad (4{\cdot}54)$$

The quantities P, K, X and ε are all measures of the extent of the polarisation produced within a material by given applied field.

Gauss's law in terms of $\vec{D}$

The Gauss's law in the presence of dielectric as given by equation (4·14) i.e., $K\oint \vec{E}_0.\hat{n}\, dS = \frac{q}{\varepsilon_0}$ can now be written as in terms of $\vec{D}$ as

$$\varepsilon_0 K \oint \vec{E}_0 . \hat{n}\, dS = q$$

or

$$\oint_S \varepsilon_0 K \vec{E}_0 . \hat{n}\, dS = q$$

$$\oint_S \vec{D} \,.\, \hat{n}\, dS = q \tag{4·55}$$

[$\because \varepsilon_0 K \vec{E}_0 = \vec{D}$ by Equation (4.47)]

i,e., flux of $\vec{D}$ over any closed surface is equal to free charge only within the surface and not the net charge (q–q') enclosed by the surface. $\nabla . \vec{D} = P$ is the differential form of gauss's law in terms of $\vec{D}$. Here P is free charge density.

Energy Stored in Dielectrics

In previous chapters, we have studies that the stored energy per unit volume in vaecuum $=1/2\ \varepsilon_0 E^2$ (4·56)
and it depends upon the field energy stored in a capacitor in the absense of dielectric is as follows:

$$U = \frac{1}{2} CV^2 \tag{4·57}$$

Consider a parallel plate capacitor within which the field is uniform. If C_{vac} is the capacitance in vacuum, KC_{vac} is the capacitance with dielectric filling. Since the stord energy is $\frac{1}{2}CV^2$, the change in energy when a dielectric is insterted with V held constant (by giving more charge to the capacitor) is from $\frac{1}{2}C_{vac}V^2$ to $\frac{1}{2}KC_{vac}V^2$. Now, the expression of stored energy ofter the introduction of dielectric constant K is given by:

$$\tfrac{1}{2}\varepsilon_0 KE^2 \text{ or } \frac{1}{2} ED \text{ joule/m}^2$$

The differeence between the two cases arise from the extra work required to polarise the dielectric.

This comparison is on the basis of identical fields in vacuum and (on the average) in the dielectric medium.

If on the other hand, we charge a capacitor with a fixed charge Q and hold this charge Q constant while we insert a dielectric medium, the stored energy becomes less. The argument follows using

$$U_{vac} = \frac{1}{2}\frac{Q^2}{C}$$

$$U_{dia} = \frac{1}{2}\frac{Q^2}{KC}$$

or
$$U_{dia} = \frac{1}{K}U \tag{4·58}$$

for Q=constant.

This constrast is due to the fact that if Q is kept fixed, then insertion of a dielectric lowers the field.

When the dielectric is introduced, what changes occure the stored field energy. The stored energy increases in the case of constant E and it decreases in the case of constant Q. Where does the energy come from or where does it go? The principle of virtual dispacement can be used to determine this, starting with the case of the isolated capacitor, keeping V (or E) constant.

The necessary equation which keeps account of all the energy and work terms is

$$dU = -Fdx - VdQ \tag{4·59}$$

where F is the force exerted by the field on to slab. The term VdQ is required because we must add charge in order to hold V constant. We can imagine that the capacitor is connected to a battery at voltage V. Then a displacement dX of the dielecdtric slab-increases the capacitance and at the same time the battery delivers a charge dQ at the voltage V.

In terms of displacement dx, dU and VdQ are expressed in following ways. First of all we calculate Q in terms of vand the capacitance C and take the expression U=1/2VQ.

$$C = \frac{\varepsilon_0 A}{d} + \frac{K\varepsilon_0 A'}{d}$$

where A is that part of the plate area without dielectric material inserted; A' is the plate area with dielectric inserted and d is the plate separation as shown in (Fig. 4·15), we find

$$C = -\frac{\varepsilon_0 b}{d}(Kx + l - x)$$

where D is the width of the plates, and l the length.

Now, since Q=CV. we have

$$Q = \frac{V\varepsilon_0 b}{d}(Kx + l - x)$$

or
$$dQ = \frac{V\varepsilon_0 b}{d}(K - l)dx$$

and since
$$U = -\frac{1}{2}VQ \text{ or } dU = \frac{1}{2}VdQ$$

$$dU = \frac{1}{2}\frac{V^2\varepsilon_0 b}{d}(K - 1)dx$$

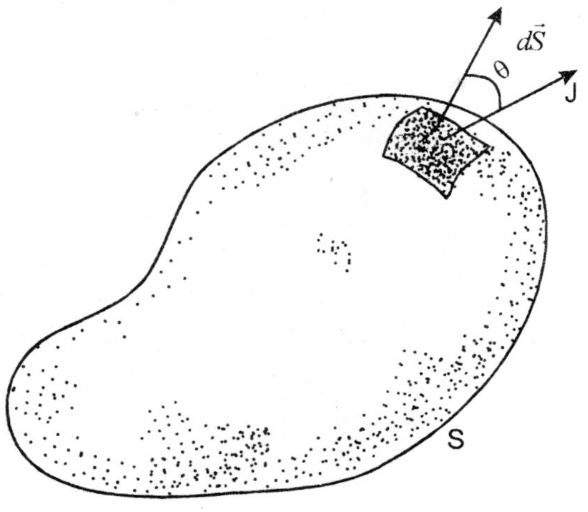

Fig. 4·1

Therefore from eqn. (4·35)

$$Fdx = -dU + VdQ$$

$$= -\frac{1}{2}\frac{V^2 \varepsilon_0 b}{d}(K-1)dx + \frac{V^2 \varepsilon_0 b}{d}(K-1)dx$$

or $$F = \frac{1}{2}\frac{V^2 \varepsilon_0 b}{d}(K-1) \text{ newtons.}$$

Since F is positive, it is in the direction of increasing x. Hence the slab is being pulled into the capacitor by electrostatic forces. Half the work done by the battery goes into increasing the stored energy in the capacitor, and half appearsd as work pulling the dielectric slab into into the space between the plates.

Equation of Continuity.

Under steady-state conditions the charge density in any given region will remain constant. We now relax the requirement of steady-state conditions and allow the charge density to become a function of time. It is experimentaly verified that the net amount of electric charge in a closed system reamins constant. Therefore if the net charge within a certain region decreases with time, this impl;ies that a like amount of charge must appear in some other region. This transport of charge constitutes a current i.e.,

$$I = -(dq/dt) \tag{4·60}$$

-ive sign here indicates that charge contained in a specified volume decreases with time.

However by definition of current density

$$I = \oint J.ds \tag{4·61}$$

So from equation (4·60) and (4·61) we have–

$$\oint_S J.ds = -\frac{dq}{dt}$$

i.e. $$\oint_S J.ds = -\frac{d}{dt}\int_\tau p\, d\tau \qquad \left[as\ q = \int_\tau p\, d\tau\right]$$

If we hold the surface S fixed in space, the time variation of the volume intgral must be solely due to the time variation of p. Thus

$$\oint_S J.ds = \int \frac{\partial p}{\partial t}\, d\tau. \tag{4·62}$$

But from Gauss's theorem

$$\oint_S J.ds = -\int_\tau (div\, J)\, d\tau \qquad (4·63)$$

So comparing expressions (4·62) and (4·63) we get

$$\int_\tau (div\, J)\, d\tau = -\int_\tau \frac{\partial p}{\partial t} d\tau$$

i.e. $$\int_\tau \left(div\, J + \frac{\partial p}{\partial t}\right) d\tau = 0 \tag{4·64}$$

Since equation (4·64) is true for any arbitrary finite volume, the integrand must vanish i.e.

$$div\, J + \frac{\partial p}{\partial t} = 0. \tag{4·65}$$

Equation (4·64) is called the equation of continuity and is an expression of the experimental fact that electric charge is conserved.

$$div\, J + \frac{\partial p}{\partial t} = 0.$$

Exercise 1

In a Van de Graaff generator, w =0.1m, u = 10m/s, and the leakage paths have resistance $10^{14}\,\Omega$. If the belt carries charge 0.5 μ C/m^2, find the potential difference bnetween the dome and the base.

Answer : 50 MV.

Miscelleneous Examples

Example 1. A wire of diameter 1 mm and conductivity 5×10^7 S/m has 10^{29} free electrons/ ^{3} when an electric field of 10 m V/m is applied. Determine

(a) The charge density of free electrons

(b) The current density

(c) The current in the wire

(d) the drift velocity of the electrons, Take the electronic charge as $e = -1.6\times10^{-19}$ C.

Solution : (In this particular problem, convection and conduction currents are the same.)

(a) $\rho v = ne\ (10^{29})\ (-1.6\times10^{-19}) = -1.6\times10^{10}$C/m3

(b) $J = \sigma E = (5\times10^7)\ (10\times10^{-3})$ 500kA/m^2

(c) $I = JS = (5\times105)\left(\frac{\pi d^2}{4}\right) = \frac{5\pi}{4}\ .\ 10^{-6}.\ 10^5 = 0.393$ A

(d) Since $J = \rho v \mu$, $u = \frac{J}{\rho v} = \frac{5\times10^5}{1.6\times10^{10}} = 3.125\times10^{-5}$m/s.

Exercise 2

The Free Charge Density In Copper Is 1.81×10^{10} C/M^3. For A Current Density of 8×10^6 A/m^2, find the electric field intensity and the drift velocity

Answer : 0.138 V/m, 4.42×10^{-4}m/s.

Example 2. A lead ($\sigma = 5\times106$ S/m) bar of square cross section has a hold borded along its length of 4 m so that its cross section becomes that of Figure 5.5. Find the resistance between the square ends.

Solution : Since the cross section of the bar is uniform, we may apply eq. (5·16); that is,

$$R = \frac{\partial}{\sigma S}$$

where $S = d^2 - \pi r^2 = 32 - \pi\left(\frac{1}{2}\right)^2 = 9 - \frac{\pi}{4} cm^2.$

Hence,

$$R = \frac{4}{5\times10^6(9-\pi/4)\times10^4} = 974\,\mu\Omega$$

Exercise 3

If the hole in the lead bar of Example 5.4 is completely filled with copper (σ = 5.8 $\times 10^6$ mhos/m), determine the resistance of the composite bar.

Answer : 876.7 $\mu\Omega$

Example 3. In example I, if 100 V battery remains connected during the time that the dielectric slab is being introduced, calculate (a) the charge on capacitor plates, (b) electric field in the gap, (c) the electric field in the slab, (d) the capacitance.

Solution.

(a) Free charge on plate

$q=CV=61.95\times10^{-12}\times100=61.95\times10^{-10}C$

(b) Electric field strength in gap

$$E_0=\frac{q}{\varepsilon_0 A}=\frac{61.95\times10^{10}}{8.85\times10^{-12}\times10^{-2}}=7\times10^4\ Vm^{-1}$$

(c) Electric field strength in dielectric

$$E=\frac{E_0}{K}=\frac{7\times10^4}{5}=1\times10^4 Vm^{-1}$$

(d) capacity of the capacitor

$$C=\frac{K\varepsilon_0 A}{d}=\frac{7\times\left(8.85\times10^{-12}\right)\times10^2}{10^{-2}}$$

$= 61.95\ \mu\mu f$

Example 4. Two parallel plates having equal and opposite charges are separate by a dielectric slab 1 cm thick and having dielectric constant 5. If the electric field strength inside the dielectric is 106 Vm-1, calculate (a) free charge surface density on the conducting plate, (b) bound charge surface density on the surface of dielectric and (c) P and D in the dielectric.

Solution. Given dielectric constant K=5

Thickness of dielectric slab=separation of plates

=d=1cm =.01m

Electric intensity inside the dielectric

E = 106 Vm-1

∴ Polarisation $P=(K-1)\ \varepsilon_0 E$

$=(5-1)\times(8.85\times10^{-12})\times10^6$

$=35.4\times10^{-6}\ Cm^{-2}=35.4\ \mu\ Cm^{-2}$

Displacement D = k $\varepsilon_0 E = 5\times(8.85\times10^{-12})\,(10^6)$

$= 44.25\times10^{-6}$ Cm^{-2}

$= 44.25$ μCm^{-2}.

As polarisation vector is in the same direction as $\vec{E}$, as it is normal to the plates. Hence bound charge per unit area on the surface of dielectric is

$$|\sigma\rho| = \left|\frac{q'}{A}\right| = P = 35.4\mu\, Cm^{-2}$$

Similarly free charge per unit area

$$|\sigma| = \left|\frac{q}{A}\right| = D$$

Example 5. Prove that the induced charge varies with the dielectric as $K=(1-\sigma_p/\sigma)^{-1}$ when a dielectric is placed in between the plates of capacitor. Also prove that $K \to \infty$ for a metal dielectric, where σp and σ are the induced and free surface charge densities respectively.

Solution. According to equation

$P' = q\left(1-\frac{1}{K}\right)$ where q' is the induced charge and q is the free charge

Dividing both sides by A (area of capacitor plate)

$$\frac{q'}{A} = \frac{q}{A}\left(1-\frac{1}{K}\right)$$

or $$\sigma p = \sigma\left(1-\frac{1}{K}\right) \quad \left[\because \frac{q'}{A} = \sigma p \text{ and } \frac{q}{A} = \sigma\right]$$

or $$\frac{\sigma\rho}{\sigma} = \left(1-\frac{1}{K}\right)$$

or $$\frac{1}{K} = \left(1-\frac{\sigma\rho}{\sigma}\right)$$

or $$K = \left(1-\frac{\sigma\rho}{\sigma}\right)^{-1}$$

In metal dielectric, magnitude of the induced charge is the same as that of free charge, i.e. q'=q or $\sigma\rho = \sigma$

Therefore, $$K = \frac{1}{\left(1 - \frac{\sigma\rho}{\sigma}\right)} = \frac{1}{(1-1)} = \frac{1}{0} = \infty$$

Example 6. A dielectric block such as shown in fig. 4.16 is uniformly polarised. The polarsation is P. find the polarisation charge density σ_v on the faces 1, 2 and 3. (Find both magnitude and sisgn of the charge).

Solution. Polarisation charge density on any surface is equal to the component of P normal to the surface i.e.

$\sigma_v = \vec{P}.\hat{n},$ where $\hat{n}$ is unit vector noraml to surface.

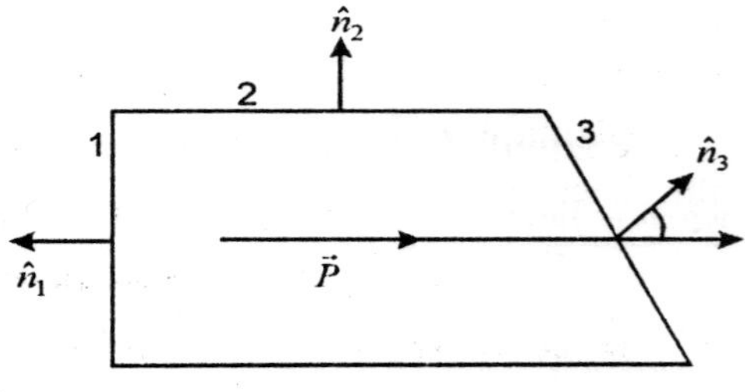

Fig. 4·16

$\therefore$ σ_v on surface 3 = $\vec{P}.\hat{n}_3$

$=P\,(1)\cos\theta$

$= P\cos\theta$

σ_v on surface 2 = $\vec{P}.\hat{n}_2$

$= P\,(1)\cos 90^\circ = 0$

σ_v on surface 1= $\vec{P}.\hat{n}_1$ = P (1) cos 180°= – P

Example 7. The dielectric constant of He at N.T.P. is 1.000074. Calculate the dipole moment of each He atom when the gas is subjected to an electric field E = 3×10^4 Vm^{-1}

(M.D.U. 1982)

Solution. Given dielectric constant K = 1.000074

Electric field intensity E=3×104 Vm-1

If N be the no. He atoms per m^3 and $\vec{p}$ be the dipole moment of each He atom, then

$\vec{P} = N\vec{p}$

But $\vec{p}$= (K-1) $\varepsilon_0\vec{E}$

$$\therefore \qquad \vec{p} = \frac{(K-1)\varepsilon_0\vec{E}}{N}$$

Now 1 gm atom of a gas at N.T.P. occupies a volume 22400 c.c. and has 6.02×10^{23}atoms.

$$\therefore \quad \text{No. of atoms per c.c.} = \frac{6.02\times10^{23}}{22400}$$

$$\therefore \quad \text{No. of atoms/m}^3 = \text{N} = \frac{6.02\times10^{23}}{22400}\times10^6$$

Hence dipole moment of each He atom = q

$$=\frac{(1.00074-1)8.85\times10^{-12}\times3\times10^{4}\times22400}{6.02\times10^{29}}$$

$=.73\times10^{36}$ coulomb metre.

Exercise 4.

A thin rod of cross section A extends along the x-axis from $x=0$ *to* $x=L$. The polarization of the rod is along its legth and is given by $Px = ax2 + b$. Calculate ρ_{qv} and ρ_{qs} at each end. Show explicity that the total bound charge vanishes in this case.

Answer : $0, -2aL, -b, aL^2$ + proof.

Example 8. The electric field intensity in polystyrene $(\varepsilon_r = 2.55)$ filling the space between the plates of a parallet-plate capacitor is 10kV/m. The distance between the plates is 1.5 mm. Calculate:

(a) D
(b) P
(c) The surface charge density of free charge on plates
(d) The surface density of polarization charge
(e) The potential difference betweenthe plates

Solution :

(a) $D=\varepsilon_0\varepsilon_r E\frac{10^{-9}}{36\pi}(2.55).10^4=225.4\ n\,C/m^2$

(b) $D=x_e\varepsilon_0 E(1.55).\frac{10^4}{36\pi}.10^4=137\ n\,C/m^2$

(c) $\rho s = D.a_n = D_n = 225.4\ nC/m^2$

(d) $\rho_{ps} = P.a_n = P_n = 137_n C/m^2$

(e) $V = Ed\ 10^4\ (1.5\times10^{-3}) = 15V$

Exerices 5

A parallel-plate capacior with plate sepaertation of 2 mm has a l-kV voltage applied to its plates. If the space between its plates is filled with polystyrene ($\varepsilon_r = 2.25$), find E.P, and ρ_{ps}.

Answer: 500_x kV/m, $6.853a_x\ \mu$ C/m^2, $6.853\ \mu$ C/m^2.

Example 9. A dielectric sphere ($\varepsilon_r = 5.7$) of radius 10 cm has a point charge 2 pC placed at its center.

Calculate:

(a) We apply Coulomb's or Gauss's law to obtain

$$E\frac{Q}{4\pi\varepsilon_o\varepsilon_r r^2}a_r$$

$$P = \chi_e\varepsilon_o E = \frac{\chi_e Q}{4\pi\varepsilon_r r^2}a_r$$

$$\rho_{ps} = P.a_r = \frac{(\varepsilon_r - 1)Q}{4\pi\varepsilon_r r^2} = \frac{(4.7)2\times10^{-12}}{4\pi(5.7)100\times10^{-4}}$$

$=13.12\text{pC/m}^2$

(b) Using Coulomb's law, we have

$$F = \frac{Q_1Q_2}{4\pi\varepsilon_o\varepsilon_r r^2}a_r = \frac{(-4)(2)\times10^{-24}}{10^{-9}}a_r$$

$= -1.263a_r$ pN

Exercise 6

In a dielectric material, Ex = 5 V/m and p $= \frac{1}{10\pi}(3a_x - a_y + 4a_z)$ nC/m².

Canculate:

(a) χ_e

(b) E

(c) D

Example 10. A metallic wire having area of cross-section equal to 10-6 carries a current of 1 ampere when connected across the terminals of a battery. Calculate average drift velocity of free electrons. Given that number of free electrons per. unit volume is 8.0×10^{22} per cm³.

Solution. Current i = 1 ampere

Area of cross-dection a=10-6m2

Number of density of free eleclrons

n = 8×10^{22} per cm³

= $8\times10^{22}\times10^4$ per m³

∴ Current density,

$$J = \frac{i}{a} = \frac{1}{10^{-6}} = 10^6 \text{ A m}^{-6}$$

But $\vec{J} = ne\,\vec{v}_{av}$

∴ $$v_{av} = \frac{J}{ne} = \frac{10^6}{-8 \times 10^{28} \times 1.6 \times 10^{-19}}$$

$$\frac{10-3}{-1.28} = -7.8 \times 10^{-4}\, ms^{-1}.$$

Example 11. Calulate drift velocity for elctron and also current density in a silver wire having eara if cross-10-5 m2, when a current of 2 ampere is flowing throught it. Given that on an average silver contains one free electron per atom, charge on electron e = 1·6×10-19 C, Avogadro number N= 6·023×1023, density of silver =10·5 gm cm-3 and atomic weight of silver =108.

(M.D.U. 1982 A)

Solution. Given atomic weight of silver = 108

Avogadro's number N = 6·023 ×10-23

density of Silver, d = 10.5 gm cm-2

= 6·03×1023 free electron

(∴ one atom contains one free electron)

Volume occupied by 108 gm of silver

$$\frac{108}{density} = \frac{108}{10.5} = 10.28cm^2$$

Thus 10.28 cm^2 of ag contains $=6.023\times10^{23}$ free electrons

or 1 cm^2 of ag contains = $\frac{6.023 \times 10^{23}}{10.28}$ free electrons

= 585×10^{23} free electrons

Number density of charge carrier

or number of free electrons in 1 m3 of ag

$= n = .585\times10^{23}\times10^{6}$

$=.585\times10^{29}$ free electrons

charge on an electron $=1.6\times10^{-19}$C

area of cross-section $a=10^{-6}$ m^2

current i=3 A

Now $\quad i$=nae v_d where v_d is drift velocity

$$\therefore \quad v_d = \frac{1}{nae} = \frac{2}{585 \times 10^{29} \times 10^{-6} \times 1.6 \times 10^{-19}}$$

$$= .\frac{2}{585 \times 1.6} \times 10^{-4} = 2.13 \times 10^{-10}\, ms^{-1}$$

and $\quad$ current density $J = \frac{i}{d} = \frac{2}{10-6} = 2 \times 10^6\, Am^{-2}$.

Example 12. In a certain region 5×10^{19} triply charged positive ions per cubic metre are miving with a speed fo 10^5 ms-1 towards west in the same region, there are 10^{11} electorns per cubic metre moving with a speed of 10^6 ms along north east. What is the magnitude and direction of $\vec{J}$.

Solution. Number density of triply ionsed ions $n_1=5\times10^{10}$

charge on triply ionised ion=q_1 =3e

Number density of electrons $n_2 = 10^{11}$

charge on electron n = –e

velocity of triply ionised ion toward north west $\vec{v}_1 = -\hat{i}(10^5)$

(as shown in Fig. 5.10)

velocity of electron towards north east = 10^6 ms^{-1} can be written as

$$\vec{v}_2 = 10^6\left[\frac{1}{\sqrt{2}}\hat{i} + \frac{1}{\sqrt{2}}\hat{j}\right]$$

(as shown in Fig. 5.11)

Now $\vec{J} = n_1 q_1 \vec{v}_1 + n_2 q_2 \vec{v}_2$

$$= 5\times10^{10}\times3\times1.6\times10^{-19}\times10^5(-\hat{i})$$

$$+10^{11}(-1.6\times10^{-19})(10^6)\left(\frac{1}{\sqrt{2}}\hat{i} + \frac{1}{\sqrt{2}}\hat{j}\right)$$

$$=5\times10^{15}\times1.6\times10^{-19}\left[-3\hat{i} - 20\left(\frac{1}{\sqrt{2}}\hat{i} + \frac{1}{\sqrt{2}}\hat{j}\right)\right]$$

$$= -8\times10^{-4}\times(17.14\hat{i} + 14.14\,\hat{j})$$

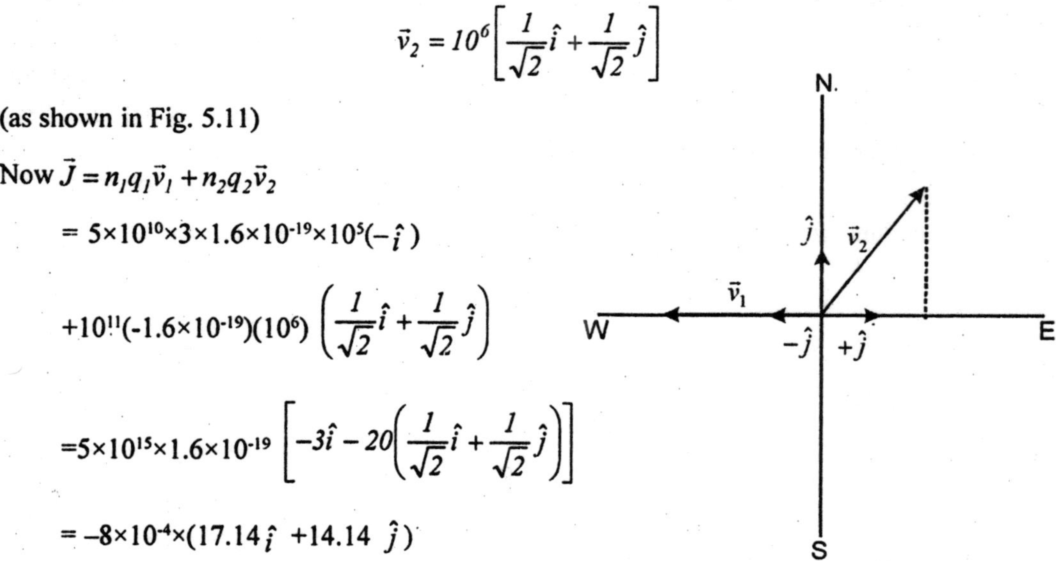

Fig. 4.17

Magnitude of $\vec{j} = |\vec{j}| = 8\times10^{-4}$ [(17.14)^2 + (14.14)2]$^{1/2}$

$=8\times10^{-4}\times22.22$

$=177.76\times10^{-4}$ Cs^{-1} m^{-2}

and direction $\theta = \tan -1\left(\frac{14.14}{17.14}\right) = \tan -1$ (8.2) from west as both $\hat{i}$ and $\hat{j}$ are negative

Example 13. An electric field intensity of 0.0088 V/m exists between two point on a conductor. find the current density J in the conductor. Also calculate the current in conductor if the cross-sectional area is 1 mm2. Given resistivity for the material of conductor is 1.72×10-8 ohm metre.

Solution. Resistivity ρ=1.72×10-8 ohm×m

∴ conductivity $\sigma = \frac{1}{\sigma} = \frac{1}{1.72\times10^{-8}} mho/m$

Electric field E=0.0686 V/m

area of cross-section a=1 mm2=10-6m2

Now from relation J=σE

current density $J = \frac{1}{1.72\times10^{-8}}\times.0686$

$=4\times10^{6}$ Am^{-2}

Also Current i=Ja

$= 4\times10^{6}\times10^{-6}=4$A.

Example 14. A steel wire whose radius is 0.05 cm is welded end to end to a copper wire with a radius of 0.032 cm. The composite wire carries a steady current of 10 amperes. What is the current density within each wire ?

Solution. The current is uniformaly distributed over the cross section of each wire except near the junction, which means that current density may be considered to be constant for all point within each wire.

Now radius of steel wire r_1=.05×10-2m

area of cross-section of steel wire a1=πr12

$= \pi\,(.05)^2\times10^{-4}$ m^2

current in composite wire i = 10A

current density in steel wire

$$=J_{steel} = \frac{i}{a_1} = \frac{10}{\pi(.05)^2\times10^{-4}}$$

$$= \frac{10 \times 10^{4}}{3.14 \times (.05)^2}$$

$$= 127.277 \times 10^5 \text{ Am}^{-2}$$

Radius of copper wire r2 = .032 cm=.032×10-2m

Area of cross-section of copper wire

$$a_2 = \pi r_2^2$$

$$= \pi (.032)^2 \times 10^{-4} m^2$$

∴ current density in copper wire

$$= \text{Jau} = \frac{i}{a_2} = \frac{10}{\pi \times (0.32)^2 \times 10^{-4}}$$

$$= \frac{10 \times 10^4}{3.14 \times (.032)^2}$$

$$= 310.735 \times 10^5 \text{Am}^{-2}$$

Example15. A current of a ampere is passed through a wire of length 1 m and radius 1 mm. If the resistvity of wire is 2.4×10-8 ohm×m, what will be the potential difference between the ends of wire?

Solution. Current i = 2A

length of wire l = 1m

resistvity ρ = 2.4×10–8 ohm×m

radius r=1 mm = 1×10-3m

area of cross-section = a= $\pi r^2 = \pi \times 10^{-6} m^2$

∴ resistance of wire $R = \rho \frac{l}{a}$

$$\frac{2.4 \times 10^{-8} \times 1}{\pi \times 10^{-6}}$$

$$= \frac{2.4 \times 10^{-2}}{\pi} \Omega$$

Hence potential difference across the wire

$$V = iR$$

$$= \frac{2 \times 2.4 \times 10^{-2}}{\pi} = 1.52 \times 10^{-2} \text{ volt.}$$

Example 16. At what temperature would the resistivity of a copper conductor be thrice its resistivity at 0°C. Temperature coefficient of resistvity α = 3.9×10-3/°C.

Solution. Let ρ_0 be the resistivity at 0°C

ρ_t be the resistvity at t°C

$\rho_t=3\rho_0$ (Given)

Now ∴ $\rho t=p_0(1+\alpha t)$

or $$3\rho_0=\rho_0(1+\alpha t)$$

or $$3=1+\alpha t$$

or $$\alpha t=2$$

or $$t=\frac{2}{\alpha}=\frac{2}{3.9\times 10^{-3}}$$

$$=\frac{2000}{3.9}=512.8^{\circ}C.$$

Example 17. A vacuum diode consists of a plane cathode and a plane anode parallet to each other and separated by a distance d and each having area A. when a potential V is applied to the anode w.r.t. cathode, a current i flows through the diode. Assuming that the electrons are emitted with zero velocity and do not change the field between electrodes, calculate electron velocity v and charge density p at any point between the electrodes.

Solution. if we neglect the effect of space charge, then for steady current

Flow of electrons is from cathode top anode only in one direction. If we take it as *x*-direction (Fig. 5.12), then $J_y=0$, $j_z=0$ and $\frac{\partial Jx}{\partial x}=0$ or Jx= constant or independent of *x*. know that $j_x=en_xv_x$ where v_x is the velocity of electrons at a distance *x* from cthode and n_x is the number of electrons per unit volume at this distance.

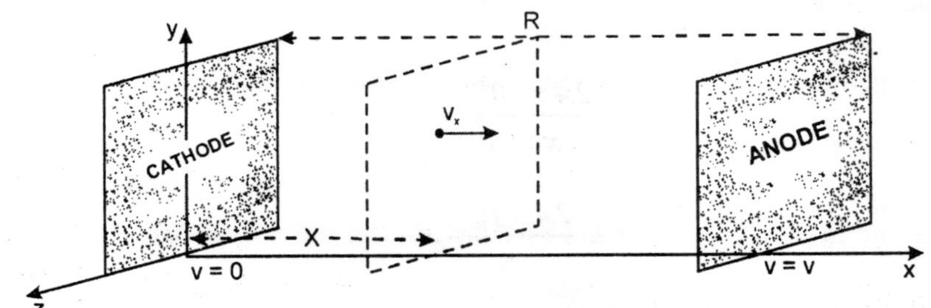

Fig. 5·12

Force acting on each electron F= eE=$e\frac{V}{d}$

∴ Acceleration on the electron $a = \frac{F}{m} = \frac{eV}{md}$

So, the velocity of electrons at a distance x from the cathode is given by

$$v_x^2 = 0 + 2ax = \frac{2eVx}{md} \text{ or } v_x = \sqrt{2eVx / md}$$

∴ currengt $i = J_x A = en_x v_x A = en_x A \sqrt{2eVx / md}$

and space charge density px =enx $= i = \sqrt{md / 2eVxA^2}$

Long Answer Question

1. Define the terms current and current density. Derive the relation

$$\vec{J} = nc\,\vec{v}_d$$

where terms have their usual meanings.

2. Define current density vector. derive an expression for it.

(K.U.1986A)

3. Explain the terms resistance, resistivity and conductivity and deduce the microscopic form of Ohm's law.

(K.U.1980)

4. What do you understand by steady current? Show that for steady current $Div\,\vec{J} = 0$

(K.U.1992A)

5. Establish the equation of continuity and discuss the physical significance of each term. **(K.U.1986A, 88S)**

6. Derive the equation of continuity. What physical fact does it. Explain.

7. Explain clearly the term drift velocity? What role this velocity play in the conduction phenomenon.

8. Prove that Div $\vec{J} = \frac{\partial \rho}{\partial t}$ =0. Give the name of this equation. Given $\vec{J}$ is current density, ρ is charge density.

(K.U.1985,88A, 92 S, 94A, 95A)

9. State and explain Ohm's law. On what factors does the resistance of a conductor depend?

10. $\vec{J} = \sigma\vec{E}$ is one form of Ohm law, Another from of the same is V=iR. Show that these are equivalent. Which form is applicable to practical problems? **(P.U.1982)**

11. Define electric current density $\vec{J}$ in a medium. Prove that $\vec{J}$ is a solenoidal vector in a conductor carrying a steady current.

(K.U.1995 S)

12. What is the difference between a conductor and a dielectric? Explain the term dielectric constant.
13. What are polar ad non-polar dielectrics? explain the mechanism of polarisation of the two types of dielectric in the presence of an external electric field.
14. Explain what do you understand by the polarisation of a dielectric. How does it explain the effect inserting dielectric slab between the plates of a given capacitor?
15. What do you understand by electric intensity $\vec{E}$, electric polarisation vector P and electric displacement D? Derive a relation between them.

    (K.U.1986 S, 1993 S, 1994A)
16. Name the three electric vectors. Define each of them and establish a relation between them.

    (K.U.1987A, 95 S)
17. Derive the relation $\vec{D} = \varepsilon_0 \vec{E} + \vec{P}$ where $\vec{D}, \vec{E}$ and $\vec{p}$ have their usual meanings.

    (K.U.1993S,1994 S)
18. Define electric polarisation $\vec{p}$ and displacement vector $\vec{D}$. Express them in terms of electric intensity $\vec{E}$.
19. Derive an expression for Gauss's law in the presence of dielectric media and derive the expression for energy stored in diclectric. **(K.U.1994S,1995 A)**
20. Derive the relation, $q' = q\left(1 - \frac{1}{K}\right)$ in case of a capacitor with dielectric in between its plates, where the letters have their usual meaning. **(K.U. 1988)**
21. Defined dielectric constant of a medium. How does electric field intensity decrease in a dielectric medium.

    (K.U. 1989A)
22. Prove $\varepsilon = 1 + \chi$ where χ is electrical susceptibility and ε is permittivity.

    (K.U. 1995)

Short Answer Questions

1. What are polar molecules?
2. Does the dielectric constant for subtance containing permanent dipoles very with temperature?
3. Molecules consisting of dissimilar atoms are usually polar. why?
4. Will a molecule having a spherically symmetric charge distribution be polar or non-polar? Explain.
5. What happens to an individual non-polar molecule when it is put in an external electric field?

6. What happens to an individual polar molacule when it is put in an external electric field?
7. How do non-polar molecules behave in bulk matter in the presence of an external field.
8. Does the dielectric constant of non-polar substances depend on temperature? If so, how?
9. Show that the dielectric constant of a conductor can be taken to be infinitely large. **(M.D.U. 1979, 1986)**
10. How bound charge are different from free charges?
11. Explain why the introduction of a dielectric slab between the plates of a capacitor changes its capacity. **(K.U. 1988A)**
12. A dielectric slab fills the space in between the two plates of a capacitor. What will be its capacity compared to vacuum filled capacitor? **(K.U. 1986)**
13. For a constant potential difference across the plates of a capacitor, does a capacitor store more or less charge with a dielectric than it does without a dielectric.
14. Is the behaviour of dielectric the same in an electrostatic field and in very electric field? **(M.D.U. 1983 S)**
15. What is volume charge density for bound charge for a uniformly polarised dielectric.

Check Yourself

1. The formula R = $\partial l(\sigma S)$ is for thin wires.

   (a) True (b) False (c) Not necessarily

2. Sea water has ε_r =80. Its permittivity is

   (a) 81 (b) 79
   (c) 5.162×10^{-10}F/m (d) 7.074×10^{-10}F/m

3. Both ε_o and χ_e are dimensionless.

   (a) True (b) False

4. If $\nabla . D = \varepsilon\nabla . E$ and $\nabla . J = \sigma\nabla . E$ in a given material, the material is said to be

   (a) Linear
   (b) Homogeneous
   (c) Isotropic
   (d) Linear and homogeneous
   (e) Linear and isotropic
   (f) Isotropic and homegeneous

5. The relaxation time of mica ($\sigma = 10^{-15}$ mho/m, ε_r =6) is

(a) 5×10-10s (b) 10-6s

(c) 5 hours (d) 10 hours

(e) 15 hours

6. The uniform fields shown in Figure 5.17 are near a dielectric-dielectric boundary but on opposite sides of is it. Which configurations are correct? Assume that the boundary is charge free and that $\varepsilon_2 > \varepsilon_1$.

7. Which of the following statements are incorrect?

(a) The conductivities of conductors and insulators very with temperature and frequency.

(b) A conductor is an equipotential body and E is always tangential to the conductor.

(c) Nonpolar molecules have no permanent dipoles.

(d) In a linear dielectric, *P* varies linearly with E.

8. The electric conditions (charge and potential) inside and outside an electric screening are completely independent of one another.

(a) True (b) False

Answer : 5.1d, 5.2c, 5.3c, 5.4c, 5.5b, 5.6d, 5.7e, 5.8e, 5.9b, 5.10a.

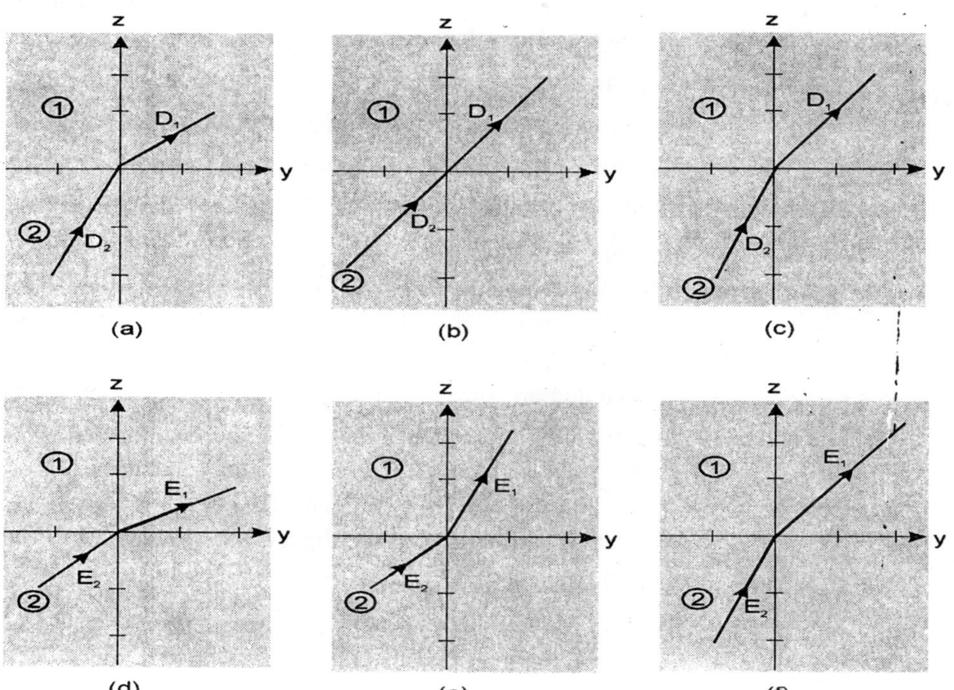

Figure 4·17 For Review Question 6

Problems

1. The charge 10-4e-3t C is removed from a sphere throught a wire. Find the current in the wire at t=0 and t=2.5s.

2. (a) Let V=x2y2z in a region ($\varepsilon = 2\varepsilon_o$ defined by -1< x, y, z < 1. Find the charge density ρ_v in the region.

   (b) If the charge travels at 104 yay m/s, determine the current crossing surface 0<x, z < 0.5, y =1.

3. If the ends of a cylindrical bar of carbon (=3×104) of radius 5 mm and length 8 cm are maintained at a potential difference of 9 V, find: (a) the resistance of the bar, (b) the current through the bar, (c) the power dissipated in the bar.

4. The resistance of round long wire of diameter 3 mm is 4.04 $\Omega / km.$ If a current of 40 A flows through the wire, find
   (a) The conductivity of the wire and identify the material of the wire
   (b) The electric current density in the wire

5. A coil is made of 150 turns of copper wire wound on a cylindrical core. If the mean redius of the turns is 6.5 mm and the diameter of the wire is 0.4 mm, calculate the resistance of the coil.

6. A composite conductor 10m long consists of an inner core of steel of radius 1.5 cm and an outer sheath of copper whose thickness is 0.5cm.
   (a) Determine the resistance of the conductor.
   (b) If the total current in the conductor is 60A, what current flow in each metal?
   (c) find the resistance of a solid copper conductor of the same length and cross-sectional areas as the sheath. Take the resistivities of copper and steel as 1.77×10^{-8} and $11.8\times10^{-8}\ \Omega\cdot$ m, respectively.

7. A hollow cylinder of length 2m has its cross section as shown in Figure 5.18. If the cylinder is made of carbon (σ = 10^5 mhos/m), determine the resistance between the ends of the cylinder. Take a = 3cm, b = 5cm.

8. At a particular temperature and pressure, a helium gas contains 5×10^{25} atoms/m^3. If a 10-kV/m field applied to the gas causes an average electron cloud shift of 10^{-18} m, find the dielectric constant of helium.

9. A dielectric material contains 2×1019 polar molecules/m3, each of dipole moment 1.8×1027 C/m. Assuming that all the dipoles are aligned the direction of the electric field E = 105 ax V/m, find P and ε_r .

10. In a slab of dielectric material for which ε=2.4 ε 0 and V=300z2V, find: (a) d and $\rho_{v,}$ (b) P and $\rho_{pv,}$.

11. For *x*< 0, P=5sin (ay) ax, where a is a constant, find $\rho_{ps,}$ and $\rho_{pv.}$

12. Consider figure 5.18 as a spherical sielectric shell so that $\varepsilon = \varepsilon_o \varepsilon_r$ for a < r< b and $\varepsilon = \varepsilon_o$ 0 < r <a. If a charge Q is placed at the center of the shell, find
    (a) P for a < r< b
    (b) ρ_{pv} for a $\leq$ r < b
    (c) ρ_{ps} at r = a and r = b
13. Two point charges when located in free space exert a force of 4.5 μN on each other. When the space between them is filled with a dielectric matectric material material, the force changes to 2μN. Find the dielectric constant of the material and identiful the material.
14. A conducting sphere of radius 10cm is centered at the origin and embedded in a dielectric material with ε=2.5 ε_o. If the sphere carries a surface charge of 4n C/m_2, find E at (–3cm, 4cm, 12cm).

    At the center of a hollow dielectric sphere $(\varepsilon = \varepsilon_o \varepsilon_r)$ is placed a point charge Q. If the sphere has inner radius a and outer radius b, calculate D, E, and P.
15. A sphere of radius a and dielectric constant ε_r has a uniform charge density of ρ_o.
    (a) At the center of the sphere, show that

$$V = \frac{\rho_o a}{6\varepsilon_o \varepsilon_r}(2\varepsilon_r + 1)$$

    (b) Find the potential at the surface of the sphere.
16. For static (time-independent) field, which of the following current densities are possible?
    (a) $J = 2x^3 y a_x + 4x^2 z^2 a_y - 6x^2 yz a_z$
    (b) J = xya_x +y(z+1)a_y + 2ya_z
    (c) J = $\frac{z}{\rho} a\rho + \cos\phi \, a_z$
    (d) $J = \frac{\sin\theta}{r^2} a_r$
17. For an anisotropic medium

$$\begin{bmatrix} D_x \\ D_y \\ D_z \end{bmatrix} = \varepsilon_o \begin{bmatrix} 4 & 1 & 1 \\ 1 & 4 & 1 \\ 1 & 1 & 4 \end{bmatrix} \begin{bmatrix} E_x \\ E_y \\ E_z \end{bmatrix}$$

Obtain d for: (a) E = 10a_x + 10a_y V/m, (b) E=10a_x + 20a_y - 30a_z V/m.

18. Two parallel conducting plates separated by a distance of 2cm. are given an equal and opposition surface charge density 30 μCm². Two slabs of dielectric, one 8 mm thick having dielectric constant K_1=2 and the other 1·2 cm thick having dielectric constant K_2=3, fill the space between the plates. Calculric constant K_2=3, fill the space between the plates. Calculate the electric intensity and the electric displacement in each dielectric as well as induced charge density on the surface dielectric as well as induced charge density on the surface of each dielectric.

$$\left[\begin{array}{l} Ans.\ D_1 = D_2 = 30 \times 10^{-6}\, Cm^{-2}, E_1 = 1.68 \times 10^6\, Vm^{-1} \\ E_2 = 1.2 \times 10^6\, Vm^{-1}, \sigma_{1p} = 15\mu Cm^{-1}, \sigma 2p = 20\mu Cm^{-2} \end{array}\right]$$

19. A potential difference of 100V is applied across a parallel plate capacitor (in the absence of dielectric) having plate area 10cm2 and separation of plates 1.0 cm. The bettery is then disconnant K=70 is inserted in between the plates of the capacitor. Calculate the magnitude of three electric vectors (a) in dielectric and (b) in air gap.

Ans. (a) $E=1.43\times10^3 Vm^{-1}$, $D=89\times10^{-2}$, 7.5×10^{-8} Cm^{-2}

(b) $E=1.00\times10^4 V\ m^{-1}$, $D=8.9\times10^{-8}$ Cm^{-2}, P=0

20. A potential difference of 100 V is applied between two aimilar parallel plates each of area 01 m2. The plates are separted by a distance of 10-3 m. If the space in between the plates is filled with some dielectric material having K=6, calculate the charge on each plate and induced surface charge at the surface of the dielectric.

[**Ans.** 53.1×10^{-9}C, 44.2×10^{-9} C]

21. A parallel plate capacitor having a plate area 100cm2and a mica dielectric K=6, has capacitance of 100. Calculate (a) E in mica (b) free charge on the plate and (c) the induced surface charge.

[**Ans.** 9.4×103 Vm-1, 5×10-9 C, 4.16×109 C]

22. Two parallel conducting conducting plates of area 100 cm2 are each given equal and opposite charges of 8.9×10-7 c. The electric field strength within the dielectric material filling the space between the plates if 1.4×105 Vm-1. Find dielectric constant of the material and also the magnitude of the charge induced on each dielectric surface.

[**Ans.**7.1, 7.6×10-6 C]

23. The dielectic constant of a gas at N.T.P. is 1.0001. Calculate the dipole moment of each atom of the gas when it is kept in an electric field E=6×104 Vm-1.

[**Ans.** 0.19×10^{-35}Cm]

24. For the atomsof diamond, the polarisation per unit volume is 6 Cm-3. Find the distance by which the centre of the cloud of six electrons is shifted from the nucleus. Given that the density of diamond is 3.5 gm cm-3.

[**Ans.** 3.57×10^{-17}m]

25. The electric field intensity between the plates of a parallel plate capacitor is E0 without dielectric. if a dielectric material having dielectric K is put between the plates, what will be the electric field then?

$$\begin{bmatrix} Hint & \vec{D} = \varepsilon_0 \vec{E} + \vec{P} \\ & \varepsilon_0 E = \varepsilon_0 \vec{E} + \varepsilon_0 (K-1)\vec{E} \end{bmatrix}$$

[Ans. E_0/K]

26. Dielectric constant of heilum at N.T.P. is 1.00074. Calculate dipole moment of each helium atom when the gas is subjected to an electric field $E=15\times10^4$Vm-1. Take Avogadro's number as 6.02×10^{23} per g.mole.

[Ans. 365×10^{-35} C×m]

27. If $J = \dfrac{100}{\rho^2} a\rho\, A/m^2$, find: (a) the rate of increase in the volume charge density, (b) the total current passing through surface definded by ρ=2, 0<z<1, 0<ϕ<2π.

28. Given that $J \dfrac{5e^{-10^4}}{r} a_r$ A/m2, at t=0.1 ms, find: (a) the amount of current passing surface r =2m, (b) the charge density ρ_v on that surface.

29. Determine the relaxation time for each of the following medium:
   (a) Hand rubber ($\sigma = 10^{-15}$S/m, $\varepsilon = 3.1\varepsilon_0$)
   (b) Mica (10^{-15} S/m, $\varepsilon = 6\varepsilon_0$)
   (c) Distilled water ($\sigma = 10^{-4}$ S/m, $\varepsilon = 80\varepsilon_0$)

30. The excess charge in a certain medium decreases to one-third of its initial value in 20 Hs. (a) If the conductivity of the medium is 10^{-4} S/m, what is the dielectric constant of the medium? (b) What is the relaxation time? (c) After 30 μs, what fraction of the charge will remain?

31. Lighting strikes a dielectric sphere of radius 20 mm for which $\varepsilon_r = 2.5$, $\sigma = 5\times10^{-6}$ mhos/m and deposits uniformaly a charge of 10μC. Determine the initial charge density and the charge density 2 μs later.

32. Region 1 (z < 0) contains a dielctric for which $r = 2.5$, while region 2 (z > 0) is characterized by $\varepsilon_r = 4$. Let e1 = – $30a_x + 50a_y + 70a_z$ V/m and find: (a) D_2, (b) P_2, (c) the angle between E_1 and the normal to the surface.

33. Given that $\mathbf{E}_1 = 10a_x - 6a_y + 12a_z$ V/m in Figure 5.19, find: (a) $\mathbf{P}_1$, (b) $\mathbf{E}_2$, and the angle $\mathbf{E}_2$ makes with the *y*-axis, (c) the energy density in each region.

34. Two homogeneous dielectric regions 1 $(\rho \le 4cm)$ and 2 $(\rho \ge 4cm)$ have dielectric constants 3.5 and 1.5 respectively. $\mathbf{D}_2 = 12a_\rho - 6a_\phi + 9a_z$ nC/m², calculate : (a) $\mathbf{E}_1$ and $\mathbf{D}_1$, (b) $\mathbf{P}_2$ and $\rho_{\rho v2,}$ (c) the energy density for each region.

35. A conducting sphere of radius a is half-embebbed in a liquid dielectric medium of permittivity ε_1 as in Figure 5·20. The region above the liquid is a gas of permittivity ε_2. If the total free charge on the sphere is Q, determine the electric field intensity everywhere.

36. The parallel sheets of glass (r = 8.5) mounted vertically are sepearated by a uniform air gap betweem their inner surface. The sheets, properly sealed, are immersed in oil (r = 3.0) as shown in Figure 5.21. A uniform electric field of strength 2000 V/m in the horizonted direction exists in the oil. Calculate the magnitude and direction of the electric field in the glass and in the enclosed air gap when (a) the field is normal to the glass surfaces, and (b) the field in the oil makes an angle of 75° with a normal to the glass surfaces. Ignore edge effects.

37. (a) Given that $E = 15a_x - 8a_z$ V/m at a point on a conductor surface, what is the surface charge density at that point? Assume $\varepsilon = \varepsilon_0$.

(b) region $y \geq 2$ is occupied by a conductor. If the surface charge on the conductor is – 20 nC/m^2, find D just outside the conductor.

38. The current in a conductor of circular cross-section of radius a varies radius according to

$$\left[Hint\ I = \int J.dS.S = \pi r^2 \right.$$
$$dS = 2\pi r\, dr$$
$$\left. = \int_0^a J_0 r\ 2\pi r\, dr \right]$$

[**Ans.** $\frac{2}{3}\pi J_0 a^3$]

39. A silver whose redius is 0.122 cm is weldedend to end to a opper wire with radius 0.0813cm. The composite wire carries a steady current of 10 A.

(a) Find the current density within each wire.

(b) Find the drift velocity of the electrons in the wire. Assume one free election per atom in copper. Given avogadro's number =6×1023 atrom/gm mole, molecular weight of Cu = 64, density of Cu = 9 gm cm^{-3}.

[**Ans.** (a) 2.121×10^6 Am^{-2}, 4.818×10^6 Am^{-2}
(b) 2.834×10^{-4} ms^{-1}]

40. A current of 5 ampere flows through a wire of 1 mm2 cross-section. If the density of charge carrier is 1027 per m3 calculate the mean drift velocity of electron.

[**Ans.** 0.0312 ms-1]

41. In a neon gas dicharge tobe 2.9×1016 Ne+ ions move to the east through a cross-section of the tube each second, 1.2×1018 electrons move to the west in the same time. Find the magnitude and direction of the current.

[**Ans.** 2.384 A from west to east]

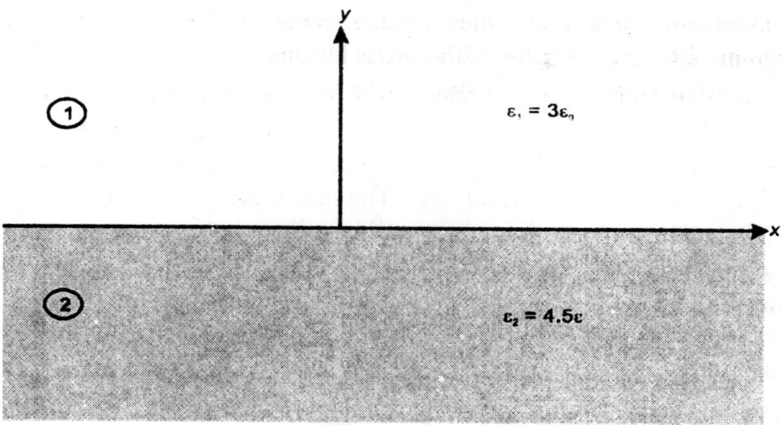

Figure 4·19 For Problem 33.

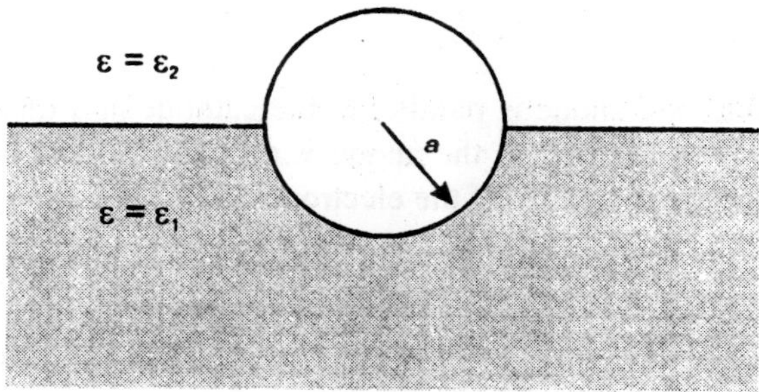

Figure 4·20 For Problem 35.

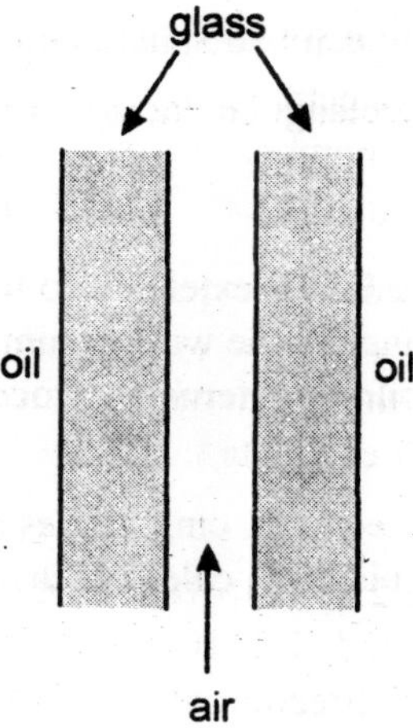

Figrue 4·21 For Problem 36

42. A silver wire 1 mm in diameter carries a charge of 9 C in 1 hr. and 15 minutes. Silver contains 5.8×1022 free electrons/cm3.Calculate

    (i) current in the wire in ampere (ii) the drift velocity of electrons in wire.

    [Ans. (i) .02 A (ii) 2.74×10^{-6} ms^{-1}]

$$\left[Hint.\quad i=\frac{Q}{t},\quad J\frac{i}{a}=nev\right]$$

43. In copper there are 10^{22} free electrons per cm^2 all of which contribute to current of 1 ampere in a wire of copper of 0.01 cm^2 cross-section area.

    (a) Calculate average drift velocity of the electrons in copper.

    (b) What is the electric field in wire? Take resistvity of Cu = 1.6×10-8 ohm ×m.

    [Ans. (a) 6.25×10^{-4} ms^{-1} (b) $1.6\times10^{-2}Vm^{-1}$]

$$\left[\begin{array}{l}Hint.\quad J\frac{I}{a}=ne\,V_{av}\\ E={}_{\rho}\,J\end{array}\right]$$

44. In a vacuum diode, the anode is parallel to the cathode at a distance of 0.2cm. If a potential of 100 V is applied to the anode w.r.t. cathode, calculate the velocity of electron at a point at the middly of the electrodes.

45. A current of 0.1 A is flowing through a Cu wire 10m in length and diameter 0.08 mm. If the resistivity of Cu wire is 1.65×10-6 ohm cm, calculate

    (i) the resistance of Cu wire (ii) p.d. between the ends of wire

    [Ans. (i) 32.85 ohm (ii) 3.285 V]

46. A rectangular carbon block has dimentsion 1.00cm×1.00 cm ×50 cm.

    (a) what is the resistance between two spuare ends?

    (b) between two opposing rectangular faces? The resistvity of carbon at 20°C is 3.5×10-6 ohm m.

    [Ans. (a) 0.157Ω (b) 7.0 ×10-5Ω]

47. A wire having a resistance of 6.0 Ω is extended so that its new lenfth is four times its orginal length. Find the resistance of the wire, assuming that resistance and density of the material are not changed during externsion process.

    [Ans. 96 Ω]

48. A copper wire having a cross-section 1 cm2 carries a current of 1 ampere. Assuming that atom contributes one free electron, calculate drift velocity of free electron. Given Atomic weight

    of copper =63.5
    Density of copper =8.92gm/cc
    Avogadro's number $=6.02\times10^{23}$ atoms/mole.

    [Ans. 8×10^{-7} ms^{-1}]

49. A copper wire and an iron wire of the same length have same potential difference applied to them.

    (a) Find the ratio of their if the current is to be same.

    (b) Can the current density he made same by suitable choice of radii, ρ for copper is 1.7×10^{-6} ohm $\times$m and ρ for iron is 1×10^{-7} ohm$\times$m.

[**Ans.** (a) $\frac{rFe}{rCu} = 2.4,$ (*b*) *No*

50. Steel trolly car rail has a cross-sectional area of 3.0 cm^2. What is the resistance of 10m of single track? The resistivity of steel is 6.0×10^{-7} ohm$\times$.

[**Ans.** $2\times10^{-2}\Omega$]

51. A rod of a certain metal is 100m along and 0.550 cm in diameter. The resistance between its ends at (20°C) is 2.87×10-3Ω. A round disc, 2.0 cm in diameter and 1.00 mm thick, is formed of this same materal.

    (a) What is the resistane between the opposite round faces?

    (b) What is the resistivity of the metal?

[**Ans.** (a) $.22\times10^{-8}$ ohm (b) $.68\times10^{-9}$ ohm $\times$m]

52. A square aluminium rod is 1.0 m long and 1 mm on edge.

    (a) What must be the diameter between its ends?

    (b) What must be the diameter of a circular 1.0m copper rod if its resistance to be the same?

$$\rho Al=2.8\times10^{-8}\ \Omega\text{-m}$$
$$\rho Cu=1.7\times10^{-8}\ \Omega\text{-m}$$

[**Ans.** 0.88×10-3m]

53. Determine the resistance of cubic centimeter of copper

    (a) when drawn out into a wire of diameter 0.32 mm and

    (b) when hammered into a flat sheet of thickness 1.2mm.

    The current is flowing normal through the sheet from one face to the other. Given that the resistvity of conductor is $1.59\times\ 10^{-6}$ ohm cm.

[**Ans.** (a) 2.549Ω (b) 2.29×10^{-8}ohm]

54. A current of 5 amperes exists in a 10 ohm resistance for 4 minutes. Calculate the number of electrons of passing in this time through any cross-section of the resistor.

[**Ans.** 7.5×10^{21}]

5

Boundary Conditions : Boundary Value Problems

The existence os electroc field in a homogeneous medium has already been discussed by us. We have alos discussed the existence of electric field in a region consisting of two different media. But now, we will analyse the conditions that the field must satisfy at the interface separating the media, this condition is called–*boundary conditions*. These conitions are helpful in determining the field on one side of the boundary if the field on the other side is known. obviously, the conditions will be dictated by the types of materical the media are made of. We shall consider the boundary conditions at an interface separating

- Dielectric (ε_{r1}) and dielectric (ε_{r2}).
- Conductor and dielectric.
- Conductor and free space.

To determine the boundary conditions, we need to use Maxwell's equations:

$$\oint E.dl = 0 \tag{5·1}$$

and

$$\oint D.dS = Q_{enc} \tag{5·2}$$

Also we need to decompose the electric field intensity E into two rothogonal components:

$$E = E_t + E_n \tag{5.3}$$

Here, the tangential and normal components of E to the inter face of interest are Et and En respectively.

A. Dielectric-Dielectric Boundary Conditions

Two different dielectrics featured by $\varepsilon_1=\varepsilon_0 \varepsilon r_1$ and $\varepsilon_2=\varepsilon_0 \varepsilon r_2$ as shown in the fig 5.1(a) have been taken in which the field E exists. $E_1$and E_2 in media 1 and 2, respectively, can be decomposed as

$$E_1 = E_{1t} + E_{1n} \tag{5·4a}$$

$$E_2 = E_{2t} + E_{2n} \tag{5·4b}$$

We apply eq. (5·1) to the closed path abcda of Figure 5.1 (a) assuming that the path is very small with respect to the veriation of E. We obtain

$$0 = E_{1t}\Delta_w - E_{1n}\frac{\Delta h}{2} - E_{2n}\frac{\Delta h}{2} - E_{2t}\Delta w + E_{2n}\frac{\Delta h}{2} + E_{1n}\frac{\Delta h}{2} \tag{5·5}$$

where $E_t = |E_t|$ and $E_n = |E_n|$. As $\Delta \to 0$, eq. (5.5) becomes

$$E_{1t} = E_{2t} \tag{5.6}$$

Hence, on the two sides of the boundary, the tangential components of E are the same. In other words, Et undergoes no change on the boundary and it is said to be continuous across the boundary. Since $D = \varepsilon E = D_t + D_n$, eq. (5.6) can be written as

$$\frac{D_{t1}}{\varepsilon_1} = E_{1t} = E_{2t} = \frac{D_{2t}}{\varepsilon_2}$$

Or

$$\frac{D_{1t}}{\varepsilon} = \frac{D_{2t}}{\varepsilon_2} \tag{5.7}$$

that is, D_t undergoes some change across the interface. Hence D_t is said to be discontinuous across the interface.

Similarly, we apply eq. (5.2) to the pillbox (Gaussian surface) of Figure 5.1 (b). Allowing $\Delta h \to 0$ gives.

$$\Delta Q = \rho s\,\Delta S = D_{1n}\,\Delta S - D_{2n}\,\Delta S$$

or

$$D_{1n} - D_{2n} = \rho s \tag{5·8}$$

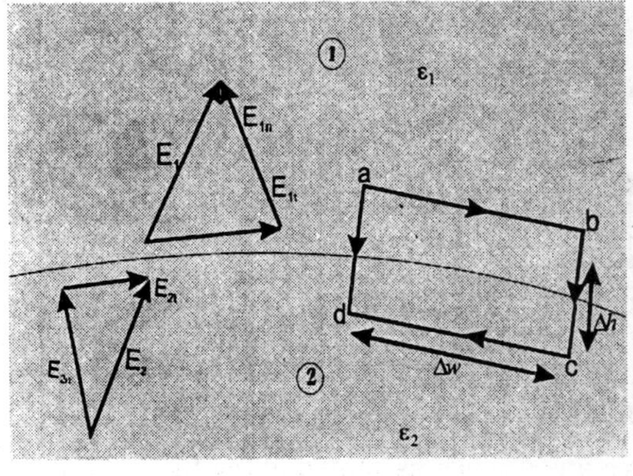

(a)

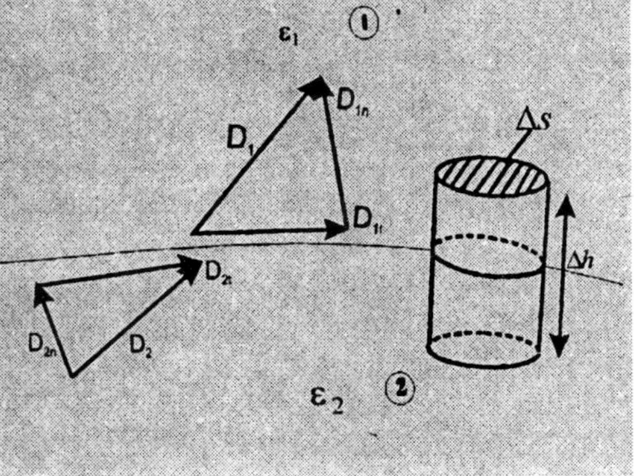

(b)

Figure (5.1) Dielectric-dielectric boundary.

Here, the deliberatily placed free charge density at the boundary is Ps and for equation (5.8), we have assumed that D is directed from region 2 to region1and must be applied accordingly. If no free charges exist at the interfce (i.e., charges are not deliberately placed there), ρs =0 and eq. (5.8) becomes

$$D_{1n} - D_{2n} \qquad (5{\cdot}9)$$

Thus the normal component of D is continuous across the interface; that is, Dn undergoes no change are at the boundary. Since D = εE, eq. (5·9) can be written as

$$\varepsilon_1 E_{1n} = \varepsilon_2 E_{2n} \qquad (5.10)$$

The equation shows the discontinuity of the normal component of E at the boundary. Equations (5·8) and (5·9), or (5·10) are collectively referred to as boundary conditions: they must be satisfied by an electric field at the boundary separating two different dielectries.

As mentioned earlier, the boundary conditions are usually applied in finding the electric field on are side of the boundary given the field on the other side. Besides this, we can use the boundary conditions to determine the "refraction" of the electric field across the interface. Consider D1 or E1 and D2 or E2 making angles θ1 and θ2 with the normal to the interface as illustrated in Figure (5.2). Using eq (5.6), we have

$$E_1 \sin\theta_1 = E_{1t} = E_{2t} = E_2 \sin\theta_2$$

or

$$E1 \sin 1 = E2 \sin 2 \qquad (5.11)$$

Similarly, by appling eq. (5·9) or (5·10), we get

$$\varepsilon_1 E_1 \cos\theta\, 1 = D_{1n} = D_{2n} = \varepsilon_2 E_2 \cos\theta_2$$

or

$$\varepsilon_1 E_1 \cos\theta_1 = \varepsilon_2 E_2 \cos\theta_2 \qquad (5{\cdot}12)$$

Dividing eq. (5.11) by eq. (5.12) gives

$$\frac{\tan\theta_1}{\varepsilon_1} = \frac{\tan\theta_2}{\varepsilon_2} \qquad (5{\cdot}13)$$

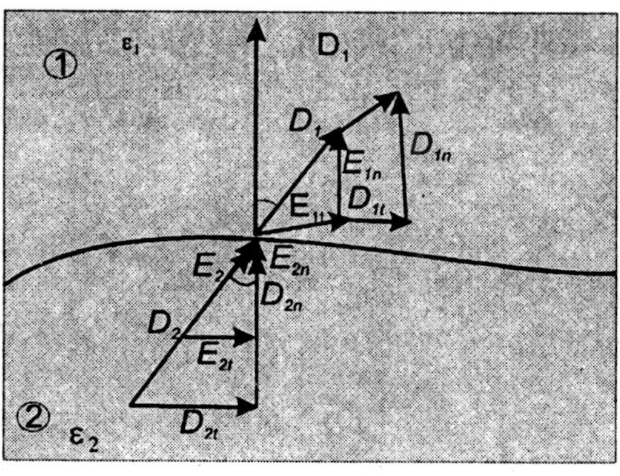

Figure 5.2. Refraction of D or E at a dielectric-dielectric boundary.

Since ε1 =ε0εr1 and ε2 = εoεr2, eq (5.13) becomes

$$\frac{tan\,\theta_1}{tan\,\theta_2} = \frac{\varepsilon_{r1}}{\varepsilon_{r2}} \tag{5.14}$$

This is the *law of reraction* of the electric field at a boundary free of charge (since ρs =0 is assumed at the interface).

B. Conductor-Dielectric Boundary Conditions:

Fig 5.3 depicts the conductor-Dielectric boundary conditions. Here σ→∞ or p_c→o the conductor is perfact. Although such a conductor is not practically realizable, we may reagard conductors such as copper and silver as though they were perfect conductors.

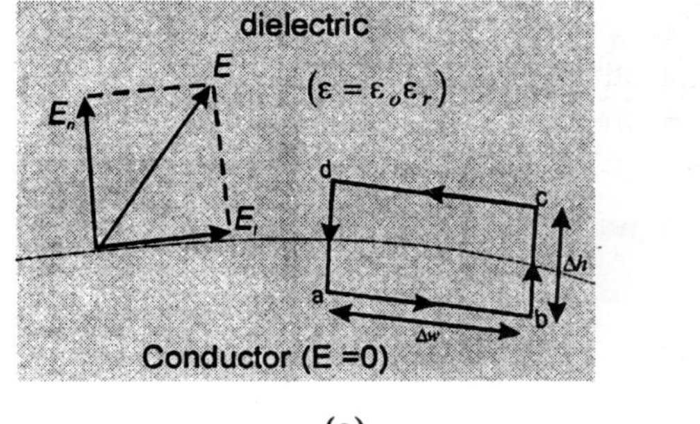

(a)

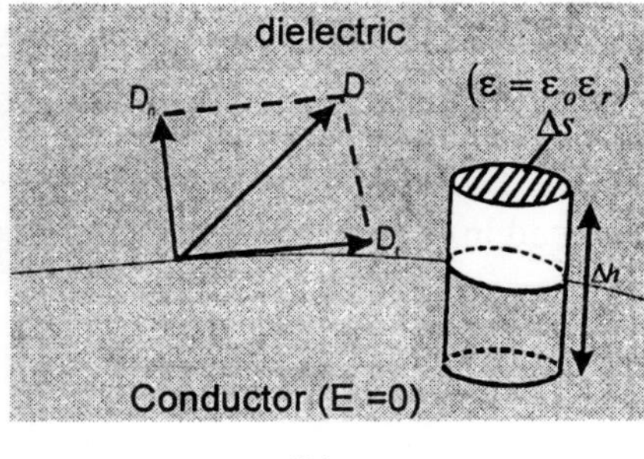

(b)

Figure 5.3. Conductor–dielectric boundary.

Almost the same, procedure is followed to determined the boundary condition for a conductor-dielectric interface which is used for dielectric-dielectric interface except that we incorporate the fact that E = 0 inside the conductor. Applying eq. (5.1) to the closed path *abcada* of Figure 5.3(a) gives

$$0 = 0.\Delta w + 0.\frac{\Delta h}{2} + E_n\frac{\Delta h}{2} - E_t.\Delta w - E_n.\frac{\Delta h}{2} - 0.\frac{\Delta h}{2} \tag{5·15}$$

As Δh→0,

$$E_t = 0 \tag{5.16}$$

Similarly, by applying eq. (5.2) to the pillbox of Figure 5.3 (b) and letting Δh→0, we get

$$\Delta Q = D_n \,.\, \Delta S - 0.\,\Delta S \tag{5.17}$$

because D = εE =0 inside the conductor. Equation (5·17) may be written as

$$D_n = \frac{\Delta Q}{\Delta S} = \rho s$$

or

$$D_n = \rho s \tag{5·18}$$

Now, we may condlude following remarks about a perfect conductor under static conditions:

(a) within a conduction no electric field may exist that is,

$$\rho_v = 0, E = 0 \tag{5·19}$$

(b) Since E = $-\Delta V$=0, there can be no potential difference between any two points in the conductor; that is, a conductor is an equipotential body.

(c) The electric field E can be external to the conductor and *normal* to its surface; that is

$$Dt = \varepsilon_0 \varepsilon_r E_t = 0, \; D_n = \varepsilon_0 \varepsilon_r E_n = \rho s \tag{5.20}$$

An important application of the fact that E = 0 inside a conductor is in electrostatic screening or shielding. if conductor a kept at zero potential surrounds conductor B as shown in Figure (5.4), B is said to be electrically sceened by A from other electric systems, such as conductor, C, outside A. Similary , conductor C outside A is dcreened by A from B.

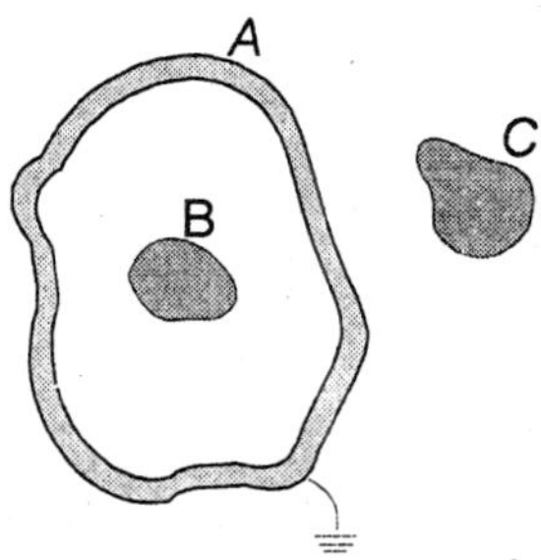

Figure 5.4. Electrostatic Screening

C. Conductor-Free Space Boundary Conditions

Here in this case of free space conductor, For free space Er is taken 1. The boundary conditions at the interface between a conductor and free space can be obtained from eq. (5·20) by replacing r by 1. We expect the electric field E to external to the conductor and normal to its surface. Thus the boundary conditions are

$$Dt = \varepsilon_0 E_t = 0, \; D_n = \varepsilon_0 E_n = \rho s \tag{5.21}$$

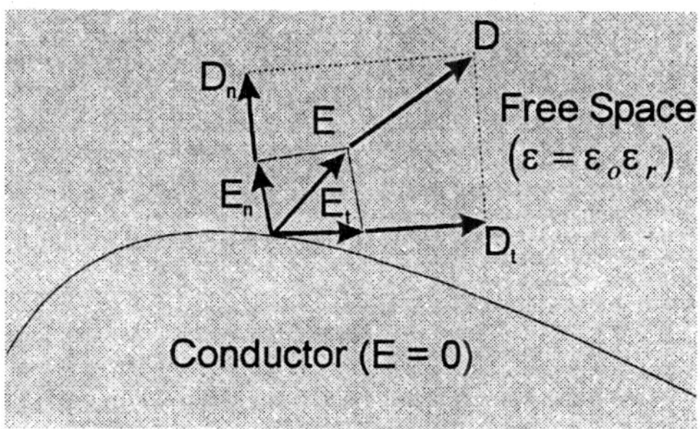

Figure 5.5. Conductor-free space boundary.

It should be noted again that eq. (5·21) implies that E field must approach a conducting surface normally.

Example 1. Two exensive homogeneous isotropic dielectrics meet on phane z = 0. For z ≥ 0, $\varepsilon_r 1 = 4$ and for z ≤ 0 $\varepsilon_r 2 = 3$. A uniform electric E1 = 5ax –2ay + 3ax KV/m exists for z ≥ 0. Find.

(a) E_2 for z < 0

(b) The angles E1 and E2 make with the interface

(c) The energy densities in J/m3 in both dielectrics

(d) The energy within a cube of side 2 m centered at (3,4,–5)

Solution: Let the problem be as illustrated in Figure 5.15.

(a) Since az is normal to the boundary plane, we obtain the normal components as

$$E_{1n} = E_1.a_n = E_1.a_z = 3$$

$$E_{1n} = 3a_z$$

$$E_{2n} = (E_2.a_z)a_z$$

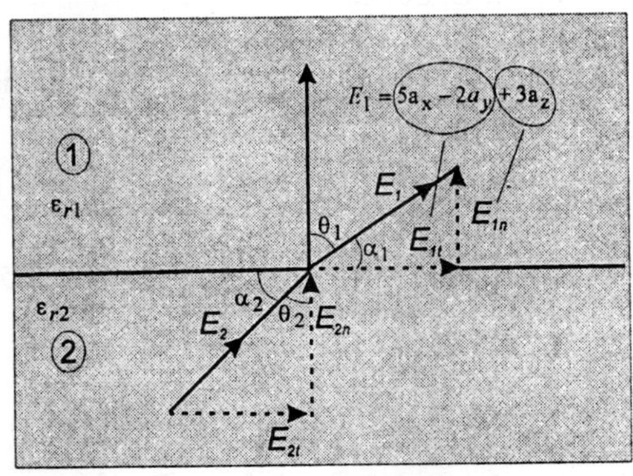

Figure 5.6. For Example1.

Also

$$E = E_n + E_t$$

Hence, $\quad E_{1t} = E_1 - E_{1n} = 5a_x - 2a_y$

Thus $\quad E_{2t} = E_{1t} = 5a_x - 2a_y$

Similary.

$$D_{2n} = D_{1n} \rightarrow \varepsilon_{r2}E_{2n} = \varepsilon_{r1}E_{1n}$$

or $\quad E_{2n} = \dfrac{\varepsilon_r 1}{\varepsilon_r 2} E_{1n} = \dfrac{4}{3}(3a_z) = 4a_z$

Thus,

$$E_2 = E_{2t} + E_{2n}$$

$$= 5a_x - 2a_y + 4a_z \text{ kV/m}$$

(b) Let α1 and α2 be the E_1 and E_2 make with the interface while θ_1 and θ_2 are the angles they make with the normal to the interface as shown in Figures 5.15; that is,

$$\alpha 1 = 90 - \theta_1$$

$$\alpha 2 = 90 - \theta_2$$

Since $\quad E_{1n} = 3$ and $E_{1t}\ \sqrt{25+4} = \sqrt{29}$

$$\tan\theta_1 \frac{E_{1t}}{E_{1n}} = \sqrt{\frac{29}{3}} = 1.795 \rightarrow \theta_1 = 60.9^o$$

Hence, $\alpha 1 = 29.1°$

Alternatively,

$$E_1 . a_n = |E_1| . 1.\cos\theta_1$$

or

$$\cos\theta_1 = \frac{3}{\sqrt{38}} = 0.4867 \rightarrow \theta_1 = 60.9°$$

Similarly,

$$E_{2n} = 4 \; E_{2t} = E_{1t} = \sqrt{29}$$

$$\tan\theta_2 \frac{E_{2t}}{E_{2n}} = \frac{\sqrt{29}}{4} = 1.346 \rightarrow \theta_2 = 53.4°$$

Hence, $\alpha 2 = 36.6°$

Note that $\dfrac{\tan\theta_1}{\tan\theta_2} = \dfrac{\varepsilon_r 1}{\varepsilon_r 2}$ is satisfied.

(c) The energy densities are given by

$$w_{E1} = \frac{1}{2}\varepsilon_1 |E_1|^2 = \frac{1}{2}.4.\frac{10^{-9}}{36\pi}.(25+4+9)\times 10^6$$

$$= 672 \;\; \mu J/m^3$$

$$w_{E2} = \frac{1}{2}\varepsilon_2 |E_2|^2 = \frac{1}{2}.3.\frac{10^{-9}}{36\pi}.(25+4+9)\times 10^6$$

(d) at the center (3, 4, –5) of the cubs of side 2m, z = –5 < 0; that is, the cubs is in region 2 with $2 \le x \le 4$, $3 \le y \le 5$, $-6 \le z \le -4$. Hence

$$W_E = \int W_{E2} dv = \int_{x=2}^{4} \int_{y=3}^{5} \int_{z=-6}^{-4} w_{E2} \, dz \, dy \, dz = w_{E2}(2)(2)(2)$$

$$= 597 \times 8 \; \mu J = 4.776 mJ$$

Exercise 1

A homogeneous dielectric (r = 2.5) fills region 1 (x < 0)while region 2 (x > 0) is free space.

(a) If $D_1 = 12a_x - 10a_y + 4a_z$ nC/m², find D_2 and θ_2.

(b) If E2 = 12 V/m and $\theta_2 = 60°$, find E_1 and θ_1. Take θ_1 and θ_2 as defined in the previous example.

Answer: (a) $12a_x - 4a_y + 1.6a_z$ nC/m², 19.75°, (b) 10.67 V/m, 77°

Example 2. Region y < 0 consists of a perfact conductor while region y > 0 is a dielectric medium (1r =2) as in Figure 5.16. If there is a surface surface of 2n C/m² on the conductor, determine E and D at

(a) A (3, –2, 2)

(b) B (–4, 1, 5)

Solution:

(a) Point A (3, –2, 2) is in the conductor since y = –2 < 0 at A. Hence.

(b) Point B (–4, 1, 5) is in the dielectric medium since y = 1>0 at B.

$$Dn = \rho s = 2\ nC/m^2$$

Hence,

$$D = 2a_y\ nC/m^2$$

and

$$E = \frac{D}{\varepsilon_o \varepsilon_r} = 2 \times 10^{-9} \times \frac{36\pi}{2} \times 10^9 a_y = 36\pi a_y$$

$$= 113.1 a_y\ V/m$$

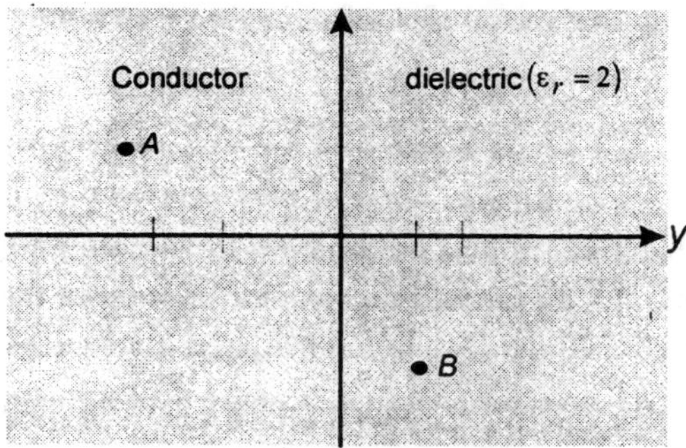

Figure 5.7. See Example 2.

Practice Exercise 2

It is found that E = 60ax + 20ay –30az mV/m at a particular point on the interface between air and a conducting surface. Find D and ρs at that point

Answer: $0.531a_x + 0.177a_y - 0.265a_z$ pC/m², 0.619 p C/m².

Boundary -Value Problems:

When the charge distribution is known then the ceulemb's law or Gauss law was used in preceding chapters for ditermining electric field E. When the potential throughout the region is know then the equation. If $E = -\nabla V$ was used for determining E. If we see practically them nither of these two is known. Now, we shall consider practical electrostatic problems where only electrostatic conditions (Charge and potential) at some boundaries are known and it is desired to find E and V throughout the region. Such problem are usually tackled using Poisson's[1] or Laplace's 2 equation or the method of images. and they are usually referred to as boundary-value problem. The concepts of resistance and capacitance will also be covered.

Equations of Poission and Laplace

Gauss's law for a linear material medium is helpful to derive Poisson's and Laplace's equation easily

$$\nabla . D = \nabla .\ \varepsilon E = \rho_v \tag{5.22}$$

and

$$E = -\ \nabla V \tag{5.23}$$

Now, putting the value of E from the equation (5.2) into equation (5.21)

$$\nabla .(-\ \varepsilon \nabla V) = \rho_V \tag{5.24}$$

Note: Equation (5.24) is for an inhomogeneous medium. Equation (5.24) changes into following equation for a homogeneous medium.

$$\nabla^2 V = -\frac{\rho_v}{\varepsilon} \tag{5.25}$$

This is known as *Poisson's equation.* for a free region $P_v = 0$ then the equation converts as follows.

$$\nabla^2 V = 0 \tag{5.26}$$

(5.26) is known as *Laplace's equation.* Equation (5.24) is Poisson's equation for an inhomogeneous medium; it becomes Laplace's equaticn for an inhomogeneous medium when $\rho_v = 0$. For the former equation, we assumed ε constant throughout the region in which V is difined but for inhomogeneous medium, ε is not constdered constant.

Laplace's equation in Cartesian, Cylindrical, or Spherical coordinates respectively is given by

$$\frac{\partial^2 V}{\partial x^2} + \frac{\partial^2 V}{\partial y^2} + \frac{\partial^2 V}{\partial z^2} = 0 \tag{5.27}$$

$$\frac{1}{\rho}\frac{\partial}{\partial \rho}\left(\rho \frac{\partial V}{\partial \rho}\right) + \frac{1}{\rho^2}\frac{\partial^2 V}{\partial \phi^2} + \frac{\partial^2 V}{\partial z^2} = 0 \tag{5.28}$$

$$\frac{1}{r^2}\frac{\partial}{\partial r}\left(r^2 \frac{\partial V}{\partial r}\right) + \frac{1}{r^2 \sin\theta}\frac{\partial}{\partial \theta}\left(\sin\theta \frac{\partial V}{\partial \theta}\right)$$

$$+ \frac{1}{r^2 \sin\theta}\frac{\partial^2 V}{\partial \phi^2} = 0 \tag{5.29}$$

depending on whether the potential is $V(x,y,z)$, or $V(\rho, \phi, z)$, or (r, θ, ϕ). Poission's equationin those corrinate may be obtained by simply replacing zero on the right-hand side of eqs. (5.27), (5·28) and (5·29) with $-\frac{pv}{E}$.

Laplace's equation is of primany importance in solving electrostatic problems involving a set of conductors maintained at different potentials. Examples of such problem .

Uniqueness Theeorem

Uniqueness theorem is peratining to laplance's and poission's theorem. If a sòlutions to Laplance's or poisson's equation can be found which satisfies the boundary conditions then the solution is considered unique. If a solution of Laplace's equation satisfies a given set of boundary conditions, is this the only possible solution? The answer is yes: there is only one solution. We say that the solution is unique. Thus any solution regardless of the method used. This is Known as the *uniqueness theorem*. The theorem applies to any solution of Poisson's or Laplace's equation in a given region or closed surface.

The theorem is proved by contradiction. We assume that there are two solutions V_1 and V_2 of Laplace's equation both of which satisfy the prescribed boundary conditions, Thus

$$\nabla^2 V_1 = 0, \qquad \nabla^2 V_2 = 0 \tag{5.30 a}$$

$$V_1 = V_2 \text{ on the boundary} \tag{5·30 b}$$

We consider their difference

$$V_d = V_2 - V_1 \tag{5·31}$$

Which obeys

$$\nabla^2 V_d = \nabla^2 V_2 - \nabla^2 V_1 = 0 \tag{5.32 a}$$

$$V_d = 0 \text{ on the boundary} \tag{5·32 b}$$

according to eq. (5·30). From the divergence theorem.

$$\int_v \nabla . A \, dv = \oint_s A . \, ds \tag{5.33}$$

We let $A = V_d \nabla v_d$ and use a vector identity

$$\nabla . A = \nabla . (V_d \nabla V_d) = V_d \nabla^2 V_d + \nabla V_d . \nabla V_d$$

But $\nabla^2 V_d = 0$ according t eq. (5·32), so

$$\nabla . A = \nabla V_d . \nabla V_d \tag{5·34}$$

Substitting eq. (5·34) into eq. (5·33) gives

$$\int_v \nabla . V_d \, \nabla V_d \, dv = \oint_s V_d \nabla V_d . \, ds \tag{5·35}$$

From eqs. (5.30) and (5.32), it is evident that the right-hand side of eq. (5.·35) vanishes.

Hence,

$$\int_v |\nabla V_d|^2 \, dv = 0$$

Since the integration is always positive.

$$\nabla V_d = 0 \quad (5.36\ a)$$

or

$$V_d = V_2 - V_1 = \text{constant everywhere in } v \quad (5{\cdot}36\ b)$$

But eq. (5·36) must be consistent with eq. (5·30b). Hence, $V_d = 0$ or $V_1 = V_2$ everywhere, showing that V_1 and V_2 cannot be different solutions of the same problem.

The theoren applies to Poisson's equation and to prove the theorem for the case where the electric field (potential gradient) is specified on the boundary; similar steps can be taken.

The unique discription of the problem is given by the following three things:

(a) The appropriate differnetial equation (Laplace's or Poisson's equation in this chapter).

(b) The solution region.

(c) The prescribed boundary conditions.

Note: If, any of the three above things is missing the problem cammet be solved complitely.

General Procedure For Solving Poisson's or Laplace's Equation

With the help of poisson's of Laplance's equation a given boundry value problem can easily be solved while following the given steps.

(a) If Pv = 0 then solve Laplance's equation and if Pv ≠ 0 then use poisson's equation in the given two ways

(b) To ditermine a unique of V, apply the boundary conditions

(i) Direct integration when V is a function of one variable, or (ii) Separation of variables if V is a function of more than one varible. The solution at this point is not unique but expressed in terms of unknown integration constant s to be determined.

(c) Imposing the given boundary conditions makes the solution unique.

(d) Having obtained V, find E using $E = -\nabla V$ and D from $D = \varepsilon E$.

(e) Find the charge Q induced on a conductor with the help of equ. $Q = \int \rho_s \, ds$ if necessary. If desired, find the charge Q inductor using $Q = \int \rho_s \, ds$ where $\rho s = D_n$ and D_n is the component of D normal to the conductor. If necessary, the capacitance between two conductors can be found using $C = Q/V$.

Example 3. Current-carrying components in high-voltage power equipment must be cooled to carry away the heat caused by ohmic losses. A means of pumping is based on the force transmitted to the cooling fluid by charges in an electric field. The electrohydrodynamic (EHD) pumping is modeled in Figure 5·8. The region between the electrodes contains a uniform charge ρ_o, which is generated at the left electrode and collected at the right electrode. Calculate the presure of the pump if $\rho_o = 25$ mC/m^3 and $V_o = 22$ kV.

Solution: Since $\rho_v \neq 0$, we apply Poisson's equation

$$\nabla^2 V = \frac{\rho_v}{\varepsilon}$$

The boundary conditions $V(z = 0) = V_o$ and $V(z = d) = 0$ show that V depends only on z (there is no ρ or ϕ dependence). Hence

$$\frac{d^2V}{dz^2} = \frac{-\rho_o}{\varepsilon}$$

Integrating once gives

$$\frac{dV}{dz} = \frac{-\rho_o z}{\varepsilon} + A$$

Integrating again yields

$$V = -\frac{\rho_o z^2}{2\varepsilon} + Az + B$$

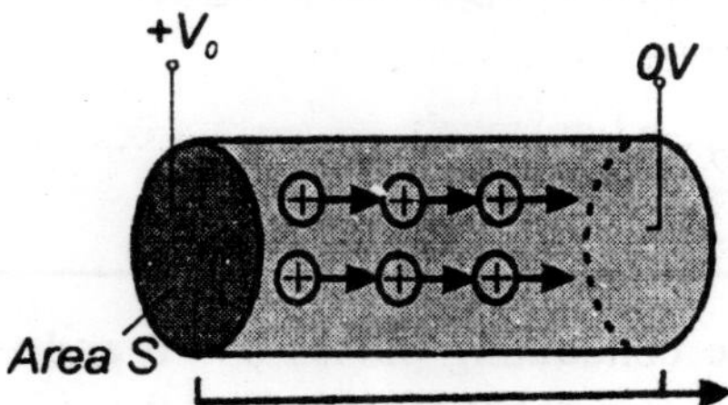

Figure 5.8. An electrohydrodynamic Pump; for Example 3.

where A and B are integration constants to be determined by applying the boundary conditions. when $z = 0$, $V = V_0$,

$$V_o = -0 + 0 + B \rightarrow B = V_o$$

When $z = d$, $V = 0$,

$$0 = -\frac{\rho_o d^2}{2\varepsilon} + Ad + V_o$$

or
$$A = \frac{\rho_0 d}{2\varepsilon} - \frac{V_0}{d}$$

The electric field is given by

$$E = -\nabla V = -\frac{dV}{dz}a_z = \left(\frac{\rho_o z}{\varepsilon} - A\right)a_z$$

$$= \left[\frac{V_o}{d} + \frac{\rho_o}{\varepsilon}\left(z - \frac{d}{2}\right)\right]a_z$$

The electric field is given by

$$F = \int \rho_v E\, dv = \rho_o \int dS \int_{z=0}^{d} Edz$$

$$= \rho_o S\left[\frac{V_o z}{d} + \frac{\rho_o}{2\varepsilon}\left(z^2 - dz\right)\right]_0^d a_z$$

$$F = \rho_o S V_o a_z$$

The force per unit area or pressure is

$$p = \frac{F}{S} = \rho_o V_o = 25\times10^{-3} \times 22 \times 10^3 = 550 \text{ N/m}^2$$

Exercise 3

In a one-dimensial device, the charge density is given by $\rho_v = \rho_o x/a$. If E = 0 at $x = 0$ and $V = 0$ at $x = a$, find V and E.

Answer: $\frac{\rho_0}{6\varepsilon a}\left(a^3 - x^3\right), \frac{\rho_o x^2}{2a\varepsilon}a_x$

Example 4. Seminifinite conducting planes $\phi = 0$ and $= \pi/6$ are separated by an infinitesimal insutlating gap as in Figure 5.9. If $V(\phi = 0) = 0$ and $V(\phi = \pi/6) = 100$ V, calculate V and E in the region between the planes.

Solution: As V depends only on ϕ, Laplace's equation in cylindrical coordinates becomes.

$$\nabla^2 V = \frac{1}{\rho^2}\frac{d^2V}{\rho^2 d\phi^2} = 0$$

Since $\rho = 0$ is excluded due to the isulating gap, we can multiply by ρ^2 to obtain.

$$\frac{d^2V}{d\phi^2} = 0$$

which is integrated twice to give.

$$V = A\phi + B$$

We apply the boundary conditions to determie A and B. When $\phi = 0$, $V = 0$,

$$0 = 0 + B \rightarrow B = 0$$

When $\phi = \phi_o$, $V = V_o$,

$$V_0 = A\phi_0 \rightarrow A = \frac{V_o}{\phi_o}$$

Hence,

$$V = \frac{V_o}{\phi_o}\phi$$

and

$$E = -\nabla V = -\frac{1dV}{\rho\, d\phi}a_\phi = -\frac{V_0}{\rho\phi_o}a_\phi$$

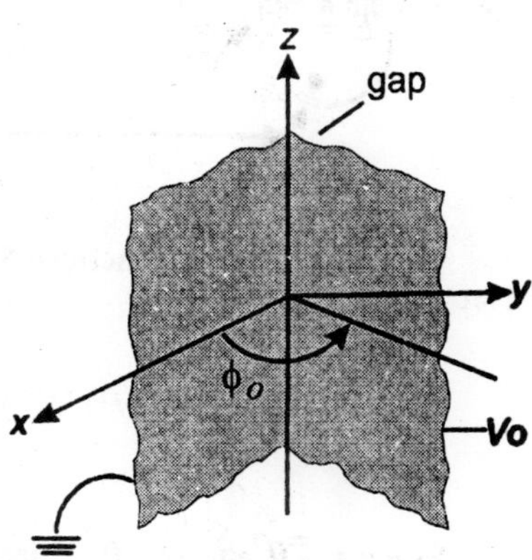

Figure 5·9 Potential $V(\phi)$ due to seminifinite condcuting planes.

Substituting $V_o = 100$ and $\phi_o = \pi/6$ gives

$$V = \frac{600}{\pi}\phi \quad \text{and} \quad E = \frac{600}{\pi\rho}a\phi$$

Check: $\nabla^2 V = 0, V(\phi = 0) = 0, V(\phi = \pi/6) = 100.$

Exercise 4

Two conducting plates of size 1×5m are inclined at 45° to each other with a gap of width 4 mm separating them as shown in Figure 6.4. Detetmine an approximate value of the charge per plate if the plates are maintained at a potential difference of 50V. Assume that he medium between them has $\varepsilon_r = 1.5$.

Answer: 22.2nC.

Example 5. Two conducting cones ($\theta = \pi/10$ and $\theta = \pi/6$) of infinite extent are separated by an infitesimal gap at $r = 0$. If $V(\theta = \pi/10) = 0$ and $V(\theta = \pi/6) = 50$ V, find V and E between the cones.

Solution: Consider the coaxial cone of Figure 5·10, where the gap, serves as an insulator between the two conducting cones. V depends only on θ, so Laplace's equation in spherical coordinates becomes

$$\nabla^2 V = \frac{1}{r^2 \sin\theta}\frac{d}{d\theta}\left[\sin\theta \frac{dV}{d\theta}\right] = 0$$

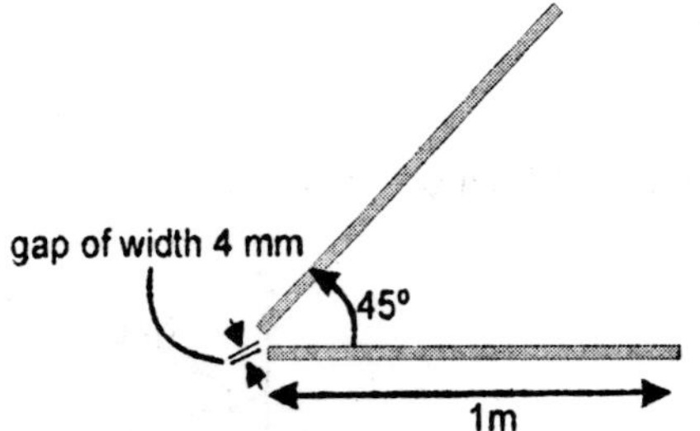

Figure 5·10. For Practicse 5.

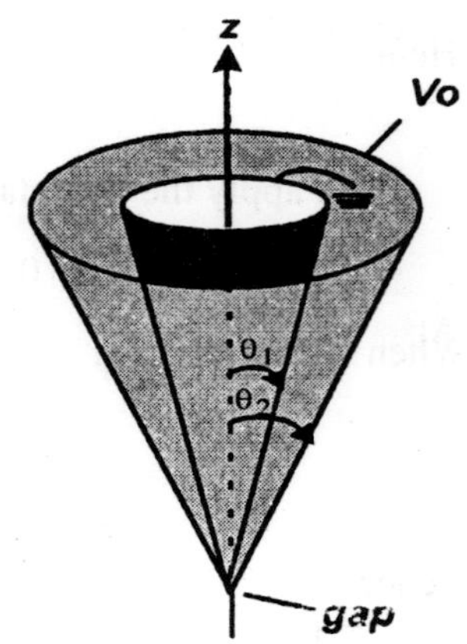

Figure 5.11. Potential V(φ) due to conducting cones.

Since $r = 0$ and $\theta = 0, \pi$ are excluded, we can multiply by $r^2 \sin\theta$ to get

$$\frac{d}{d\theta}\left[\sin\theta \frac{dV}{d\theta}\right] = 0$$

Integrating once gives

$$\sin\theta \frac{dV}{d\theta} = A$$

or

$$\frac{dV}{d\theta} = \frac{A}{\sin\theta}$$

Integrating this results in

$$V = A\int \frac{d\theta}{\sin\theta} = A\int \frac{d\theta}{2\cos\theta/2\ \sin\theta/2}$$

$$= A\int \frac{1/2\sec^2\theta/2 d\theta}{\tan\theta}$$

$$= A\int \frac{d(\tan\theta/2)}{\tan\theta/2} = \text{A In} (\tan\theta/2) + B$$

We now apply the boundary conditions to determie the integration constants A and B.

$$V(\theta = \theta_1) = 0 \rightarrow 0 = \text{A In} (\tan\theta_1/2) + B$$

or

$$B = -\text{A In} (\tan\theta_1/2)$$

Hence

$$V = A\ In\left[\frac{tan\,\theta/2}{tan\,\theta_1/2}\right]$$

Also

$$V(\theta=\theta_2) = V_o \rightarrow V_o = A\ In\left[\frac{tan\,\theta_2/2}{tan\,\theta_1/2}\right]$$

or

$$V = \frac{V_o}{In\left[\frac{tan\,\theta_2/2}{tan\,\theta_1/2}\right]}$$

Thus

$$V = \frac{V_o\ In\left[\frac{tan\,\theta/2}{tan\,\theta_1/2}\right]}{In\left[\frac{tan\,\theta_2/2}{tan\,\theta_1/2}\right]}$$

$$E = -\nabla V = -\frac{1}{r}\frac{dV}{d\theta}a_\theta = -\frac{A}{r\,sin\,\theta}a_\theta$$

$$= -\frac{V_o}{r\,sin\,\theta\ In\left[\frac{tan\,\theta_2/2}{tan\,\theta_1/2}\right]}a_\theta$$

Taking $\theta_1 = \pi/10$, $\theta_2 = \pi/6$, and $V_o = 50$ gives

$$V = \frac{50\,In\left[\frac{tan\,\theta/2}{tan\,\pi/20}\right]}{In\left[\frac{tan\,\pi/12}{tan\,\pi/20}\right]} = 95.1\ In\left[\frac{tan\,\theta/2}{0.1584}\right]V$$

and

$$E = -\frac{95.1}{r\,sin\,\theta}a_\theta V/m$$

Check: $\nabla^2 V = 0$, $V(\theta = \pi/10) = 0$, $V(\theta = \pi/6) = V_o$.

Resistance and Capcitance

We have already discussed about resistance and now to find out resistance for uniform cross- section on and also for non-uniform cross-secton in preceding chapters. To recall see equ. (4·17) and equ (4·18). We arerewriting equ. (4.18) here; Boundary-volue Problem is applied to find the resistance of a conduction of non-cross-section.

$$R = \frac{V}{I} = \frac{\int E.dI}{\oint \sigma E.dS} \tag{5·37}$$

Following are some steps which can be opted to find the resistance of a given conducting material with the help of equ. (5·37).

(a) Choose a suitable coordinate system.

(b) Assume Vo as the potential difference between conductor terminals.

(c) Solve Laplace's equation $\nabla^2 V$ to obain V. Then determine E from E = $-\nabla V$ and I from $I = \int \sigma E.dS$.

(d) Finally, obtain R as V_o/I.

For determining $R = \frac{V_o}{I}$, we assume Vo and find I if necessary. Alternatively, it is possible to assume current I_o, find the corresponding potential difference V, and determine R from R = V/I_o. As will be discussed shortly, the capacitance of capacitor is obtained using a similar technique.

We know that the two plates of a conductor carry equal and opposite charges. Therefore, we must say that the fluxline coming from one plate finishes at the other plate. The conductors are sometimes referred to as the *plates* of the capacitor. The plates may be separated by free space or a dielectric.

Consider the two-conductor capacitor of Figure 5.12. The conductors are maintained at a potential difference V given by

$$V = V_1 - V_2 = -\int_2^1 E.dI \tag{5·38}$$

where E is the electric field existing between the onductors and conductor is assumed to carry a positive charge and the E field is always normal to the conducting surfaces. 'the ratio of magnitude of charge on the of the plates to the potential difference between the two plates is called capacitance of a capacitor; this is,

$$C = \frac{Q}{V} = \frac{\varepsilon \oint E.dS}{\int E.dI} \tag{5.39}$$

The capacitance of a capacitor is denoted by C and it is measured in farads (F). We are interested in the absolute value of V that's why we left out the negative sign of equation

$V = -\int E.dl$, C can be obtained for any eiven two-conductor capacitance by following either of these methods with the help of equ (5·39).

1. Assuming Q and determining V in terms of Q (involving Gauss's law).
2. Assuming V and determining Q in terms of V (involving solving Laplace's equation).

The former method involves taking the following steps:

1. Choose a suitable coordinate system.
2. Let the two conducting plates carry charges +Q and –Q.

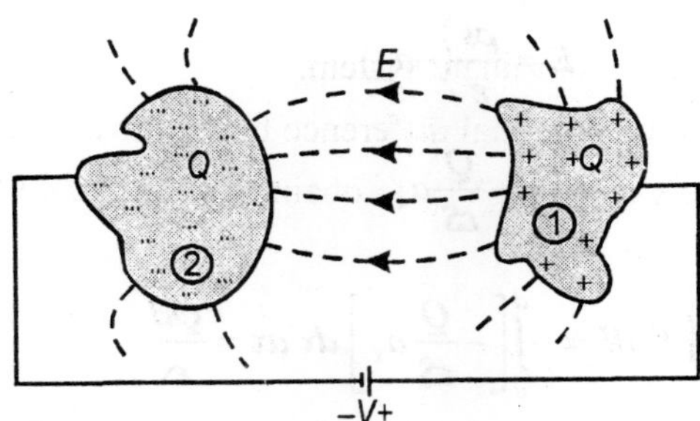

Figure 5·12 A two-conductor capacitor.

3. Now, Gauss's law and coulomb's law must be used to determine E and equation $V = -\int E.dl$ should be used to find the value of V. The nagative sign may be ignored in this case becasue we are interested in the obsolute value of *V*.
4. Finally, obtain *C* from *C* = *Q*/*V*.

A. Parallel-Plate Capacitor

We now take a parallel plate capacitor as shown in Fig.5.13 (a). The two plates of capacitor, carry charges +*Q* and –*Q* uniformly distributed on them and has a surface area s. The plates of capacitor are separated by a distanced.

therefore,
$$\rho s = \frac{Q}{S} \tag{5·40}$$

If the distance between the two plates d is very small in comparison with the dimensions of the plate than the parallel-plate capacitor is considered ideal one. Assuming such an idea case, the fringing field at the edge of the plates, as illustrated in Fingure 5.13 (b), can be ignored so that the field between them is considered uniform. If the space between the plates is filled with a homogeneous dielectric with permittivity ε and we ignore flux fringing at the edges of the plates,

D= $-\rho_s a_x$ as we have studies in preseding chepter;

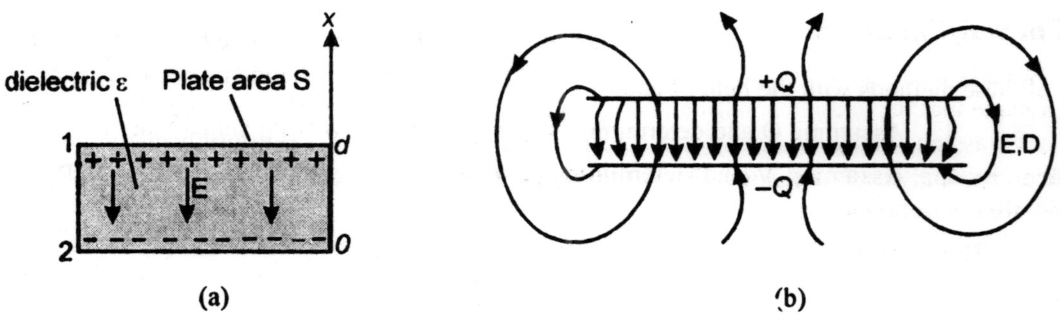

Figure 5.13 (a) parallel-plate capacitor, (b) fringing effect ude to a parallel-plate capacitor.

or,
$$E = \frac{\rho s}{\varepsilon}(-a_x) \tag{5.41}$$
$$= -\frac{Q}{\varepsilon S}a_x$$

Hence
$$V = -\int_2^1 E.dl = -\int_0^d \left[-\frac{Q}{\varepsilon S}a_x\right].dx\,ax = \frac{Qd}{\varepsilon s} \tag{5·42}$$

and thus for a parallel-plate capacitor.

$$C = \frac{Q}{V} = \frac{\varepsilon S}{d} \tag{5.43}$$

The dielectric constant *Er* of a given dielectric can be measured by the formula offered in the equation (5.43). By measuring the capacitance C of a parallel-plate capacitor with the space between the plates filled with the dielectric and the capacitance C^{o} with air between the plates, we find ε_r from

$$\varepsilon_r \frac{C}{C_{\circ}} \tag{5.44}$$

Now, the energy stored in a capacitor is given by

$$W_E = \frac{1}{2}CV^2 = \frac{1}{2}QV = \frac{Q^2}{2C} \tag{5.45}$$

To verify this for a parallel-plate capacitor. If we substitute eq. (5.41) into the equation of energy stored in a capacitor then we get:

$$W_E = \frac{1}{2}\int_v \varepsilon \frac{Q^2}{\varepsilon^2 S^2}dv = \frac{\varepsilon Q^2 Sd}{2\varepsilon^2 S^2}$$

$$= \frac{Q^2}{2}\left(\frac{d}{\varepsilon S}\right) = \frac{Q^2}{2C} = \frac{1}{2}QV$$

B. Coaxial Capacitor

We take a conductor of inner radius, outer radius and Length of a, bond l respectively and also assume that the homogeneous dielectric with permitivity ε between free space of the two plates. We assume that conductors 1 and 2, respectively, carry $+Q$ and $-Q$ uniformly distributed on them. By applying Gauss's law to an arbitrary Gaussian cylindrical surface of radius $\rho(a < \rho < b)$, we obtain

$$Q = \varepsilon\oint E.dS = \varepsilon E_\rho 2\pi\rho L \tag{5.46}$$

Hence
$$E = \frac{Q}{2\pi\varepsilon\rho L} a_\rho \tag{5.47}$$

Neglecting flux fringing at the cylinder ends.

$$V = -\int_2^1 E.dl = -\int_b^a \left[\frac{Q}{2\pi\varepsilon\rho L} a_\rho\right].d\rho\, a\rho \tag{5.48 a}$$

$$= \frac{Q}{2\pi\varepsilon L} In\frac{b}{a}$$

Thus the capacitance of a coaxiat cylinder is given by

$$C = \frac{Q}{V} = \frac{2\pi\varepsilon L}{In\dfrac{b}{a}} \tag{5.49}$$

C. Spherical Capacitor

Take the two concentric spherical conductors with radius a and b as outer and inner respectively. Separated by a dielectric medium with permittvity ε as shown in Figure (5.14). We assure charges $+Q$ and $-Q$ on the inner and outer spheres respectively. By applying Gauss's law to an arbitrary Gaussion Spherical surface of redius r $(a < r < b)$,

$$Q = \varepsilon\oint E.dS = \varepsilon E_r 4\pi r^2 \tag{5.50}$$

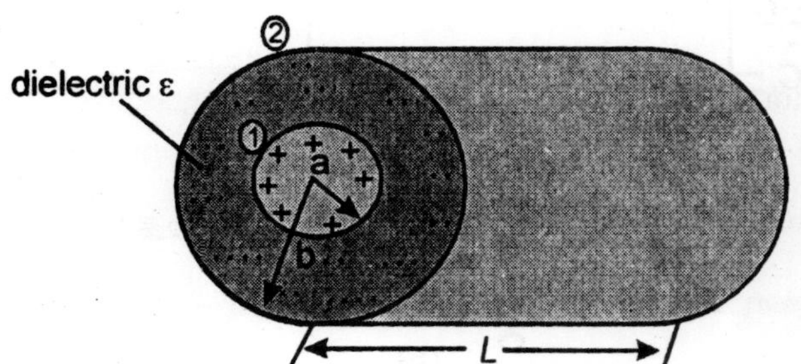

Figure 5.14. Coaxial Capacitor

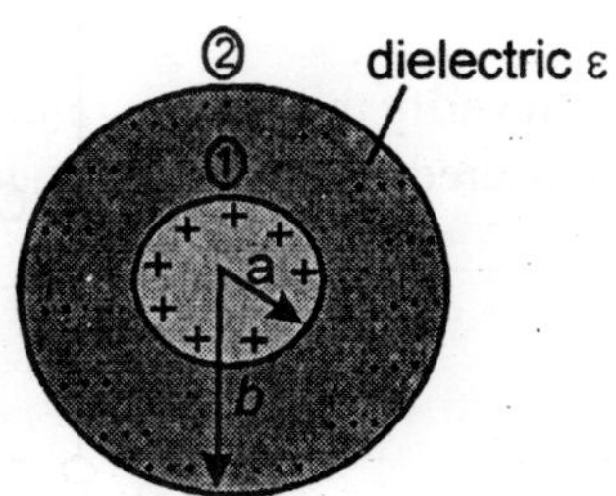

Figure 5.15. Spherical Capacitor

this is,
$$E = \frac{Q}{4\pi\varepsilon r^2} a_r \tag{5.51}$$

The potential difference between the conductors is

$$V = -\int_2^1 E.dl = -\int_b^a \left[\frac{Q}{4\pi\varepsilon r^2}\right].dr\, a_r$$

$$= \frac{Q}{4\pi\varepsilon}\left[\frac{1}{a} - \frac{1}{b}\right] \tag{5.52}$$

Thus the capacitance of the spherical capacitor is

$$C = \frac{Q}{V} = \frac{4\pi\varepsilon}{\frac{1}{a} - \frac{1}{b}} \tag{5.53}$$

By letting $b \to \infty$, $C = 4\pi\varepsilon a$, which is the capacitance of a shpherical capacitor whose outer plate is infinitely large. Such is the case of a spherical conductor at a large distance from other conducting bodies–the *Isolate shpere*. Even an irregularly shaped object of about the same size as the sphere will have nearly the same capacitance. This fact is useful in estimating the stray capacitance of an isolated body or piece of equipment.

Recall from network theory that if two capacitors with capacitance C_1 and C_2 are in series (i.e., they have the same charge on them) as shown in Figure 5.16 (a) the total capacitance is

$$\frac{1}{C} = \frac{1}{C_1} + \frac{1}{C_2}$$

or

$$C = \frac{C_1 C_2}{C_1 + C_2} \tag{5.54}$$

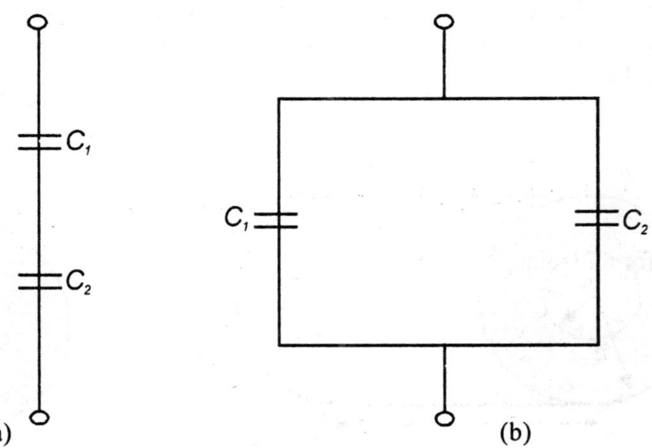

Figure 5.16. Capacitors in (a) series, and (b) parallel.

If the capacitors are in parallel (i.e., they have the same voltage acroos their plates) as shown in figure 5.16 (b), the total capacitances is

$$C = C1 + C2 \tag{5.55}$$

Let us reconsider the expressions for finding the resistance R and the capacitance C of an electrial system.

$$R = \frac{V}{1} = \frac{\int E.dI}{\oint \sigma E.dS}$$

$$C = \frac{Q}{V} = \frac{\varepsilon \oint E.dS}{\int E.dI}$$

The product of these expressions yields

$$RC = \frac{\varepsilon}{\sigma} \tag{5.56}$$

which is the relaxation time *Tr* of the medium separating the conductors. It should be remarked that eq. (5.56) is valid only when the medium is homogeneous. Assuming homogeneous media, the resistance of various capacitors mentioned earlier can be readily obtained using eq. (5.56). The following examples are provided to illustrate this idea. For a parallel-plate capacitor.

$$C = \frac{\varepsilon S}{d}, \qquad R = \frac{d}{\sigma S} \tag{5.57}$$

For a cylindrical capacitor,

$$C = \frac{2\pi\varepsilon L}{In\frac{b}{a}}, \qquad R = \frac{In\frac{b}{a}}{2\pi\sigma L} \tag{5.58}$$

For a spherical capacitor.

$$C = \frac{4\pi\varepsilon}{\frac{1}{a} - \frac{1}{b}}, \qquad R = \frac{\frac{1}{a} - \frac{1}{b}}{4\pi\sigma} \tag{5.59}$$

And finally for an isolated spherical conductor,

$$C = 4\pi\varepsilon a, \qquad R = \frac{1}{4\pi\sigma a} \tag{5.60}$$

It should be noted that the resistance *R* in each of eqs. (5.56) to (5.60) is not the resistance of the capacitor plate but the leakage resistance between the plates; therefore, σ in those equations is the conductivity of the dielectric medium separating the plates.

Example 3. A metal bar of conductivity σ is bent to form a flat 90°sector of inner redius a, outer radius b, and thickness t as shown in Figure 5.17. Show that (a) the resistance of the bar between the vartical curved surfaces at $\rho = a$ and $\rho = b$ is

$$R = \frac{2\ In\frac{b}{a}}{\sigma \pi t}$$

and (b) the resistance between the two horintal surface at $z = 0$ and $z = t$ is

$$R' = \frac{4t}{\sigma\pi\left(b^2 - a^2\right)}$$

Solution: (a) Between the vertical curved ends located at $\rho = a$ and $\rho = b$, the bar has a nonuni form cross section. Let a potential difference V_o be maintained between the surved surfaces at $\rho = a$ and $\rho = b$ so that

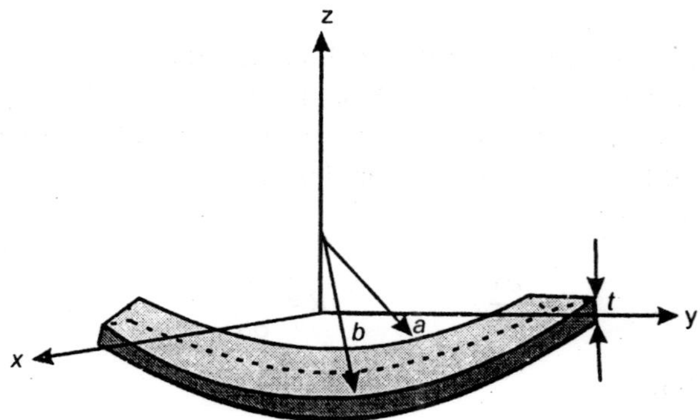

Figure 5.17. Metal bar of Example 3.

$V(\rho = a) = 0$ and $V(\rho = b)$ V_o. We solve for V in Laplace's equation $\nabla^2 V = 0$ in cylindrical coordinates. Since $V = V(\rho)$,

$$\nabla^2 V = \frac{1}{\rho}\frac{d}{d\rho}\left(\rho\frac{dV}{d\rho}\right) = 0$$

As $\rho = 0$ is excluded, upon multiplying by ρ and integrating once, this becomes

$$\rho\frac{dV}{dp} = A$$

or

$$\frac{dV}{d\rho} = \frac{A}{\rho}$$

Integrating once again yields

$$V = A \ln \rho + B$$

where A and B constants of integration to be determined from the boundary conditions.

$$V(\rho = a) = 0 \rightarrow 0 = A \ln a + B \qquad \text{or} \qquad B = -A \ln a$$

$$V(\rho = b) = V_o \rightarrow V_o = A \ln b + B = A \ln b - A \ln a = A \ln \frac{b}{a}$$

or

$$A = \frac{V_o}{\ln \frac{b}{a}}$$

Hence,

$$V = A \ln - A \ln a = A \ln \frac{b}{a} = \frac{\frac{V_o}{\ln \frac{b}{a}}}{} \ln \frac{\rho}{a}$$

$$E = -\nabla V = -\frac{dV}{d\rho} a_\rho = -\frac{A}{\rho} a_\rho = -\frac{V_o}{\rho \ln \frac{b}{a}} a_\rho$$

$$J = \sigma E, \qquad dS = -\rho \, d\phi \, dz \, a_\rho$$

$$I = \int J.ds = \int_{\phi=0}^{\pi/2} \int_{z=0}^{t} \frac{V_o \sigma}{\ln \frac{b}{a}} dz \, \rho \, d\phi = \frac{\pi}{2} \frac{t V_o \sigma}{\ln \frac{b}{a}}$$

Thus

$$R = \frac{V_o}{I} = \frac{2 \ln \frac{b}{a}}{\sigma \pi t}$$

As required.

(b) Let V_o be the potential difference between the two horizontal surfaces so that $V(z = 0) = 0$ and $V(z = t)$ V_o. $V = V(z)$, so Laplace's equation $\nabla^2 V = 0$ becomes

$$\frac{d^2 V}{dz^2} = 0$$

Integrating twice gives

$$V = Az + B$$

We apply the boundary conditions to determine A and B.

$$V(z = 0) = 0 \rightarrow 0 = 0 + B \qquad \text{or} \quad B = 0$$

$$V(z = t) = V_o \rightarrow V_o = At \qquad \text{or} \quad A = \frac{V_o}{t}$$

Hence,

$$V = \frac{V_o}{t} z$$

$$E = -\nabla V = -\frac{dV}{dz} a_z = -\frac{V_o}{t} a_z$$

$$J = \sigma E = -\frac{\sigma V_o}{t} a_z, \quad dS = -\rho\, d\phi\, d\rho\, a_z$$

$$I = \int J.dS = \int_{\rho=a}^{b} \int_{\phi=0}^{\pi} \frac{2V_o\sigma}{t} \rho\, d\phi\, d\rho$$

$$= \frac{V_o\sigma}{t} \cdot \frac{\pi}{2} \frac{\rho^2}{2} \Big|_a^b \frac{V_o\sigma\pi\left(b^2 - a^2\right)}{4t}$$

Thus

$$R' = \frac{V_o}{I} = \frac{4t}{\sigma\pi\left(b^2 - a^2\right)}$$

Alternatively, for this case, the cross section of the bar is uniform between the borizontal surfaces at $z = 0$ and $z = t$ and eq. (5.16) holds. Hence, as reuired.

$$R' = \frac{\partial}{\sigma S} = \frac{t}{\sigma \frac{\pi}{4}\left(b^2 - a^2\right)}$$

$$= \frac{4t}{\sigma\pi\left(b^2 - a^2\right)}$$

Exercise 3.

A disc of thickness t has radius b and a central hole of radius a. Taking the conductivity of the disc as , find the resistance between

(a) The hole and the rim of the disc

(b) The two flat sides of the disc

Answer: (a) $\frac{Inbla}{2\pi t\sigma}$, (b) $\frac{t}{\sigma\pi\left(b^2 - a^2\right)}$

Example 4. A coaxial cable contains an insulating material of conductiveity σ. If the radius of the central wire is a and that of the sheath is b, show that the conductance of the cable per unit length is

$$G = \frac{2\pi\sigma}{In\ b/a}$$

Solution: Consider length L of the coaxial cable as shown in Figure 6.14. Let V_o be the potential difference between the inner and outer conductors so that $V(\rho = a) = 0$ and $V(\rho = b) = V_o$. V and E can be found just as in part (a) of the last example. Hence:

$$J = \sigma E = \frac{-\sigma V_o}{\rho\ In\ bla} a_\rho, \qquad dS = -\rho\ dz\ d\phi$$

$$I = \int J.dS = \int_{\phi=0}^{2\pi}\int_{z=0}^{L} \frac{V_o\sigma}{\rho\ In\ b/a}\rho\ dz\ d\phi$$

$$= \frac{2\pi L\sigma V_o}{In\ b/a}$$

The resistance per unit length is

$$R = \frac{V_o}{I}.\frac{1}{L} = \frac{In\ b/a}{2\pi\sigma}$$

and the conductance per unit length is as required.

$$G = \frac{1}{R} = \frac{2\pi\sigma}{In\ b/a}$$

Exercise 4

A coaxial cable contains an insulating material of conductivity σ_1 in its upper half and another material of conductivity σ_2 in its lower half (similar to the situation in Figure 5.19 b). If the radius of the central wire is a and that of the sheath is b, show that the leakage resistance of length ℓ of the cable is

$$R = \frac{1}{\pi\ell(\sigma_1 + \sigma_2)} In\frac{b}{a}$$

Answer: Proof.

Example 5. Conducing spherical shells with radii $a = 10$cm and $b = 30$cm are maintained at a potential difference of $100V$ such that $V(r = b) = 0$ $V(r = a) = 100$V. Determine V and E in the region between the shells. If $\varepsilon_r = 2.5$ in the region, determine the total charge induced on the shells and the capacitance of the capacitor.

Solution: Consider the spherical sheels shown in Figure 5.18. V depends only on r and hance Laplace's equation becomes

$$\nabla^2 V = \frac{1}{r^2}\frac{d}{dr}\left[r^2\frac{dV}{dr}\right] = 0$$

Since $r \neq 0$ in the region of interest, we multiply through by r^2 to obtain

$$\frac{d}{dr}\left[r^2\frac{dV}{dr}\right] = 0$$

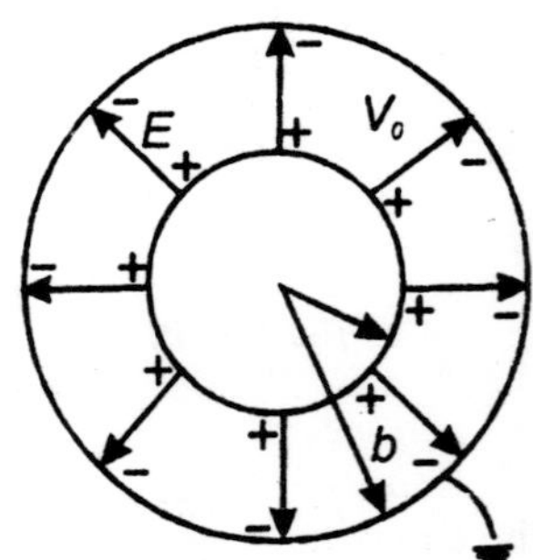

Figure 5.18. Potential V(r) due to conducting spherical shells.

Integrating once gives

$$r^2\frac{dV}{dr} = A$$

or

$$\frac{dV}{dr} = \frac{A}{r^2}$$

Integrating again gives

$$V = -\frac{A}{r} + B$$

As usual, constants A and B are determined from the boundary conditions.

When $\quad r = b,\ V = 0 \rightarrow 0 = -\frac{A}{b} + B \quad \text{or } B = \frac{A}{b}$

Hence $\quad V = A\left[\frac{1}{b} - \frac{1}{r}\right]$

Also when $\quad r = a,\ V = Vo \rightarrow Vo = A\left[\frac{1}{b} - \frac{1}{a}\right]$

or

$$A = \frac{V_o}{\frac{1}{b} - \frac{1}{a}}$$

Thus

$$V = V_o\frac{\left[\frac{1}{r} - \frac{1}{b}\right]}{\frac{1}{b} - \frac{1}{a}}$$

$$E = -\nabla V = -\frac{dV}{dr}a_r = -\frac{A}{r^2}a_r$$

$$= \frac{V_o}{r^2\left[\frac{1}{a} - \frac{1}{b}\right]}a_r$$

$$Q = \int \varepsilon E.dS = \int_{\theta=0}^{\pi} \int_{\phi=0}^{2\pi} \frac{\varepsilon_o \varepsilon_r V_o}{r^2\left[\frac{1}{a} - \frac{1}{b}\right]} r^2 \sin\theta \, d\phi \, d\theta$$

$$= \frac{4\pi\varepsilon_o \varepsilon_r V_o}{\frac{1}{a} - \frac{1}{b}}$$

The capacitance is easily determined as

$$C = \frac{Q}{V_o} = \frac{4\pi\varepsilon}{\frac{1}{a} - \frac{1}{b}}$$

which is the same as we obtained in eq. (6.32); there in section 6.5, we assumed Q and found the corresponding V_o, but here we assumed V_o and found the corresponding Q to detremine C. Substituting $a = 0.1$ m, $b = 0.3$m, $V_o = 100V$ yields

$$V = 100\frac{\left[\frac{1}{r} - \frac{10}{3}\right]}{10 - 10/3} = 15\left[\frac{1}{r} - \frac{10}{3}\right]V$$

Check: $\nabla^2 V = 0$, $V(r = 0.3\text{m}) = 0$, $V(r = 0.1\text{ m})$ 100.

$$E = \frac{100}{r^2[10 - 10/3]}a_r = \frac{15}{r^2}a_r \; V/m$$

$$Q = \pm 4\pi.\frac{10^{-9}}{36\pi}.\frac{(25).(100)}{10 - 10/3}$$

$$= \pm 4.167nC$$

The positive charge is induced on the inner shell; the negative charge is induced on the outer shell. Also

$$C = \frac{|Q|}{V_o} = \frac{4.167 \times 10^{-9}}{100} = 41.47pF$$

Exercise 5

If Figure 5.19 represents the cross sections of two spherical capacitors, determine their capacitances. Let a = 1mm, b = 3 mm, c = 2mm, ε_{r1} = 2.5, and ε_{r2} = 3.5.

Answer: (a) 0.53 pF, (b) 0.5 pF

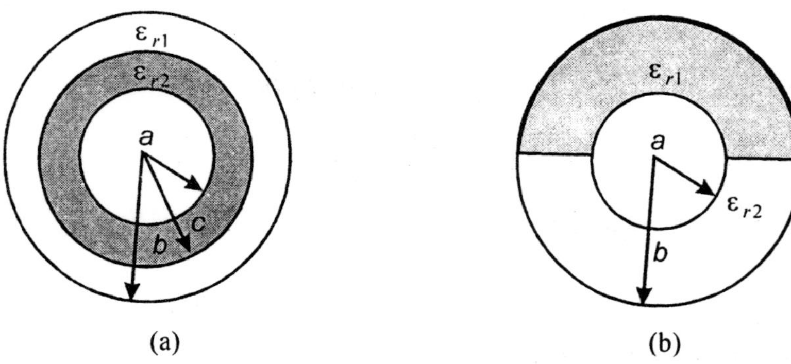

Figure 5.19. For Practice Exercises,

Check Yourself

1. Equation $\nabla.(-\varepsilon\nabla V) = \rho_v$ may be regarded as Poisson's equation for an inhomogeous medium.
   (a) True
   (b) False
2. In cylinrical coordinates, equation

$$\frac{\partial^2 \psi}{\partial \rho^2} + \frac{1 \partial \psi}{\rho \partial \rho} + \frac{\partial^2 \psi}{\partial z^2} + 10 = 0$$

   is called
   (a) Maxwell's equation
   (b) Laplace's equation
   (c) Poisoon's equation
   (d) Helmholtz's equation
   (e) Lorentz's equation
3. Two potential functions V_1 and V_2 satisfy Laplace's equation within a closed region and assume the same values on its surface. V_1 must be equal to V_2.
   (a) True
   (b) False
   (c) Not Necessarily

4. Which of the following potentials does not satisfy Laplace's equation?

   (a) V = 2x + 5 (b) V = 10xy

   (c) $V = r\cos\phi$ (d) $V = \frac{10}{r}$

   (e) $V = \rho \cos \phi + 10$

5. Which of the following is not true?

   (a) –5 cos 3*x* is a solution to $\phi''(x) + 9\phi(x) = 0$

   (b) 10 sin 2*x* is a solution to $\phi''(x) - 4\phi(x) = 0$

   (c) –4 cosh 3*y* is a solution to $R''(y) - 9R(y) = 0$

   (d) Sinh 2*y* is a solution to $R''(y) - 4R(y) = 0$

   (e) $\frac{g''(x)}{g(x)} = -\frac{h''(y)}{h(y)} = f(z) = -1$ where g(*x*) = sin *x* and *h*(*y*) = sinh *y*

6. If $V_1 = X_1 Y_1$ is a product solution of Laplace's equation, which of these are not solutions of Laplace's equation?

   (a) $-10X_1Y_1$

   (b) $X_1Y_1 + 2xy$

   (c) $X_1Y_1 - x + y$

   (d) $X_1 + Y_1$

   (e) $(X_1 - 2)(Y_1 + 3)$

7. The capacitance of a capacitor filled by a linear dielectric is independent of the charge on the plates and the potential difference between the plates.

   (a) True (b) False

8. A parallel-plate capacitor connected to a battery stores twice as much charge with a given dielectric as it does with air as dielectric, the susceptibility of the dielectric is,

   (a) 0 (b) 1 (c) 2

   (d) 3 (e) 4

9. A potential difference V_0 is applied to a mercury column in a cylindrical container. The mercury is now poured into another cylindrical container of half the radius and the same potential difference V_0 applied across the ends. As a result of this change of space, the resistance will be increased.

   (a) 2 times (b) 4 times

   (c) 8 times (d) 16 times

10. Two conducting plates are inclined at an angle 30° to each other with a point charge between them. The number of image charges is

    (a) 12 (b) 11

    (c) 6 (d) 5

    (e) 3

Answer: 6.1a, 6.2c, 6.3a, 6.4c, 6.5b, 6.6d,e, 6.7a, 6.8b, 6.9d, 6.10b.

Problems

1. In free space, $V = 6xy^2z + 8$. At point $P(1, 2, -5)$, find E and ρ_v.
2. Two infinitely large conducting plates are located at $x = 1$ and $x = 4$. The space between them is free space with charge distribution $\frac{x}{6\pi}$ nC/m³. Find V at $x = 2$ if $V(1) = -50V$ and $V(4) = 50$ V.
3. The region between $x = 0$ and $x = d$ is free space and has $\rho_v = \rho_o(x - d)/d$. If $V(x = 0) = 0$ and $V(x = d) = V_o$, find: (a) V and E, (b) the surface charge densities at $x = 0$ and $x = d$.
4. Show that the exact solution of the equation

$$\frac{d^2V}{dx^2} = f(x) \quad 0 < x < L$$

subject to

$$V(x = 0)\ V_1 \quad V((x = L)\ V_2$$

is

$$V(x) = \left[V_2 - V_1 - \int_0^L f(\mu)\, d\mu\, d\lambda\right]\frac{x}{L} + V_1 + \int_0^x \int_0^\lambda f(\mu)\, d\mu\, d\lambda$$

5. A certain material occupies the space between two conducting slabs located at $y = \pm$ 2cm. When heated, the material emits electrons such that $\rho_v = 50(1 - y^2)$ μC/m³. If the slabs are both held at 30 kV, find the potential distribution within the slabs. Take $\varepsilon = 3\varepsilon_o$.
6. Determine which of the following potential field distributions satisfy Laplace's equation.

    (a) $V_1 = x^2 + y^2 - 2z^2 + 10$ (b) $V_2 = \frac{1}{(x^2 + y^2 + z^2)^{1/2}}$

    (c) $V_3 = \rho z \sin\phi + \rho^2$ (d) $V_4 = \frac{10 \sin\theta \sin\phi}{r^2}$

7. Show that the following potentials satisfy Laplace's equation.
   (a) $V = e^{-5x} \cos 13y \sinh 12z$

   (b) $V = \dfrac{z \cos \phi}{\rho}$

   (c) $V = \dfrac{30 \cos \theta}{r^2}$

8. Show that $E = (E_x, E_y, E_z)$ satisfies Laplace's equation.

9. Let $V = (A \cos nx + B \sin nx)(Ce^{ny} + de^{-ny})$, when A, B, C, and D are constants. show that V satisfies Laplace's equation.

10. The potential field $V = 2x^2yz - y^3z$ exists in a dielectric medium having $\varepsilon = 2\varepsilon_0$. (a) Does V satisfy Laplace's equation? (b) Calculate the total charge within the unit cube $0 < x,y,z < 1$m.

11. Consider the conducting plates shown in figure 5.26. If $V(z=0) = 0$ and $V(z = 2\text{mm}) = 50$V, determine V, E, and D in the dielectric region ($\varepsilon_r = 1.5$) between the plates and ρ_s on the plates.

12. The cylindrical capacitor whose cross section is in Figure 5.27 has inner and outer radii of 5 mm and 15 mm, respectively. If $V(\rho = 5\text{mm}) = 100V$ and $V(\rho = 15\text{mm}) = 0$ V, calculate V, E, and D at $\rho = 10$ mm and ρs on each plate. Take $\varepsilon_r = 2.0$.

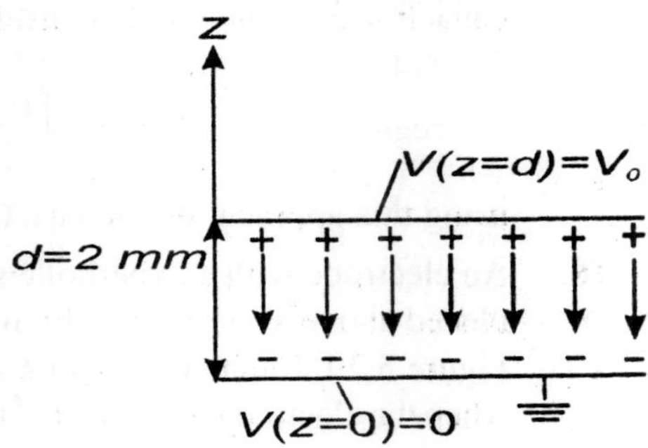

Figure 5.26. For Problem 11.

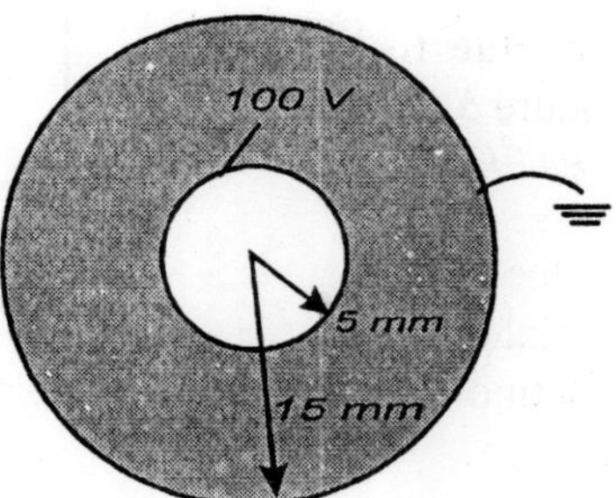

Figure 5.27 cylindrical capacitor of Problem 12.

13. Concentric cylinders $\rho = 2$ cm and $\rho = 6$cm are maintained at $V = 60$ V and $V = -20$ V, respectively. Calculate V, E, and D at $\rho = 4$ cm.

14. The region between concentric spherical conducting shells r = 0.5m and r = 1m is charge free. If $V(r = 0.5) = -50V$ and $V(r = 1) = 50V$, determine the potential distribution and the electric field strength in the region between the shells.

15. Find V and E at (3, 0, 4) due to the two conducting cones of infinite extent shown in Figure 5.28.

16. The inner and outer electrodes of a diode are coaxial of radii a = 0.6m and b = 30 mm, respectively. The inner electrodes is maintained at 70 V while the outer electrode is grounded. (a) Assuming that the length of the electrodes $\ell >> a$, b and ignoring the effects of space charge, calculate the potential at ρ = 15mm. (b) If an electron is injected radially through a small hole in the inner electrode with velocity 10^7 m/s. find its velocity at ρ = 15 mm.

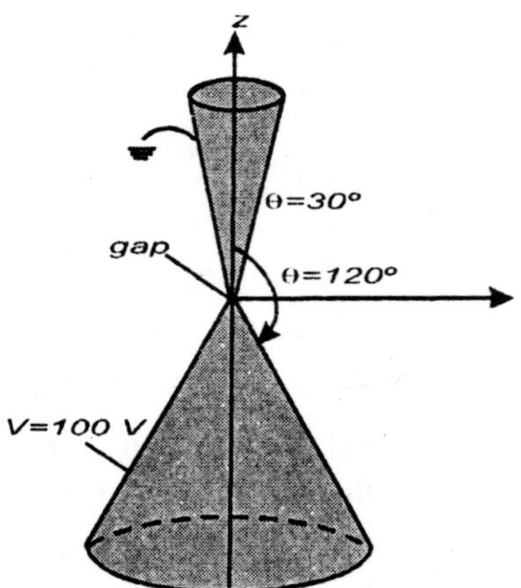

Figure 5.28 Conducting cones of Problem 15.

17. Another method of finding the capacitance of a capacitor is using energy considerations, that is

$$C = \frac{2W_E}{V_o^2} = \frac{1}{V_o^2}\int \varepsilon |E|^2 \, dv$$

using this apprach, derive eqs. (5.22), (5.28), and (5.32).

18. An electrode with a hyperbolic shape ($xy = 4$) is placed above an earthed right-angle corner as in Figure 5.29. Calculate V and E at point (1,2, 0) when the electrode is connected to a 20-V source.

19. Solve Laplace's equation for the two-dimensional electrostatic systems of Figure 5.30 and find the potential $V(x, y)$.

20. Find the potential $V(x, y)$ due to the two-dimensional esystems of Figure 5.31.

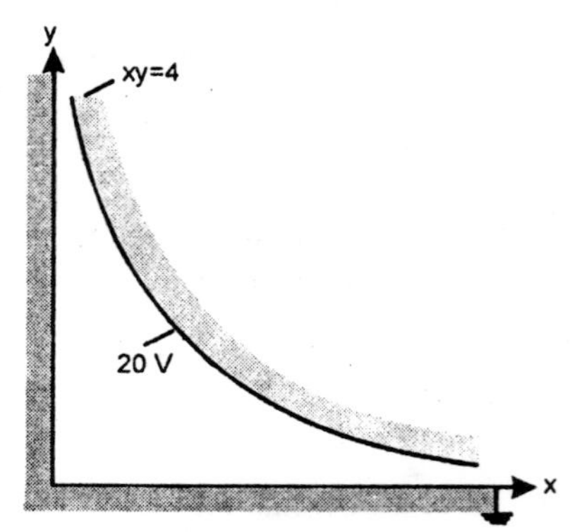

Figure 5.29. For Problem 18.

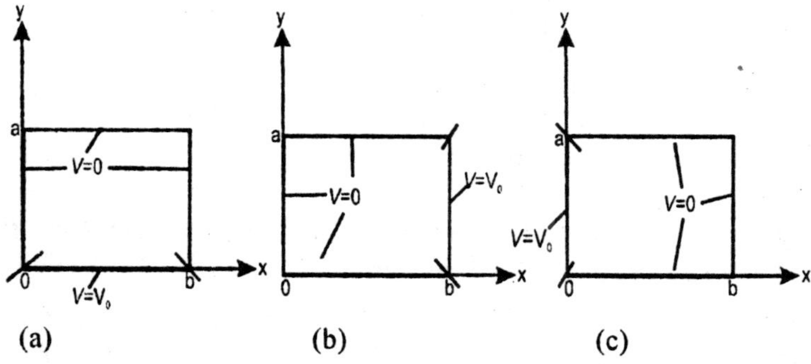

Figure 5.30. For Problem 19.

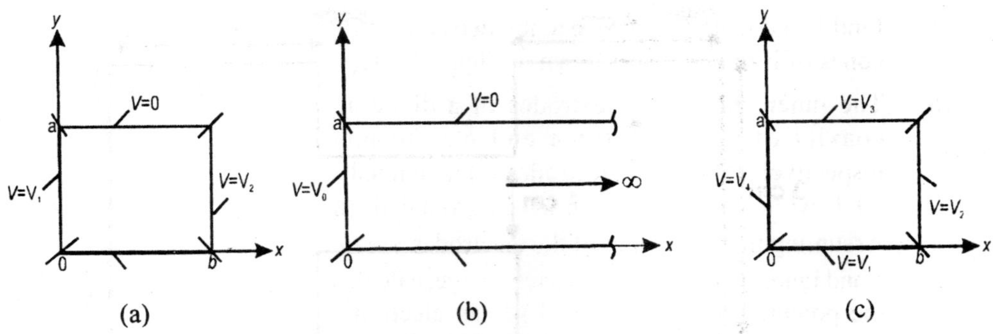

Figure 5.31. For Problem 20.

21. By letting $V(\rho, \phi) = R(\rho)\, \Phi(\phi)$ be the solution of Laplace's equation in a region where $\rho \neq 0$, show that the separated differential equations for R and Φ are,

$$R'' + \frac{R'}{\rho} - \frac{\lambda}{\rho^2} R = 0$$

$$\Phi'' + \lambda \Phi = 0$$

where λ is the seqaration constant.

22. A potetial in spherical coordinates is a function of r and ϕ. Assuming that $V(r, \theta) = R(r)\,F(\theta)$, obtain the separated differential equations for R and F in a region for which $\rho_v = 0$.

23. Show that the resistance of the bar of figure 5.17 between the etical ends located at $\phi = 0$ and $\phi = \pi/2$ is.

$$R = \frac{\pi}{2\sigma t\, In\, b/a}$$

24. Show that the resistance of the sector of a spherical shell of conductivity σ, with cross section shown in Figure 5. 32 (where $0 < \phi < 2\pi$), between its base is

$$R = \frac{1}{2\pi\sigma(1 - cos\,\alpha)}\left[\frac{1}{a} - \frac{1}{b}\right]$$

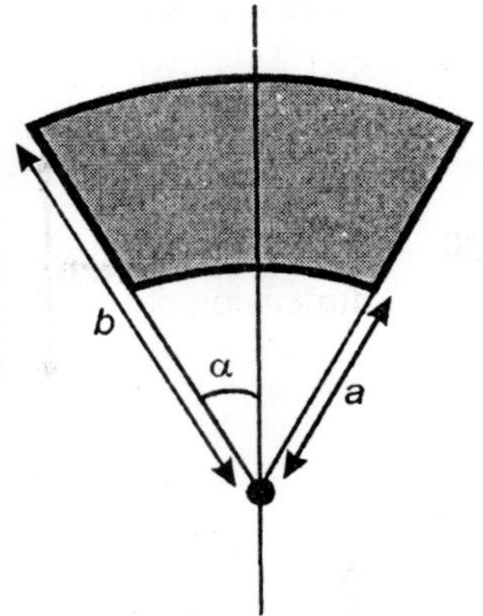

Figure 5.32. For Problem 24.

25. A hollow conducting hemisphere of radius a is buried with its flat face lying flush with the earth surface thereby serving as an earthing electrode. If the conductivity of earth is σ, show that the leakage cobnductance between the electrode and earth is $2\pi a\sigma$.

26. The cross section of an electric fuse is shown in Figure 5.33. If the fuse is made of copper and of thickness 1.5 mm, calculate its resistance.

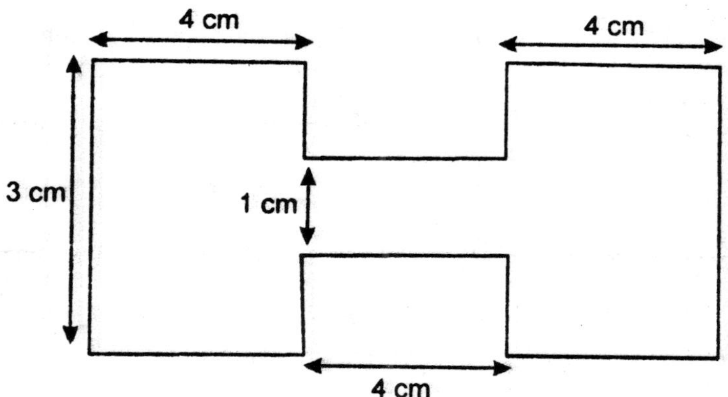

Figure 5.33. For Problem 26.

27. In an integrated circuit, a capacitor is formed by growing a silicon dioxide layer (ε_r = 4) of thickness 1 μm over the condiucting silicon substrate and covering it with a metal electrode of area S. Determine S if a capacitance of 2 nF is desired.

28. The parallel-plate capacitor of Figure 5.34 is quarted-filled with mica (ε_r = 6). Find the capacitance of the capacitor.

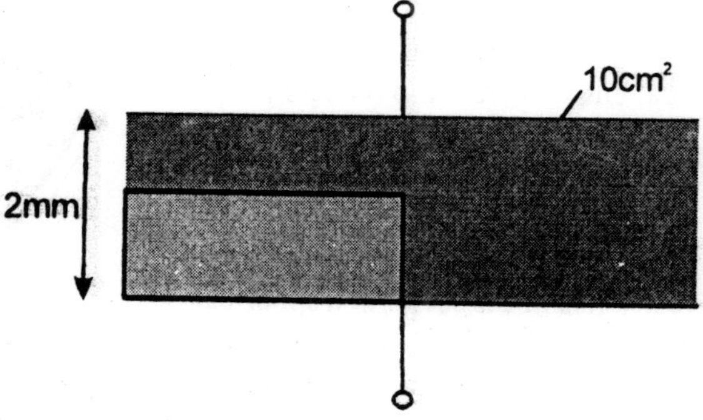

Figure 5.34. For Problem 28.

29. An air-filled parallel plate capacitor length L, width a, and plate separation d has its plates maintained at constant potential difference V_o. If a dielectric slab of dielectric constant ε_r is slid between the plates and is withdrawn until only a length x remains between the plates as in figure 5.35, show that the force tending to restore the slab to its orginal position is.

$$F = \frac{\varepsilon_o(\varepsilon_r - 1)aV^2{}_o}{2d}$$

30. A parallel-plate capacitor has plate area 200 cm^2 and plate separation 3mm. The charge density is 1 $\mu C/m^2$ with air as dielectric. Find

    (a) The capacitance of the capacitor

    (b) The voltage between the plates

    (c) The force with which the plates attract each other

31. Two conducing plates are placed at $z = -2$cm and $z = 2$ cm and are, respectively, maintained at potentials 0 and 200 V. Assuming that the plates are separated by a polypropylene ($\varepsilon = 2.25\varepsilon_o$). Calculate: (a) the potential at the middle of the plates, (b) the surface charge densities at the plates.

32. The conducting parallel plates are separated by a dielectric material with $\varepsilon = 5.6\varepsilon_o$ and thickness 0.64 mm. Assume that each plate has an area of 80 cm^2. If the potential field distribution between the plates is $V = 3x + 4y - 12z + 6$kV, determine: (a) the capacitance of the capacitor, (b) the potential difference between the plates.

6

Moving Charges

Electromagnetic Phenomenon

You may have studied about the evolution of magnets. But let us recall; Magnatites were naturally. Occured stones which were found in the magnesia district in Asia Minor, Greece. These stones possesses the property of attracting small bits of iron and when a piece of magnetite is suspended freely, it stablized in North-South direction.

After the name of the district where it was found the stone was named magnete and the property of attraction magnetism.

Oersted first noticed the magnetic effect of current by observing that a pivoted magnetic needle gets deflected when a steady surrent is passed through a wire kept above or below and parallel to it (Fig. 6.1). It was the first ever experiment to establish the relationship between electrical and magnetism. Then after the electromagnetic phenemence was developed. The deflection of the needle must essentially be due to the magnetic field round the wire carrying current. Oersted thus showed a connection between electricity and magnetism. He said that all magnetic effects (temporary or permanent)

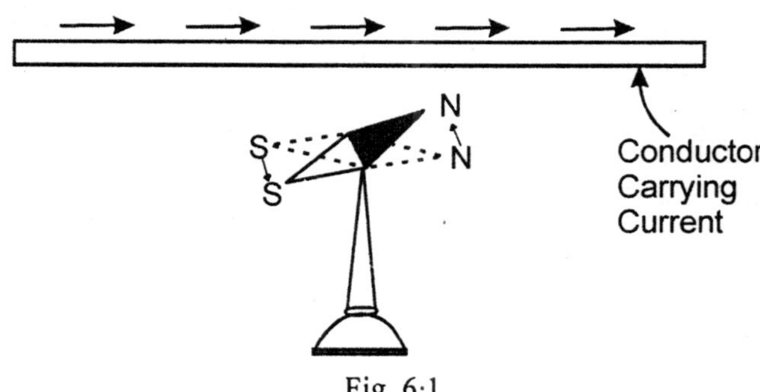

Fig. 6·1

A mometary, current is produced in circuit when the current in nearby circuits was being started or stopped. It was observed by Faraday after tweleve years. The same effect was observed by moving a magnet towards or away from the circuit. Thus we see that the work of Oersted demonstrated the magnetic effect of moving electric charges and that tof Faraday, the electric effect (production of current) by moving magnets. Now a-day it is believed that all magnetic phenomena result from forces between electrostatic charges in motion and the magnetism cannot be considered as a separate science.

The Magnetic Field and Magnetic Induction

Magnetic field is the space around the current carrying conductor or magnet.

The magnetic field around a moving charge exists in addition to the electrostatic field which exists around the charge whether the charge is in motion or not. If a second charged particle (text charge) is placed near the moving charge, it will experience a force due to the electric field whether it is in motion or at rest and a force due to magnetic field only if it is in motion. Like electric field, magnetic field is also a vector field whose magnitude and direction at any point are specified by a vector $\vec{B}$. It can be represented by the lines called lines of magnetic induction or magnetic flux lines, just as the electric field was represented by lines of force. About magnetic induction, we will be discussing in subs equent chapters. Following are few point which signified the relationship between magetic field $\vec{B}$ and lines of induction.

(a) The tangent to the line of induction at any point gives the direction of magnetic induction vector $\vec{B}$ at that point.

(b) The lines are drawn in such a way that the *number of lines per unit ared of cross-section at a certain point taken normal to the lines, is equal to the magnitude of* $\vec{B}$ *at that point.*

Magnitude of Magnetic Flux

The magnetic flux through a surface is the total number of magnetic lines of induction threading that surface. If the magnetic induction varie from point to point on a surface and the surface is not normal to the lines of induction everywhere, then magnetic flux through the surface is calculated as follows:

Consider an infinitesimally small area element ds, normal to which makes an θ angle with the direction of line of induction $\vec{B}$ (Fig. 6.2).

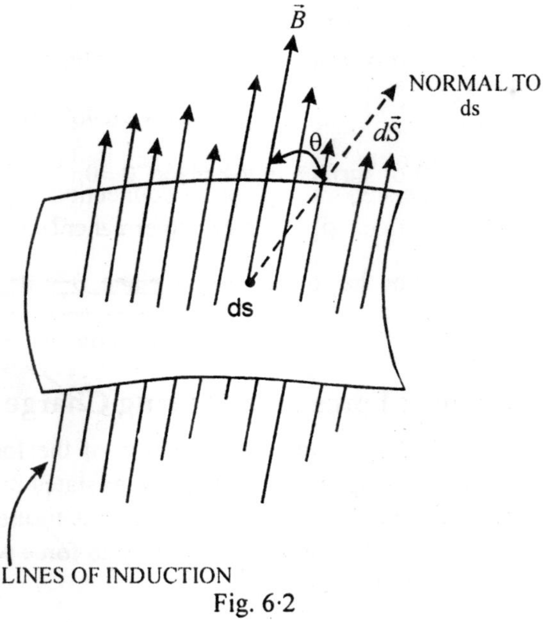

Now, across the area ds the flux $d\theta_B$ is given;

$$d\phi_B = \vec{B} \cdot \vec{d}\,s = B\,ds\cos\theta$$

$\vec{B}$

NORMAL TO ds

θ $d\vec{S}$

ds

LINES OF INDUCTION

Fig. 6·2

∴ Total flux across a finite area S is given by

$$\phi_B = \int d\phi = \int_S \vec{B}.\vec{ds} = \int_S B\,ds\cos\theta \tag{6.1}$$

The flux over the finite area S is expressed by the following equation if B is uniform and normal to the area.

$$\phi_B = \int_S B\,ds = B\int_S ds$$

($\because$ B is uniform, so it is taken out of integral sign)

or $$\phi_B = BS \tag{6.2}$$

Thus, we must conclude here that the magnetic flux density or magnetic induction $\vec{B}$ at a point is equal to the flux passing through a unit area held perpendicular to the lines of induction at that point.

Magnetic flux may be positive or negative depending on the angle θ between the vector. $\vec{B}$ and vector area $\vec{ds}$ is acute or obtuse.

Example 1. Calculate the magnetic flux across the surfaces each of area 0.1 sqm in (i) xy plane (ii) y-z plane and (iii) z-x plane in a region specified by the magnetic field vector $\vec{B} = 2\hat{i}$ tesla.

Solution. Magnetic flux $\phi_B = \vec{B}.\vec{A} = BA\cos\theta$ where θ is the angle between vectors $\vec{B}$ and $\vec{A}$ (i.e. normal to the plane of area).

Given $\vec{B} = 2\hat{i}$ tesla.

$A = 0.1$ m^2

(i) For surface in x-y plane, 0 = 90°

$$\therefore \quad \phi_{x-y} = \vec{B}.\vec{A} = BA\cos 90^o = 0$$

(ii) For surface in y-z plane, θ =0

$$\therefore \quad \phi_{y-z} = BA\cos 0^o = BA = 2\times .1 = 0.2\ weber$$

(iii) For the surface in z=x plane, θ =90°

$$\therefore \quad \phi_{z-x} = \vec{B}.\vec{A} = BA\ \cos\ 90° = 0.$$

Magnetic Force on a Moving Charge

The magnetiude and direction of the force on a moving test charge is measured for determining magnetic field, because as stated in section 6.2, the moving test charge experiences a force due to magnetic field also. Let us further assume that there is no electric field present, which means that neglecting gravity, no force will act on the test charge when placed at rest at the point say P in question.

Now, assume, a positive test charge q_0 which is moving with an arbitrary velocity $\vec{V}$ through the point in the space. If a sideways deflecting force $\vec{F}$ acts on the test charge, there must exist a magnetic field at P and magnetic induction vector $\vec{B}$ can be defined in terms of $\vec{F}$ and other measured quantities.

If, we keep the magnitude of $\vec{u}$ constant and vary the direction of $\vec{v}$ through the point p then we find that although $\vec{F}$ remains always normal to $\vec{v}$ *but its magnitude* $|\vec{F}|$ changes. By changing orientation of $\vec{v}$, we further find that there is one direction in which the magnetic field exerts no force (i.e force $\vec{F}$ becomes zero). we define this direction as the direction of $\vec{B}$ vector.

Once the direction of $\vec{B}$ has been found, the magnitude can easily be found. We now keep the velocity of the moving charge perpendicular to the magnetic field $\vec{B}$ as shown in fig. 6.3 where at P, $\vec{B}$ acts along y-axis and + q_0 charge moving with velocity $\vec{v}$ along z-axis. It is also found that the force $\vec{F}$ is now maximum. If the speed v is increased along with the magnitude of test charge q_0 then the maximum value of $\vec{F}$ also increases as per the experiment. x-axis. $\vec{F}$ is found to act along z-axis (Fig. 6.3). The expression of above explanation is given follows

$$F_{max} \propto q_0$$

$$\propto v$$

or $$F_{max} \propto q_0 v$$

or $$F_{max} = B\, q_0 v$$

i.e. $$B = \frac{F_{max}}{q_0 v}$$

where constant B is magnetude of the field.

If we put $q_0 = 1$, $v = 1$ then, $B = F_{max}$. *The magnitude of magnetic induction at any point in a magnetic field is the maximum force experienced by a unit positive charge moving with unit velocity perpendicular to the direction of magnetic field.*

Now, for giving a complete definition of $\vec{B}$, combine the magnitude and direction of $\vec{B}$. If a positive test charge q_0 is moving with a velocity $\vec{v}$ through the point *P* in any direction and if a force $\vec{F}$ acts on the moving charge, a magnetic induction $\vec{B}$ is present at the the point *P*, where $\vec{B}$ is vector which satisfies the relation.

$$\vec{F} = q_0\left(\vec{v} \times \vec{B}\right) \tag{6.4}$$

where $\vec{v}$, q_0 and $\vec{F}$ are measured quantities. The magnitude F of the magnetic deflecting force $\vec{F}$, Now, we aply the principle of vector product.

$$F = q_0 v\, B \sin\phi \tag{6.5}$$

where ϕ is the angle between $\vec{v}$ and $\vec{B}$. it is clear from equation (6.4) that the direction of $\vec{F}$ is perpendicular to both $\vec{v}$ and $\vec{B}$ in accordance with the experimental observation. Unlike the electric force, magnetic force is velocity dependent.

Fig 6.4. depicts the relation among the three vectors and $\vec{B}, \vec{F}$ and $\vec{v}$. Vector $\vec{B}$ is acting along y-axis, $\vec{v}$ lies in x-y plane, making an angle with direction of $\vec{B}$ which is working along y-axis. The force $\vec{F}$ is directed along z-axis and it is at right angles to x-y plane.

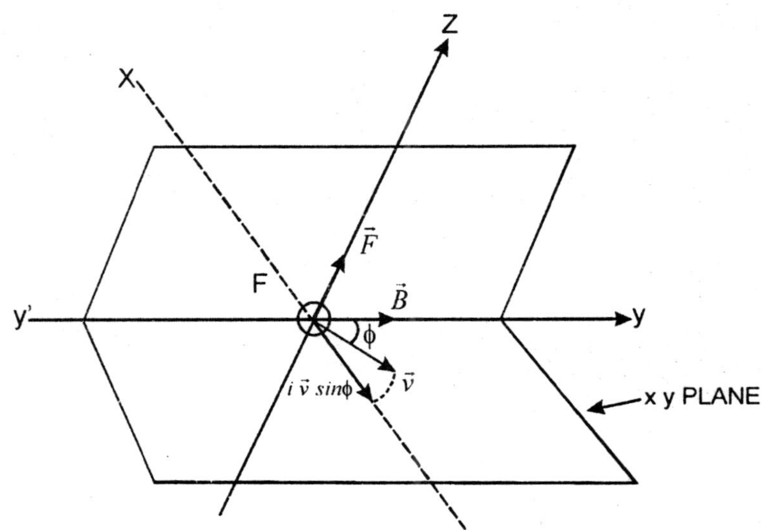

Fig. 6.4. Illustration of general reaction between $\vec{F}$, $\vec{v}$ and $\vec{B}$

We may now conclude following points from equations (6.4) and (6.5).

(i) According to the above experimental observations, $\vec{F}$ will always be a side-ways deflecting force and at right angles to $\vec{v}$ and also to $\vec{B}$. $\vec{F}$ is normal to the plane formed by $\vec{v}$ and $\vec{B}$.

(ii) The deflecting force $\vec{F}$ vanishes when the charge is at rest i.e. v = 0. This means that a charge at rest in a magnetic field does not experience any magnetic force.

(iii) If the test charge moves in the direction of $\vec{B}$ or opposite to it *i.e.*, if $\vec{v}$ and $\vec{B}$ are either parallel or anti-parallel ($\vec{v} \uparrow\uparrow \vec{B}$ or $\vec{v} \uparrow\downarrow \vec{B}$), the deflecting force $\vec{F}$ vanishes, because in these cases $\phi = 0°$ or $180°$ so v B sin ϕ = 0.

(iv) The deflecting force is maximum when the direction of motion of charge is perpendicular to the magnetic field.

$$F_{max} = q_0\, vB,\ if\, \vec{u} \perp \vec{B}\ i.e\ \phi = 90°.$$

(v) The kinetic energy of a moving charge remains unaltered in a static magnetic field. As the work done by force in a static magnetic field is zero because the direction of force and the direction of motion of charge are always at right angle to each other. The magnetic field simply produces a sideways deflection on it.

As the direction of $\vec{F}$ is perpendicular to the plane containing $\vec{v}$ and $\vec{B}$, so it can be given by fleming's left hand rule. According to this rule, the forefinger, the central finger and the thumb of the left hand are stretched in such a way that they are mutually at right angle to each other. *If the forefinger points in the direction of magnetic field vector $\vec{B}$ and the central finger points in the direction of current or velovity vector $\vec{v}$, then the thumb gives the direction of the deflecting force $\vec{F}$ on the charge* (Fig. 6.5).

The direction of $\vec{F}$ may also be given by right hand thumb rule, which is explained below:

Curl the fingers of the right hand in the direction in which the vector $\vec{v}$ sin is rotated towards $\vec{B}$, then the direction of the thumb gives the direction of the force on the positive charge.

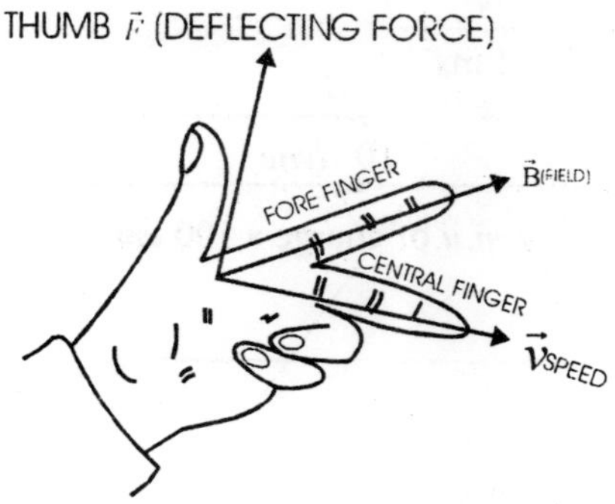

Fig. 6.5. Fleming'sleft hand rule.

Units of $\vec{B}$ and Units of Flux

S.I. unit of $\vec{B}$ is called tesla (T). If we put F=1 N, q_0 = 1 C, v = 1 ms^{-1} and $\phi = 90°$ in equation (6.5), we have

$$B = \frac{1N}{1C \times 1ms^{-1} \times sin\,90^o}$$

$$= \frac{1N}{1A \times 1m} \qquad \left(\because \frac{1\,C}{1S} = 1A\right)$$

$$= 1N\,A^{-1}\,m^{-1}$$

and it is given a special name weber/m^2 or tesla.

If a charge of one coulomb moving with a speed of 1 m/s at right angles to the magnetic induction field B at a point and experiences a force of 1N. Then the magnetic induction at that point is said to be one tesla.

Gauss is the CGS m (cgs electro magnetic unit of $\vec{B}$. If we put F = 1 dyne, q_0 = 1 e.m.u. of charge, v = 1 cm s^{-1}, $\phi = 90°$ in equation (6.5), then B is 1 gauss so

$$1Gs = \frac{1\,dyne}{1\text{ e.m.u. of charge} \times 1\text{ cm s}^{-1} \times sin\,90^o}$$

$$= \frac{1\,dyne}{1\text{ e.m.u. of current} \times 1\text{ cm}}$$

if one e.m.u. of charge moving with a velocity of 1 cms^{-1} perpendicular to the magnetic field $\vec{B}$ at a point experiences a force of one dyne. then the magnetic induction at that point is one gauss.

$$1 \text{ gauss} = 1 \text{ max well/cm}^2$$

Relation between tesla (T) and gauss (Gs)

$$1T = \frac{1N}{1C \times 1 \text{ ms}^{-1}}$$

$$= \frac{10^5 \, dyne}{\frac{1}{10} e.m.u. \text{of charge} \times 100 \, cms^{-1}}$$

$$\left(\because 1C = \frac{1}{10} \text{ e.m.u of charge}\right)$$

$$= \frac{10^4 \, dyne}{1 e.m.u. \text{ of charge} \times 1 \, cms^{-1}}$$

$$= 10^4 \text{ Gs.}$$

Hence,

1 Tesla or 1 wb m^{-2} = 10^4 gauss of 10^4 maxwell cm^{-2}

Units of flux : The S.I. unit of magnetic flux is weber and cgs m unit of magnetic flux is maxwell.

$$1 \text{ weber} = 10^8 \text{ maxwells}$$

Dimensions of Magnetic Flux

Using the formula of the force experienced by a particle having charge q moving perpendidenar to magnetic flux density B with a velocity v.

$$F = qvB$$

$$\therefore \quad B = \frac{F}{qv} = \frac{MLT^{-2}}{q\,LT^{-1}}$$

$$= \frac{MLT^{-2}}{AL}$$

[where qT^{-1} = A which represents current *i.e.* ampere A]

Now, use the formula of flux

$$\phi_B = B \times \text{area}$$

$$\therefore \qquad \phi_B = \frac{MLT^{-2}}{AL} \times L^2$$

or
$$\boxed{\phi_B = M^1 L^2 T^{-2} A^{-1}}$$

Example 2. A 5.0 MeV proton is falling vertically downwword in a uniform field of magnetic induction having magnitude 1.5 wb m^{-2}, pointing horizontally from South to North. Find the force exerted on it. Given mass of proton = 1.7×10^{-27} kg and charge on proton 1.6×10^{-19} C.

Solution. Mass of proton m = 1.7×10^{-27} kg

Charge on proton q = 1.6×10^{-19} C

Magnetic Field B = 1.5 Wb m^{-2}

K.E. of proton K = 5 MeV

$= 5 \times 1.6 \times 10^{-13}$ J $= 8 \times 10^{-13}$ J

If v be the velocity of proton, then

$$K.E. = K = \frac{1}{2} mv^2$$

or

$$v = \sqrt{\frac{2K}{m}} = \sqrt{\frac{2 \times 8 \times 10^{-13}}{1.7 \times 10^{-27}}}$$

$$= 3.1 \times 10^7 ms^{-1}$$

Magnetic force $\vec{F} = q(\vec{v} \times \vec{B})$ and its magnitude

$$F = qv\, B \sin \theta$$
$$= (1.6 \times 10^{-19})\,(3.1 \times 10^7) \times 1.5 \times \sin 90°$$
$$= 7.44 \times 10^{-12}\ N$$

(θ = 90°, $\because$ $\vec{v}$ is vertically downward direction and $\vec{B}$ is horizontal)

Relation $\vec{F} = q(\vec{v} \times \vec{B})$ indicates that the direction of $\vec{F}$ is per pendicular to $\vec{v}$ and $\vec{B}$ both. As $\vec{v}$ is in downward direction and $\vec{B}$ is from South to North horizontally, hence $\vec{F}$ will be in a horizontal plane in West East direction.

Lorentz Force Equation

The resultant force on the moving test charge q_o which moves through a region in which both the electric and magnetic field are present is found by combining equations (6.4) and (6.6) and is given by

$$\vec{F} = q_0\, \vec{E} + q_0(\vec{v} \times \vec{B}) \qquad (6.7)$$

This equation is called Lorentz force equation. In cgs m units the relation (6.7) is written as

$$\vec{F} = q_0 \vec{E} + \frac{q_0}{c}\left(\vec{v} \times \vec{B}\right) \qquad (6.8)$$

where c is the velocity of light in vacuum and is used to convert q_0 from e.s.u. to e.m.u.

Example 3. A test charge having 1.6×10^{-19}C is moving with a velocity of $2\hat{i} + 3\hat{j}\ ms^{-1}$ is an electric field of intensity $3\hat{i} + 6\hat{j} + 1\hat{k}$ and a magnetic field of $2\hat{i} + 3\hat{k}$ tesla. Determine the magnitude and direction of the Lorentz force acting on the test charge.

(K.U. 1993A)

Solution. Electric field $\vec{E} = 3\hat{i} + 6\hat{j} + 1\hat{k}$

Magnetic field $\vec{B} = 2\hat{j} + 3\hat{k}$

Charge on test charge $q_0 = 1.6 \times 10^{-19}$C

Velocity of test charge $v = \left(2\hat{i} + 3\hat{j}\right)ms^{-1}$

$$\therefore \quad \vec{F} = q_0\left(\vec{E} + \vec{v} \times \vec{B}\right)$$

$$= 1.6 \times 10^{-19}\left[3\hat{i} + 6\hat{j} + 1\hat{k} + \left(2\hat{i} + 3\hat{j}\right) \times \left(2\hat{j} + 3\hat{k}\right)\right]$$

$$= 1.6 \times 10^{-19}\left(3\hat{i} + 6\hat{j} + 1\hat{k} + 4\hat{k} - 6\hat{j} + 9\hat{i}\right)$$

$$= 1.6 \times 10^{-19}\left(12\hat{i} + 5\hat{k}\right)$$

So $$F = 1.6 \times 10^{-19}\sqrt{12^2 + 5^2} = 13 \times 1.6 \times 10^{-19}$$

$$= 20.8 \times 10^{-19}\text{ N.}$$

It is in x-z plane, making angle $\theta = \cos^{-1}\left(\frac{12}{13}\right)$ with x–axis.

Note : In this book magnetic field is denoted by Hor B.

Biot-Savart's Law

According to biot-savart's law the intensity of magnetic field d H at a point P, as shown in Figure 6.6, by the differential current element I dl is proportional to the product I dl and the sine of the angle α between the element and the line joining *P* to the element and is inversely proportional to the spuare of the distance *R* between *P* and the element.

i. e.,

$$dH \propto \frac{Idl\sin\alpha}{R^2} \qquad (6.9)$$

or

$$dH = \frac{kIdl\,sin\,\alpha}{R^2} \tag{6.10}$$

Here, the constant of proportionality is k. In SI unit, k =1/4π, so eq. (6.10) converts

$$dH = \frac{Idl\,sin\,\alpha}{4\pi R^2} \tag{6.11}$$

It is easy to notice that eq. (6.11) is better put in vector form after applying the rules of cross product;

$$dH = \frac{Idl \times a_R}{4\pi R^2} = \frac{Idl \times R}{4\pi R^3} \tag{6.12}$$

where $R = |R|$ and $a_R = R/R$. Thus the direction of *d*H can be determined by the right hand rule with the right-hand thumb pointing in the direction of the current, the right-hand fingers encircling the wire in the direction of *d*H as shown in Figure 6.7(a). Alternatively, we can use the right-handed screw rule to determine the direction of *d*H: with the screw placed along the wire and pointed in the direction of current flow, the direction of advance of the screw is the direction of *d*H as in Figure 6.7 (b).

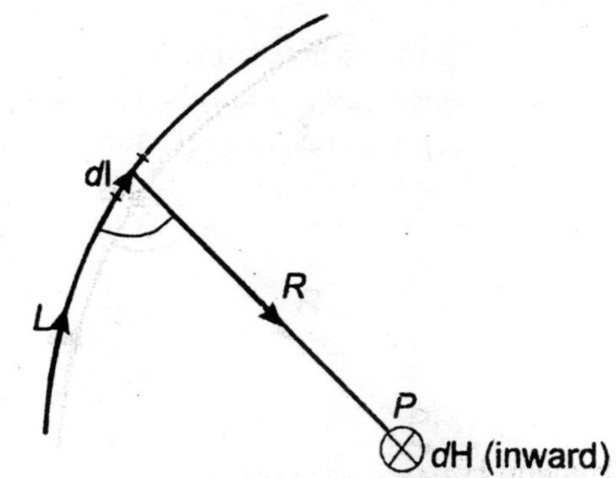

Figure 6.6 magnetic field *d*H at *P* due to current element *I dl*.

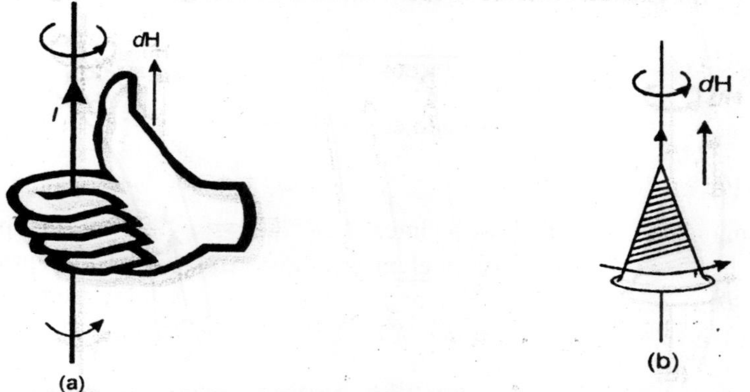

Figure 6.7 Determining the direction of *d*H using (a) the right-hand rule, or (b) the right-handed screw rule.

It is a very traditional method to represent the direction of the magnetic field intensity H (or current *I*) by a small circle with a dot or cross sign depending on whether H (or *I*) is out of, or into, the page as illustrated in figure 6.8.

We can have different current distributions: line current, surface current, and volume current as shown in Figure (6.9). If we define K as the surface current density (in amperes/meter) and J as the volume current density (in amperes/meter square.) the source element are related as

$$I\,dl \equiv K\,dS \equiv J\,dv \qquad (6.13)$$

Thus in terms of the distributed current sources, the Biot–Savart law as in eq. (6.12) becomes

$$H = \int_L \frac{Idl \times a_R}{4\pi R^2} \text{ (line current)} \qquad (6.14)$$

$$H = \int_S \frac{KdS \times a_R}{4\pi R^2} \text{ (surface current)} \qquad (6.15)$$

$$H = \int_v \frac{Jdv \times a_R}{4\pi R^2} \text{ (Volume current)} \qquad (6.16)$$

As an example, let us apply eq. (6.14) to determine the field due to a straight current carrying filamentary conductor of finite length AB as Figure 6.10. We assume that the conductor is along the z-axis with its upper and lower ends respectively subtending angles α_2 and α_1 at *P*,

Figure 6.8 Coventional representation of H (or *I*) (a) out of the page and (b) into the page.

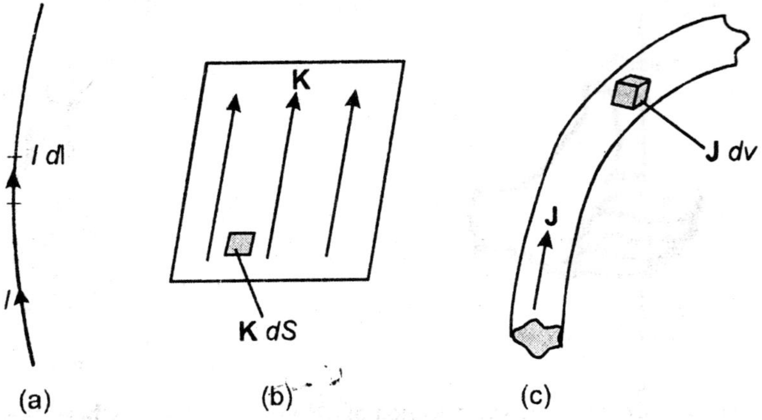

Figure 6.9. Current distributions: (a) line current, (b) surface current, (c) volume current.

the point at which **H** is to be determined. Particular note should be taken of this assumption as the formula to be derived will have to be applied accordingly. If we consider the contribution $d\mathbf{H}$ at P due to an element $d\mathbf{l}$ at (0, 0, z),

$$dH = \frac{IdI \times R}{4\pi R^3} \tag{6.17}$$

But di = dz a_z and R = $\rho a_\rho - z a_z$, so

$$dI \times R = \rho \, dz \, a_\phi \tag{6.18}$$

Hence,

$$H = \int \frac{I\rho \, dz}{4\pi[\rho^2 + z^2]^{3/2}} a_\phi \tag{6.19}$$

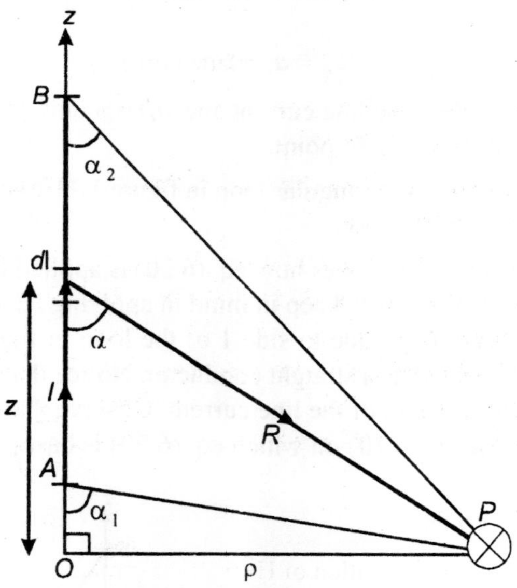

Figure 6.10. Field at point P due to a straight filamentary conductor.

Letting z = ρ cot α, dz = – ρ cosec² α dα, and eq. (6.19) becomes

$$H = -\frac{1}{4\pi} \int_{\alpha 1}^{\alpha 2} \frac{\rho^2 \operatorname{cosec}^2 \alpha \, d\alpha}{\rho^3 \operatorname{cosec}^3 \alpha}$$

$$= -\frac{1}{4\pi\rho} a\phi \int_{\alpha 1}^{\alpha 2} \sin \alpha \, d\alpha$$

or

$$\boxed{H = \frac{I}{4\pi\rho} (\cos \alpha_2 - \cos \alpha_1) a_\phi} \tag{6.20}$$

This expression is generally applicable for any straight filamentary conductor of definite length. Notice from eq. (6.20) that H is always along the unit vector α_ϕ (i. e., along concentric) circular paths) irrespective of the length of the wire or the point of interesect *P*. As a special case, when the conductor is *semiinfinite* (with respect to *P*) so that point A is now at O(0, 0, 0) while B is at (0, 0, ∞); $\alpha_1 = 90°$, $\alpha_2 = 0°$, and eq. (6.20) converts;

$$H\frac{I}{4\pi\rho}a_\phi \tag{6.21}$$

Another special case is when the conductor is *Infinite* in length. For this case, point a is at (0, 0, – ∞) while B is at (0, 0, ∞); $\alpha_1 = 180°$, $\alpha_2 = 0°$, so eq. (6.21) reduces to

$$H\frac{I}{2\pi\rho}a_\phi \tag{6.22}$$

To find unit vector a_ϕ in eqs. (6.20) to (6.22) is not always easy. A simple approach is to determine a_ϕ from

$$a_\phi = a_\ell \times a\rho \tag{6.23}$$

where a_ℓ is a unit vector along the line current and $a\rho$ is a unit vector along the perpendicular line from the line current to the field point.

Example 4. The onducting triangular loop in figure 1.11(a) carries a current of 10A. find H at (0,0, 5) due to side 1 of the loop.

Solution. This example illustrates how eq. (6.20) is applied to any straight, thin, current-carrying conductor. The key point to keep in mind in applying eq. (6.20) is figuring out α_1, α_2, ρ, and a_ϕ. the find **H** at (0, 0, 5) due to side 1 of the loop in figure 6.11(a), consider Figure 6.11(b), where side 1 is treated as a stratght conductor. Notice that we join the point of interest (0, 0, 5) to the beginning and end of the line current. Observe that α_1, α_2 and ρ are assigned in the same manner as in Figure (6.10) on which eq. (6.20) is based.

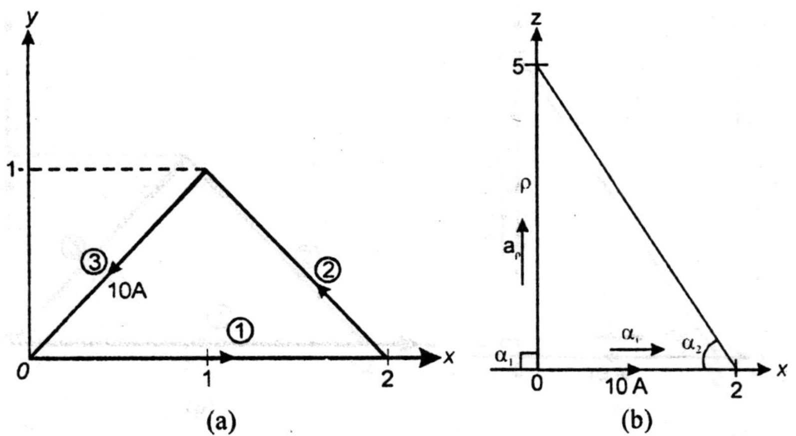

Figure 6.11. For example 4: (a) conducting triangular loop, (b) side 1 of the loop.

$\cos\alpha_1 = \cos 90° = 0$, $\quad \cos\alpha_2 = \frac{2}{\sqrt{29}}$, $\quad \rho = 5$

To determine a_ϕ is often the hardest part of applying eq. (6.20). According to eq. (6.23), $a_\ell = a_x$ and $a_\rho = a_z$, so

$$a_\phi = a_x \times a_z = -a_y$$

Hence,

$$H_1 = \frac{I}{4\pi\rho}(\cos\alpha_2 - \cos\alpha_1)a_\phi = \frac{10}{4\pi(5)}\left(\frac{2}{\sqrt{29}} - 0\right)(-a_y)$$

$$= -59.1a_y \text{ mA/m}$$

Practice Exercise 1

Find H at (0, 0, 5) due to side 3 of the triangular loop in figure 6.11(a)

Answer : $-30.63a_x + 30.63a_y$ mA/m.

Example 5. Find **H** at (–3, 4, 0) due to the current filament shown in figure (6.12a).

Solution : Let $H = H_x + H_z$ where H_x and H_z are the contributions to the magnetic field intensity at P(–3, 4, 0) due to the portions of the filament along *x* and *z*, respectively.

$$H_z = \frac{I}{4\pi\rho}(\cos\alpha_2 - \cos\alpha_1)a_\phi$$

At *P*(–3, 4, 0), p = (9 + 16)1/2 =5, $\alpha_1 = 90°$, $\alpha_2 = 0°$ and as obtained as unit vector along the circular path through P on plane z = 0 as in figure 6.12 (b). The direction of a_ϕ is determined using the right-handed screw rule or the right-hand rule. From the geometry in Figure (6.12b),

$$a_\phi = \sin\theta\, a_x + \cos\theta\, a_y = \frac{4}{5}a_x + \frac{3}{5}a_y$$

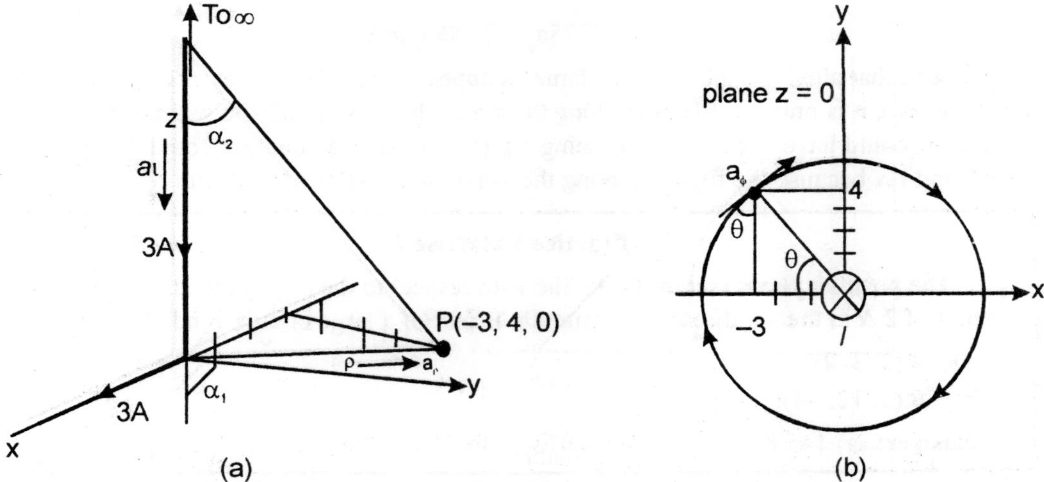

Figure 6.12. For Example 5. (a) current filament along seminfinite x-and z-axes; a_ℓ and a_ρ for H_z only; (b) determining aρ for H_z.

Alternatively, we can determine a_ϕ from eq. (6.23). At point P, a_ℓ and a_ρ are as illutstrated in Figure (6.12) (a) for H_z, Hence,

$$a_\phi = -a_z \times \left(-\frac{3}{5}a_x + \frac{4}{5}a_y\right) = \frac{4}{5}a_x + \frac{3}{5}a_y$$

$$H_z = \frac{3}{4\pi(5)}(1-0)\frac{(4a_x + 3a_y)}{5}$$

$$= 38.2a_x + 28.\ 65a_y \text{ mA/m}$$

It should be noted that in this case a_ϕ happens to be the negative of the regular a_ϕ of cylindrical coordinates. H_z could have also been obtained in cylindrical coordinates as

$$H_z = \frac{3}{4\pi(5)}(1-0)(-a_\phi)$$

$$= -47.75a_\phi \text{ mA/m}$$

Similarly, for H_x at P, $\rho = 4$, $\alpha_2 = 0°$, $\cos\alpha_1 = 3/5$, and $a_\phi = a_z$ or $a_\phi = a_\ell \times a_\rho = a_x \times a_y = a_z$. Hence,

$$H_x = \frac{3}{4\pi(4)}\left(1 - \frac{3}{5}\right)a_z$$

$$=23.88\ a_z \text{ mA/m}$$

Thus

$$H = H_x + H_z = 38.2a_x + 28.65a_y + 23.88a_z \text{ mA/m}$$

or

$$H = -\ 47.75a_\phi + 23.88a_z \text{ mA/m}$$

Notice that although the current filaments appear semiinfinite (they occupy the positive *z*-and *x*-axes.), it is only the filament along the z-axis that is semiinfinite with respect to point *P*. Thus H_z could have been found by using eq. (6.21), but the equation could not have been used to ind Hx because the filament along the x-axis is not semiinfinite with respect to *P*.

Practice Extercise 2

The positive y-axis (semiinfinite line with respect to the orgin) carries a filamentary current of 2 A in the $-a_y$ direction. Assume it is part of a large circuit. Find H at

(a) *A* (2, 3, 0)

(b) *B* (3, 12, –4)

Answer: (a) $145.8a_z$ mA/m, (b) $48.97a_x + 36.73a_z$ mA/m.

Example 6. A circular loop located on $x^2 + y^2 = 9$, $z = 0$ carries a direct current of 10 A along a_ϕ. Determine H at (0, 0, –4) and (0, 0, -4).

Solution. Consider the circular loop shown in Figure 6.13(a). The magnetic field intensity dH at point P(0, 0, h) contributed by current element I dl is given by Biot–Savart's law:

$$dH = \frac{IdI \times R}{4\pi R^3}$$

where dl = $\rho\, d\phi\, a_\varphi$, R = (0, 0, h) – (x, y, 0) = – $\rho a_\rho + h a_z$, and

$$dI \times R = \begin{vmatrix} a_\rho & a_\phi & a_z \\ 0 & \rho\, d\phi & 0 \\ -\rho & 0 & h \end{vmatrix} = \rho h\, d\phi\, a\rho + \rho^2\, d\phi\, a_z$$

Hence,

$$dH = \frac{I}{4\pi[\rho^2 + h^2]^{3/2}}\left(ph\, d\phi\, a_\rho + \rho^2 d\phi\, a_z\right) = dH_\rho\, a_\rho + dH_z\, a_z$$

By symmetry, the contributions alogns aρ add up to zero because the radial components produced by pairs of current element 180° apart cancel. This may also be shown mathematically by writing aρ in rectangular systems (i.e., $a_\rho = \cos\phi\, a_x + \sin\phi\, a_y$).

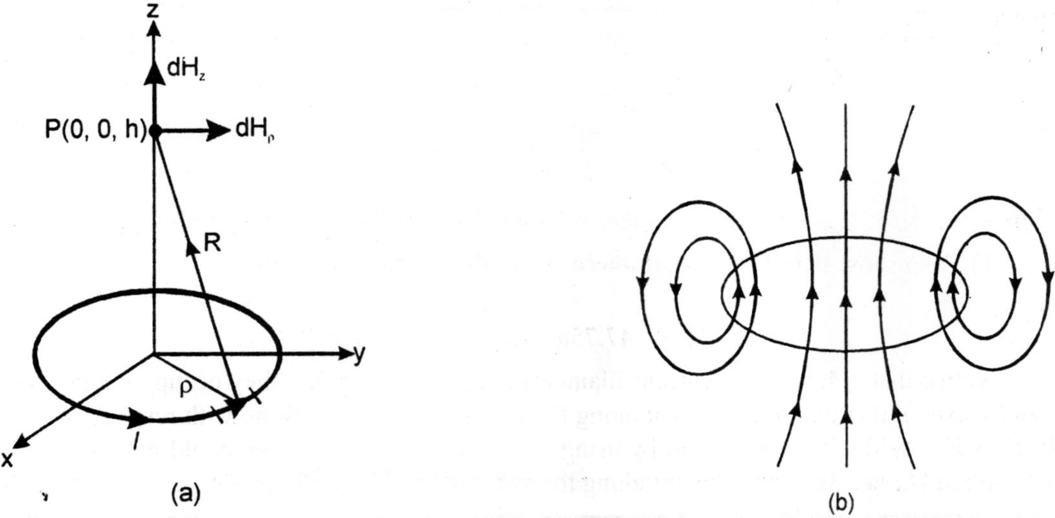

Figure 6.13. For Example 6 (a) circular current loop, (b) flux lines due Integrating cosϕ or sinϕ over $0 \le \phi \le 2\pi$ gives zero, thereby showing that $H_\rho = 0$. thus

$$H = \int dH_z a_z = \int_0^{2\pi} \frac{I\rho^2 d\phi a_z}{4\pi[\rho^2 + h^2]} = \frac{I\rho^2 2\pi a_z}{4\pi[\rho^2 + h^2]^{3/2}}$$

or

$$H = \frac{I\rho^2 a_z}{2[\rho^2 + h^2]^{3/2}}$$

(a) Substituting I = 10A, ρ= 3, h = 4 gives

$$H(0,0,4) = \frac{10(3)^2}{2[9+16]^{3/2}} = 0.36a_z A/m$$

(b) Notice from dl × R above that if H is replaced by – h, the z-component of dH remains the same while the-component still adds up to zero due to the axial symmetry of the loop. Hence

$$H(0, 0, -4) = H(0, 0, 4) = 0.36a_z \text{ A/m}$$

The flux lines due to the circular current loop are sketched in Figure 7.8 (b).

Practice Exercise 3

A thin ring of radius 5 cm is placed on plene z = 1 cm so that its center is at (0, 0, 1 cm). If the ring carries 50 mA along a_ϕ, find H at

(a) (0, 0, –1 cm)

(b) (0, 0, 10 cm)

Answer : (a) 4000a_z mA/m, (b) 57. 3a_zmA/m.

Example 7. A solenoid of length ℓ and radius a consists of N turns of wire carrying I. Show that at point P along its axis,

$$H = \frac{nI}{2}(\cos\theta_2 - \cos\theta_1)a_z$$

where n = N/ ℓ, θ_1 and θ_2 are the angles sbutended at P by the end turns as illustrated in Figure (6.14). Also show that if ℓ >> a, at the center of the solenoid,

$$H = nIa_z$$

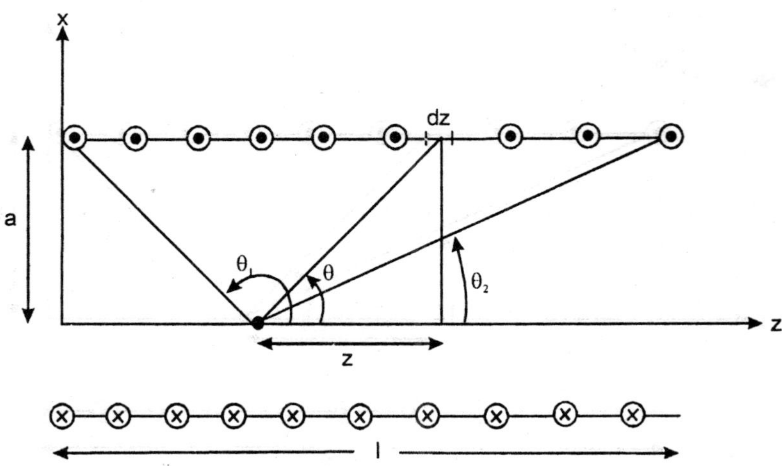

6.14. For Example, cross section of a solenoid.

Solution. Consider the cross section of the solenoid as shown in Figure 6.14. Since the solenoid consists of circular loop, we apply the result of Example. The contribution to the magnetic field H at P by an element of the solenoid of length *dz* is

$$dH_z = \frac{Idla^2}{2[a^2+z^2]^{3/2}} = \frac{Ia^2ndz}{2[a^2+z^2]^{3/2}}$$

where dl = n dz = (N/ ℓ) dz. From Figure 7.9, tanθ = a/$_z$; this is,

$$dz = -\text{a coses}^2\theta\, d\theta = -\frac{[z^2+a^2]^{3/2}}{a^2}\sin\theta\, d\theta$$

Hence.

$$dH_z = -\frac{nI}{2}\sin\theta\, d\theta$$

or

$$\text{Hz} = -\frac{nI}{2}\int_{\theta_1}^{\theta_2}\sin\theta\, d\theta$$

Thus

$$\text{H} = \frac{nI}{2}(\cos\theta_2 - \cos\theta_1)\, \text{a}_z$$

as required. Substituting n = N/ ℓ gives

$$\text{H} = \frac{NI}{2\ell}(\cos\theta_2 - \cos\theta_1)\, \text{a}_z$$

At the center of the solenoid,

$$\cos\theta_2 = \frac{\ell/2}{[a^2+\ell^2/4]^{1/2}} = -\cos\theta_1$$

and

$$\text{H} = \frac{In\ell}{2[a^2+\ell^2/4]^{1/2}}a_z$$

If

$$\ell >> \text{a or } \theta_2 \cong 0°,\ \theta_1 \cong 180°$$

$$\text{H} = nIa_z = \frac{NI}{\ell}a_z$$

Practice Exerise 4

If the solenoid of Figure 6.14 has 2.000 turns, a length of 75 cm, a radius of 5 cm, and carries a current of 50 mA along a_ϕ, find H at

(a) (0, 0, 0)

(b) (0, 0, 75 cm)

(c) (0, 0, 50 cm)

Answer: (a) 66.52a_z A/m, (b) 66.52a_z, A/m (c 131.7 a_z A/m.

Ampere's Circuit Law-maxwell's Equation

According to Ampere's circuit law, the net current I_{enc} enclosed by the path and the line integral of the tangential component of H is same

In other words, the circulation of H equals I_{enc}; that is,

$$\oint H.dI = I_{enc} \tag{6.24}$$

Ampere's law is similar to Gauss's law and it is easily applied to determine H when the current distribution is symmetrical. It should be noted that eq. (6.24) always holds whether the current distribution is symmetrical or not but we can only use the equation to determine H when symmetrical current distribution exits. Ampere's law is a special case of Biot-Savart's law: the former may be derived from the latter.

By applying Stoke's theorem to the left-hand side of eq. (6.24), we obtain

$$I_{enc} = \int_L H.dI = \int_S (\nabla \times H) \cdot dS \tag{6.25}$$

But

$$I_{enc} = \int_S J \cdot dS \tag{6.26}$$

Comparing the surface integrals in eqs. (6.25) and (6.26) clearly reveals that

$$\nabla \times H = J$$

This is the third Maxwell's equation to be derived; it is essentially Ampere's law in differential (or point) form whereas eq. (6.26) is the integral form. From eq. (6.27), we should observe that $\nabla \times H = J \neq 0$; that is, magnetostatic field s not conservative.

Properties of B

(i) **Curl of magnetic induction** $\vec{B}$. Consider a region of distributed current flow in which current density at any point (x, y, z) is given by is vector $\vec{J}$. This current density vector varies from point to point but is time independent i.e.,

$$\text{Div } \vec{J} = 0.$$

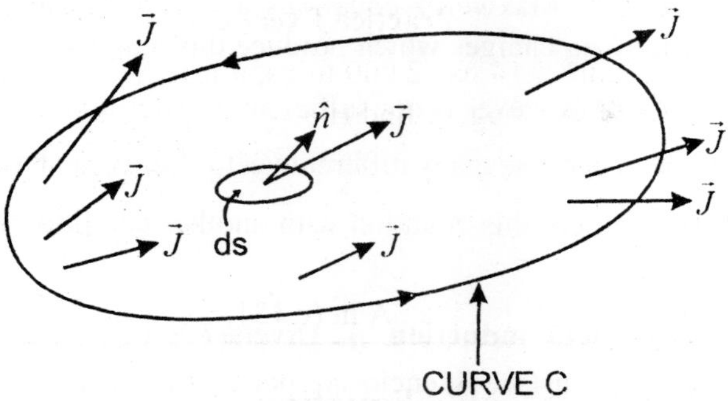

Fig. 6.15

Now consider any closed path 'C' in a region where currents are flowing as shown in Fig. 6.15. The total current enclosed by the curve C is the flux of current density $\vec{J}$ through the surface S enclosed by C. i, e.

$$i = \int_S \vec{J}.\hat{n}\, ds$$

when $\hat{n}$ is the unit vector normal to small area element ds and integral is taken over the entire surface S. Substituting this value of current in equation $\oint \vec{A}.\vec{d}l = \mu_0 i$, we get

$$\oint_C \vec{B}.\vec{d}\,l = \mu_0 \int_S \vec{J}.\,\hat{n}\, ds \tag{6.28}$$

By Stoke's theorem, we have

$$\oint_C \vec{H}.d\vec{l} = \int_S \left(\nabla \times \vec{H}\right).\,\hat{n}\, ds \tag{6.29}$$

Comparing (6.29) and (6.29), we get

$$\oint_C \left(\nabla \times \vec{B}\right).\hat{n}\, ds = \mu_0 \int_S \left(\vec{J}.\hat{n}\right) ds$$

or

$$\nabla \times \vec{B} = \mu_0 \vec{J} \tag{6.30}$$

In cgs m units equation (6.30) can be written as

$$\nabla \times \vec{B} = 4\pi \vec{J}$$

and in cgs e.s.u. units equation (6.30) can be written as

$$\nabla \times \vec{B} = \frac{4\pi}{c} \vec{J}$$

where c is the velocity of light in vacuum.

Equation (6.30) is one of Maxwell's equations and gives a simple relation between magnetic field B and the moving charges which produce this magnetic field.

Note that equation (6.30) however is not sufficient to determine the value of $\vec{B}$ (x, y, z) uniquely for given $\vec{J}$ (x, y, z) because many different vector fields can have the same curl. It is therefore necessary to supplement this equation with another condition $\nabla \times \vec{B} = 0$ which are going to derive below.

(ii) **Divergence of magnetic induction** $\vec{B}$**.** Divergence of a magnetic induction $\vec{B}$, is defined as the flux through the surface S, enclosing per unit volume i.e.

$$\text{Div. } \vec{B} = \underset{v \to 0}{\text{Limit}} \frac{flux}{v} = \underset{v \to 0}{\text{Limit}} \frac{\int_B \vec{B} \cdot \vec{ds}}{v}$$

The magnetic flux $\int_B \vec{B} \cdot \vec{ds} = 0,$ because of the fact (article 6.10) that the line of B are closed curves and do not have sources or there is magnetic monopole.

therefore $\qquad$ Div $\vec{B} = 0.$

or

$$\boxed{\nabla \cdot \vec{B} = 0.} \qquad (6.31)$$

Equations (6.31) and (6.32) help us to determine $\vec{B}$ uniquely if $\vec{J}$ is Known.

Equation (6.31) is known as Gauss's law in magnetism and states that the divergence of magnetic induction i.e. total outward flux of magnetic induction $\vec{B}$ through any closed surface is zero.

Alternative proof for ∇. B =0

(using Biot-Savart Law)

Let dB be the magnetic induction field produced at the pt. P(x, y, z) due to the current element idl at a distance 'r' from the point P [Fig. 6.15]. Then by using Biot Savart law, we get

$$d\vec{B} = \frac{\mu_0}{4\pi} \frac{id\vec{l} \times \vec{r}}{r^2}$$

The magnetic induction field at P due to the extra current isfound by integrating equation (6.32) over the entire loop, i.e.

$$\vec{B} = \frac{\mu_0}{4\pi} \oint \frac{id\vec{l} \times \vec{r}}{r^2}$$

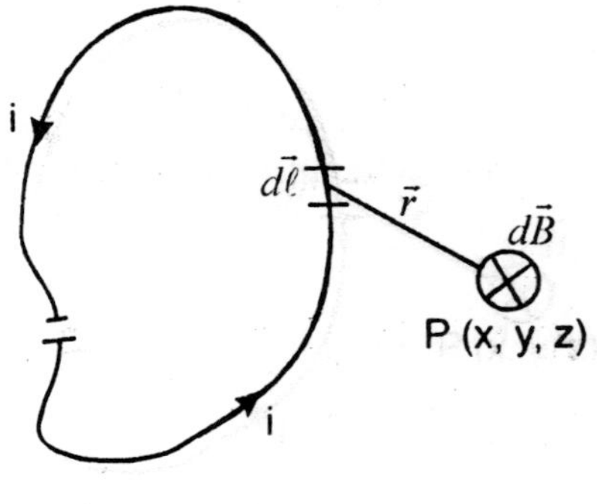

i
$d\vec{\ell}$
$\vec{r}$
$d\vec{B}$
P (x, y, z)
i

Fig 6.16

The direction of $\vec{B}$ (as given by right hand rule) is perpendicular to the plane of the paper and is directed inward.

Further taking divgence of $\vec{B}$, we have

$$\nabla.\vec{B} = \frac{\mu_0 i}{4\pi}\nabla.\oint\frac{d\vec{l}\times\vec{r}}{r^3} \tag{6.33}$$

As the integration in the equation (6.33) over the current top and divergence over $\vec{B}$ is at the point P(x, y, z) away from the curnet top, so the integration and divergence in equation (6.33) are independent of each other. Hence the integration and divergence opereations are interchangeable, so equation (6.33) can be written as

$$\nabla.\vec{B} = \frac{\mu_0 i}{4\pi}\nabla.\oint\frac{d\vec{l}\times\vec{r}}{r^3} \tag{6.34}$$

making use of the vector identity

$$\nabla.(\vec{A}\times\vec{B}) = \vec{B}.(\nabla\times\vec{A}) - \vec{A}.(\nabla\times\vec{B})$$

Equation (6.34) can be written as

$$\nabla.\vec{B} = \frac{\mu_0 I}{4\pi}\oint\left[\frac{\vec{r}}{r^3}.(\nabla\times d\vec{l}) - d\vec{l}.\left(\nabla\times\frac{\vec{r}}{r^3}\right)\right]$$

Now $\nabla\times$ $d\vec{l} = 0$ as $d\vec{l}$ is not a function of x, y, z the coordinates of point P, where we are to find $\nabla.\vec{B}$ and

$$\nabla\times\frac{\vec{r}}{r^3} = -\nabla\times\nabla\left(\frac{1}{r}\right)$$

because

$$\nabla\left(\frac{1}{r}\right) = -\frac{\vec{r}}{r^3}$$

Therefore equation (6.34) can be written as

$$\nabla.\vec{B} = -\frac{\mu_0 i}{4\pi}\oint d\vec{l}.\left(\nabla\times\nabla\left(\frac{1}{r}\right)\right)$$

$$\left(\because \nabla\times\nabla\left(\frac{1}{r}\right) = 0 \text{ i.e., curl of a gradient is always zero}\right)$$

$$\therefore \nabla.\vec{B} = 0$$

Now we can compare the relevant equation of magnetic field and electrostatic field as under

Comparrison of Electrostatic and Magnetic Fields

Electrostatic Field	*Magnetic Fields*
(i) Line integal along a closed path is zero $\oint \vec{E}.d\vec{l} = 0$	(i) Line integral along a closed path is given by $\oint \vec{B}.d\vec{l} = \mu_0 i$
(ii) Curl $\vec{E} = 0$	(ii) Curl $\vec{B} = \mu_0 \vec{j}$
(iii). Div $\vec{E} = \frac{\rho}{\varepsilon_0}$	(iii) Div $\vec{B} = 0$

Example 8. A wire shaped in the form of regular hexagon of side 1cm carries a current of 2 A. Calculate the strength of the magnetic field at the centre of the hexagon.

Solution. Let ABCDEF be a regular haxagon having each side equal to a = 1cm and the corners of the hexagon will be at a distance 'a' from the centre *O* (Fig. 6.17). If i ampere current is flowing through the hexagon, then by Biot Savart law the magnetic field at the centre O due to side AB is given by

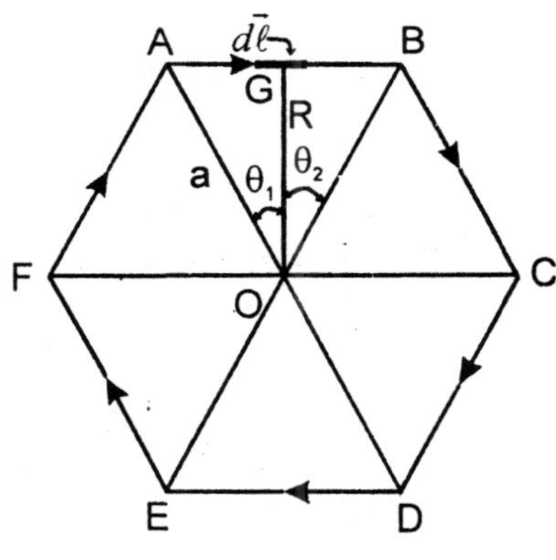

Fig. 6.17

$$\vec{B}_{AB} = \frac{\mu_0 i}{4\pi} \sum \frac{d\vec{l} \times \vec{r}}{r^3}$$

Here summation indicates the vector sum of the elements of the field $d\vec{B}$ due to the current element such as $d\vec{l}$ of AB.

The magnitude of magnetic field at O due to AB is

$$B_{AB} = \frac{\mu_0 i}{4\pi} \sum \frac{dl \sin\theta}{r^2}$$

$$= \frac{\mu_0 i}{4\pi R}(\sin\theta_1 + \sin\theta_2)$$

Here $\theta_1 = \theta_2 = 30°$

$$R = GO = OA \cos 30° = a \cos 30° = a\frac{\sqrt{3}}{2}$$

$$\therefore \quad B_{AB} = \frac{\mu_0 i}{4\pi \frac{a\sqrt{3}}{2}}(\sin 30° + \sin 30°)$$

$$= \frac{\mu_0 i}{2\pi a\sqrt{3}}\left(\frac{1}{2}+\frac{1}{2}\right) = \frac{\mu_0 i}{2\pi a\sqrt{3}}$$

Similarly the magnitude of the field intensity at O due to other side of the hexagon are

$$B_{BC} = B_{CD} = B_{DE} = B_{EF} = B_{FA} = \frac{\mu_0 i}{2\pi a\sqrt{3}}$$

With the direction of the current taken-in Fig. 6.17. we find by right hand screw rle, that the fields due to all the current element act in the same direction i.e. perpendicular to the plane of the hexagon at O.

Hence resultant field at O is $B = 6 \times B_{AB}$

$$= \frac{6\mu_0 i}{2\pi a\sqrt{3}} = \frac{\sqrt{3}\mu_0 i}{\pi a}$$

Here $i = 2A$

$a = 1 cm$

$\mu_0 = 4\pi \times 10^{-7}$ weber /m^2

$$\therefore \quad B = \frac{\sqrt{3} \times 4\pi \times 10^{-7} \times 2}{\pi \times 1 \times 10^{-2}} = 1.73 \times 8 \times 10^{-5}$$

$$= 13.84 \times 10^{-5} \text{ wb m}^{-2}$$

Example 9. A long wire has a semi-circular loop of radius r as shown in Fig. 6.18 and carries a current i. Find the magnetic induction field at the centre of curvature of the loop.

Solution. By Biot-Savart law the magnetic field at the centre O of the semi-circular loop of the wire due to a small current element idl is

$$d\vec{B} = \frac{\mu_0}{4\pi}\frac{id\vec{l} \times \hat{r}}{r^2}$$

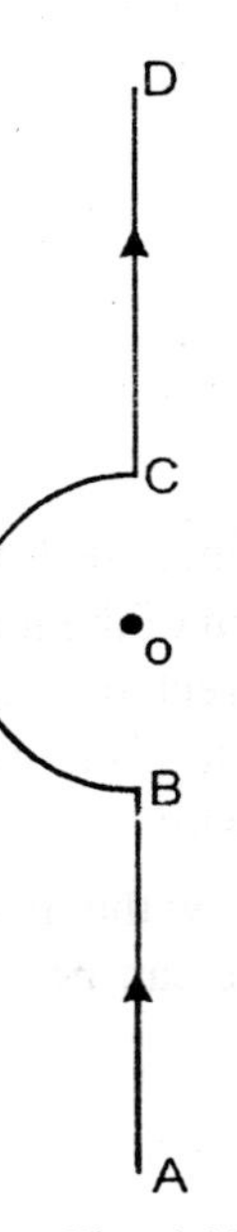

Fig. 6·18

where $\hat{r}$ is a unit vector from the element dl to the point O and r' is the distance of dl from O.

Further the angle between a current element idl of the upper straight portion of the wire and the unit vector $\hat{r}$ will be zero and the angle between a current element idl of the lower straight portion and the vector $\hat{r}$ will be 180°. Thus the cross product in the above experession for dB therefore shows that the magnetic field at O due to both the upper and lower straight portion of the the wire will be zero.

Angle between the current element idl of the semi-circular portion of the wire with $\hat{r}$ is $\frac{\pi}{2}$ and r' = r so the magnitude of the magnetic field at O due to a current element idl of the semi-circular portion will be

$$dB = \frac{\mu_0}{4\pi}\frac{idl}{r^2}$$

∴ The magnitude of the magnetic field at O due to the semi circular loop is

$$B = \int dB = \frac{\mu_0 i}{4\pi r^2}\int dl = \frac{\mu_0 i}{4\pi r^2}\pi r$$

$$= \frac{\mu_0 i}{4r} wb\ m^{-2}$$

($\because \int dl$ = circumference of semi-circle)

Example 10. Determine $\vec{B}$ at a point P due to a current i in a straight wire of length L which forms a part of a circuit.

Solution. Let CD be a straight wire of length L. P is the point of observation at a distance x from the mid-point of CD Fig. (6.19). Let dl be a small element at a distance l from O. Then by Biot Savart law the magnetic induction at P due to the element dl is

$$d\vec{B} = \frac{\mu_0 i}{4\pi}\frac{d\vec{l} \times \hat{r}}{r^2}$$

∴ Magnitude of $d\vec{B}$

$$dB = \frac{\mu_0 i}{4\pi}\frac{dl \sin\theta}{(l^2 + x^2)}$$

($\because$ from Fig. 6.12, $r^2 = x^2 + l^2$)

or $$dB = \frac{\mu_0 i}{4\pi} \frac{dl\, x}{\left(l^2 + x^2\right)^{3/2}} \qquad \text{(i)}$$

$$\left(\because \sin\theta = \frac{x}{\left(l^2 + x^2\right)^{1/2}}\right)$$

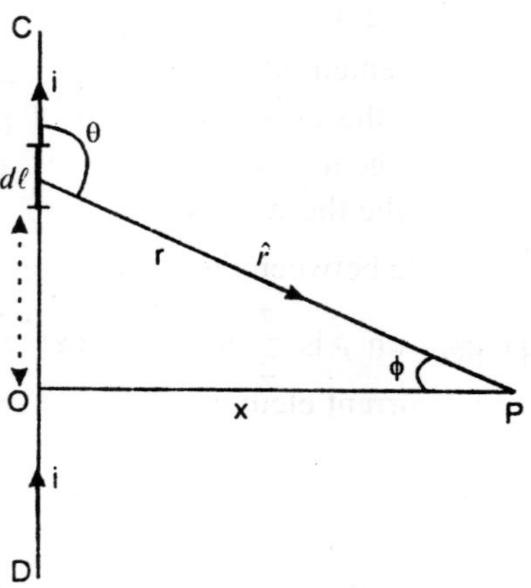

Fig. 6·19

Since vector d l × e is directed into the page, the direction of the magnetic induction due to this element is also directed into the page. Rather this would be the direction of the magnetic induction due to each element of the wire.

Now the magnetic induction due to entire length of the wire can be obtained by integrating (i) between the limits

$$l = -\frac{L}{2}, \quad l = \frac{L}{2}$$

$$\therefore \quad B = \int dB = \frac{\mu_0 i}{4\pi} \int_{-L/2}^{L/2} \frac{x\,dl}{\left(l^2 + x^2\right)^{3/2}}$$

$$= \frac{\mu_0 i}{4\pi} x \int_{-L/2}^{L/2} \frac{dl}{\left(l^2 + x^2\right)^{3/2}}$$

For solving $$\int \frac{dl}{\left(l^2 + x^2\right)^{3/2}}$$

Put $$l = x \tan\phi$$

or $$dl = x \sec^2\phi\, d\phi$$

$$\therefore \int \frac{dl}{\left(l^2 + x^2\right)^{3/2}} = \frac{x \sec^2\phi\, d\phi}{\left(x^2 \tan^2\phi + x^2\right)^{3/2}}$$

$$= \int \frac{x \sec^2\phi\, d\phi}{x^3\left(\tan^2\phi + 1\right)^{3/2}} = \int \frac{x \sec^2\phi\, d\phi}{x^3 \sec^3\phi}$$

$$= \frac{1}{x^2} \int \cos\phi\, d\phi = \frac{1}{x^2} \sin\phi = \frac{1}{x^2} \frac{l}{\left(l^2 + x^2\right)^{1/2}}$$

$$\therefore \quad \frac{1}{x^2}\left[\frac{l}{\left(l^2+x^2\right)^{1/2}}\right]_{-L/2}^{L/2} = \frac{1}{x^2}\left[\frac{L/2}{\left(\frac{L^2}{4}+x^2\right)^{1/2}} - \frac{-L/2}{\left(\frac{L^2}{4}+x^2\right)^{1/2}}\right]$$

$$= \frac{1}{x^2}\frac{L}{\left(\frac{L^2}{4}+x^2\right)^{1/2}}$$

Hence

$$B = \frac{\mu_0 i}{4\pi}x\frac{L}{x^2\left(\frac{L^2}{4}+x^2\right)^{1/2}}$$

$$= \frac{\mu_0 iL}{4\pi x\left(\frac{L^2}{4}+x^2\right)^{1/2}}$$

Magnetic Flux Density on the Axis of a Current loop and Magnetic Moment

Consider a coil of n turns and of radius r carrying current i amperes. Let P be a point at a distance x from the centre of the circular coil along its axis. Let the distance of P from any point on the circumference of the coil be 'a' as shown in Fig. 6.20. Consider two opposite current elements AB and CD each of length dl. Both these elements produce magnetic fields dB at P in directions as shown. These directions are perpendicular to the lines joining the mid points of the elements with the point P. Let θ be, the angle which the line joining P and any point on the circumference of coil makes with the axis of the coil. Resolve these fields into two

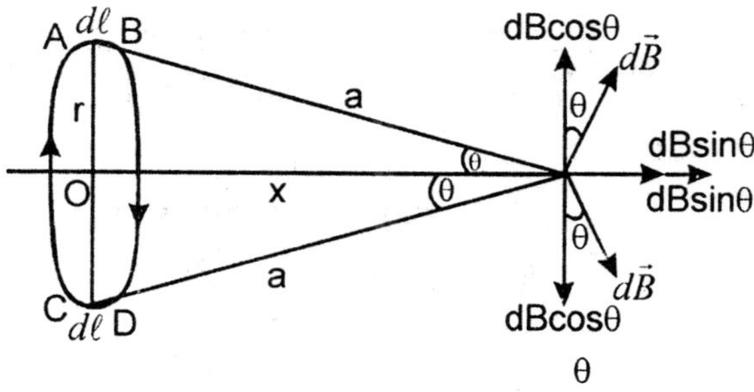

Fig. 6·20

perpendicular components. The components dB $\cos\theta$ perpendicular to the axis and due to the two opposite elements cancel each other but components dB $\sin\theta$ along the axis point out in the same direction. Therefore the magnetic field intensity at P due to each turn of coil is given by

$$B = \int dB \, Sin\,\theta$$

But $dB = \frac{\mu_0}{4\pi}\frac{idl}{a^2}$ [$\therefore$ P is at distance a from current element]

$$\therefore \quad B = \frac{\mu_0}{4\pi}\int \frac{idl \sin\theta}{a^3}$$

$$= \frac{\mu_0}{4\pi}\frac{i\sin\theta}{a^2}\int_0^{2\pi r} dl = \frac{\mu_0}{4\pi}\frac{i\,sin\theta 2\pi r}{a^2}$$

$$= \frac{\mu_0 i \sin\theta}{2a^2}$$

Now $$\sin\theta = \frac{r}{a}$$

$$B = \frac{\mu_0}{2a^2}\frac{ir}{}\frac{r}{a}$$

or $$B = \frac{\mu_0}{2a^3}\frac{ir^2}{}$$

If the coil consists of n turns, then

$$B = \frac{\mu_0}{2a^3}\frac{ir^2 n}{} wb\,m^{-2} \quad (6.35)$$

Now, from Fig (6.17), a = $(r^2 + x^2)^{1/2}$

$$\therefore \quad \boxed{B = \frac{\mu_0\, nir^2}{2\left(r^2 + x^2\right)^{3/2}} wb\,m^{-2}} \quad (6.36)$$

In cgs m units Eq. (6.51) is written as

$$B = \frac{2\pi nir^2}{\left(r^2 + x^2\right)^{3/2}}$$

or
$$B = \frac{\mu_0 niA}{2\pi\left(r^2 + x^2\right)^{3/2}} \tag{6.37}$$

where A = πr², the area of the current loop.

or
$$\boxed{B = \frac{\mu_0 M}{2\pi\left(r^2 + x^2\right)^{3/2}}} \tag{6.38}$$

where M=niA called magnetic dipole moment of the current loop. Hence a curretn loop is considered as a magnetic dipole which produces its own magnetic field.

Flux density at the centre of the current loop : At the centre of the current loop, $x = 0$. Hence the magnitude of the flux density at the centre of the loop having n turns is given by

$$B = \frac{\mu_0 nir^2}{2r^3} \text{ [from Eq. 6.36]}$$

$$\text{or } B = \frac{\mu_0 ni}{2r} \text{ wb/m}^2 \tag{6.39}$$

In cgs electromagnetic units, the magnetic flux density at the centre of the loop is given by

$$B = \frac{2\pi ni}{r} \text{ oersted} \tag{6.40}$$

Magnetic field lines due to a circular coil carrying current are shown in fig 6.21.

Flux Density Inside a Long Solenoid

A solenoid consists of a long cylindrical coil of insulated copper wire wound on a a cylindrical format as shown in Fig. 6.21. The length of the solenoid is large as compared to its radius. Let l be the length of the coil having n turns and r be its radius. Let a current i amperes flow through the coil. Due to the current flow, some magnetic field is set up. At point inside the solenoid, the magnetic field is parallel to the length of the solenoid and outside the solenoid the magnetic field is zero. The sectional view of the coil carrying current is shown in Fig. 6.22. The circles marked with dot indicate the current entering in the plane of the paper of the coil and circles marked with cross indicate the current comig out of the plane of the paper in the coil. consider a small element AB of the solenoid of length Dx which at a distance x from a point P at which the magnetic flux density is to be determined.

The number of turns per unit length of the solenoid = $\frac{n}{l}$

∴ The number of turns contained in element of length

$$dx = \frac{n}{l}dx.$$

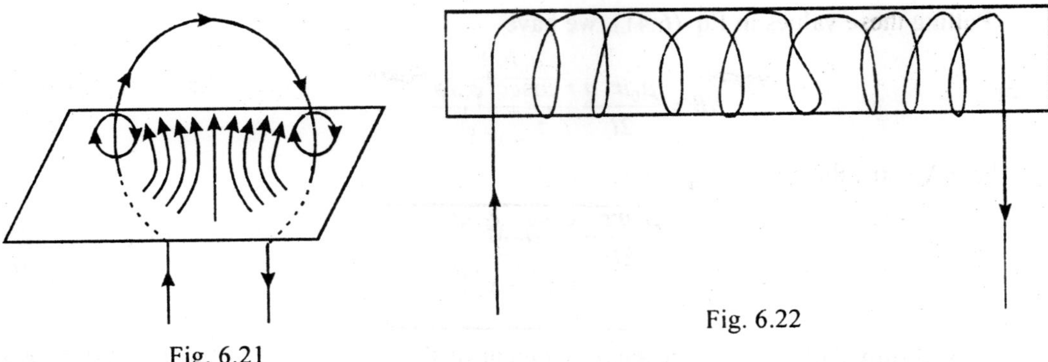

Fig. 6.21

Fig. 6.22

Let the element AB of the solenoid makes an angle ϕ with the axis of the coil. Let *PD* and *PC* are inclined at angles ϕ_1 and ϕ_2 with the axis respectively. Now the flux density at *P* due the current in the element *dx* is given by

$$dB = \frac{\mu_0 i r^2}{2} \cdot \frac{n dx}{l} \frac{1}{\left(r^2 + x^2\right)^{3/2}} \qquad \text{[Eq. 6.41]}$$

[see Eq. 6.46]

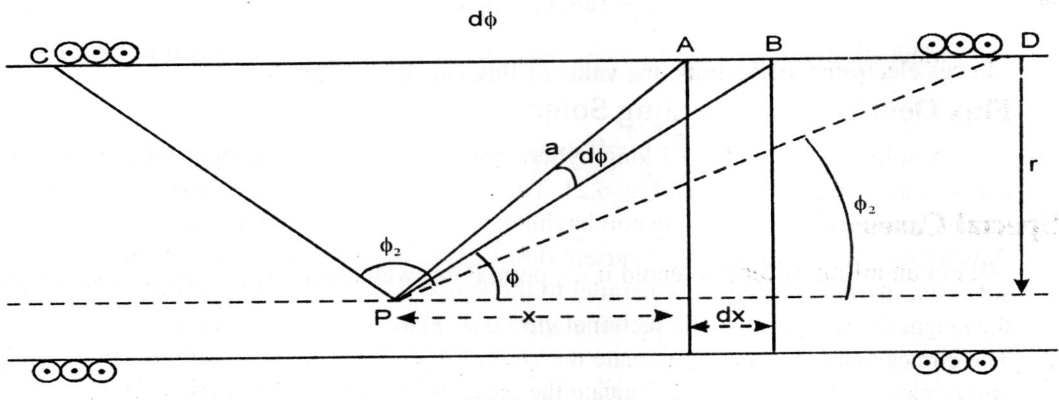

Fig. 6.23

Form Fig. 6.23. $\quad x = r \cot \phi$

$$dx = -r \operatorname{cosec}^2 \phi \, d\phi$$

Putting these values in Eq. (6.41), we have

$$dB = \frac{\mu_0 i r^2 n}{2l}\frac{r\,\mathrm{cosec}^2\phi\, d\phi}{\left(r^2+x^2\right)^{3/2}}$$

$$= -\frac{\mu_0 i n r^3}{2l}\frac{\mathrm{cosec}^2\phi\, d\phi}{a^2} \qquad [\because r^2 + x^2 = a^2]$$

$$= -\frac{\mu_0 i n}{2l}\sin^3\phi\,\mathrm{cosec}^2\phi\, d\phi \qquad \left[\because \sin\phi = \frac{r}{a}\right]$$

$$= -\frac{\mu_0\, i n}{2l}\sin\phi\, d\phi$$

The total flux density due to the entire length of the coil is given by

$$B = -\frac{\mu_0\, i n}{2l}\int_{\phi 2}^{\phi 1}\sin\phi\, d\phi$$

$$= \frac{\mu_0\, i n}{2l}\left[\cos\phi\right]_{\phi_2}^{\phi^1}$$

or

$$B = \frac{\mu_0\, i n}{2l}(\cos\phi_1 - \cos\phi_2) \qquad (6.42)$$

In cgs electromagnetic units, the value of flux density is given by

$$B = \frac{2\pi\, i n}{l}(\cos\phi_1 - \cos\phi_2) \qquad (6.43)$$

Special Cases

(i) For an infinitely long solenoid if the point P lies well inside from either end such that

$$\phi_1 = 0 \ and\ \phi_2 = \pi, \text{ than}$$

$$B = \frac{\mu_0\, i n}{2\ell}(\cos 0 - \cos\pi)$$

$$= \frac{\mu_0\, i n}{2l}(1+1)$$

or

$$\boxed{B = \frac{\mu_0\, i n}{l}} \qquad (6.44)$$

(ii) If the point P is taken near one end of a very long solenoid such that ϕ_1 =0 and ϕ_2 = 90°, then

$$B = \frac{\mu_0\, in}{2l}(\cos 0 - \cos 90^\circ)$$

$$= \frac{\mu_0\, in}{2l}(1-0)$$

or
$$B = \frac{\mu_0\, in}{2l} \qquad (6.45)$$

Hence it is clear from eqns. (6.44) and (6.45) that the magnetic flux density at a point near one end of a long solenoid is half of the flux density at its centre.

Example 11. Find the magnetic field due to a circular coil of radius 0.1m and having 200 number of turns (i) at a point at the centre from the centre. Current in the coil =5A.

Solution. Here n=200, r=0.1m,

x=.2m, i =5A,

$\mu_0 = 4\pi \times 10^{-7}$ wb/amp. Metre.

(i)
$$B = \frac{\mu_0\, ni}{2r}$$

$$= \frac{4\pi \times 10^{-7} \times 200 \times 5}{2 \times .1}$$

$$= 2\pi \times 10^{-3}$$

or
$$B = 6.28 \times 10^{-3} \text{ Tesla.}$$

(ii)
$$B = \frac{\mu_0\, ni\, r^2}{2(r^2 + x^2)^{3/2}}$$

$$= \frac{4\pi \times 10^{-7} \times 200 \times 5}{2(0.01 + 0.04)^{3/2}} \times \frac{1}{100}$$

or
$$B = 5.62 \times 10^{-4} \text{ Tesla.}$$

Example 12. A circular loop of wire has a radius of 7.cm and a source of 2.2 volts is made to send current through it. The wire has a resistance of 5 ohm per metre. Calculate the magnetic moment of the loop.

(K.U. 1976)

Solution. The length of the wire is equal to its circumference

$= 2\pi \times$ radius

$= 2\times\frac{22}{7}\times7 = 44\text{cm}$

$= \cdot 44$ m

Resistance per metre = 5 ohms

$\therefore$ Resistance of ·44m of wire

$= \cdot 44\times5$

= 2.2 ohms

Potential Difference =2.2 Volts

$\therefore$ Currnet $=\frac{2.2}{22}=1$ amp

Area of cross section of the loop

$= \pi r^2 = \frac{22}{7}\times7\times7$

=157 sq cm

$=154 \times10^{-4}$ m^2

The magnetic moment of loop M

= ni A

$= 1\times1\times154\times10^{-4}$

= 0.0154 S.I. units.

Example 13. Calculate the magnetic field along the axis of a long solenoid having 300 turns per metre when current is 10 A.

Solution. The magnetic field on the axis of a long solenoid is given by

$$B = \frac{\mu_0 i\, n}{l}$$

where n is the total number of turns and l is the length of the solenoid. The number of turns per unit length is given by $\frac{n}{l}$.

Here $\mu 0 = 4 \times 10^{-7}$ wb/amp-metre

i =10 amp

$\frac{n}{l}=300$

$\therefore$ $B = 4\pi \times 10^{-7} \times 10 \times 300$ or $B = 3.768 \times 10^{-3}$ Tesla.

Diplacement Current

We know that Ampere's circuital law in its most gneral from is given by

$$\oint_C H.dl = \int_S J.ds \qquad \text{[see equation (6.26)]}$$

i.e

$$\int_S curl\ H.ds = \int_S J.ds$$

or

$$\text{Curl H} = \text{J} \qquad (1)$$

Let us now examine the validity of this equation in the event that the fields are allowed to very with time. If we take the divergence of both sides of equation (1) then

$$\text{div (curl H)} = \text{div J}. \qquad (2)$$

Now as div of curl of any vector is zero, we get from equation (2)

$$\text{div J} = 0 \qquad (3)$$

In note we have shown that for electrostatic effects a conductor acts like a meterial of infinite dielectric constant. However in case of steady current as polarisation effects are completely overshadowed by dispersion of metals this result does not hold good. For purpose of estimation, in case of conduction through metal we usually take $Er \to 1$

Now the continuity equation in general states

$$div\ J = -\frac{\partial \rho}{\partial t} \qquad (4)$$

and will therefore vanish only in the special case that the charge density is static. Consequently we must conclude that Ampere's law as stated in equation (1) is valid only for steady state conditions and is insufficient for the case of time-dependent fields. Because of this Maxwell assumed that equation (1) is not complete but should have something else to it. Let this 'something' be denoted be J_d, then equation (1) can be rewritten as

$$\text{curl H} = J + J_d \qquad (5)$$

In order to identify Jd, we calculate the divergence of equation (2) again and get

$$\text{div curl H} = \text{div}\ (J+J_d)$$

i.e.

$$\text{div}\ (J + J_d) = 0 \qquad \text{(as div curl H = 0)}$$

or

$$\text{div J} + \text{div}\ J_d = 0$$

or

$$\text{div}\ J_d = -\ \text{div J}$$

i.e.

$$\text{div}\ J_d = \frac{\partial \rho}{\partial t} \qquad \text{[from equation (4)]}$$

i.e.

$$\text{div Jd} = \frac{\partial}{\partial t}\text{(div D)} \qquad [\text{ as div D} = \rho]$$

i.e.
$$\text{div}\left(J_d \frac{\partial D}{\partial t}\right)=0 \tag{6}$$

As equation (6) is true for any arbitrary volume

$$J_d \frac{\partial D}{\partial t} \tag{A}$$

And so the modified form of Ampere's circuital law becomes

$$\text{curl H} = J+\frac{\partial D}{\partial t} \tag{B}$$

The term which Maxwell added to Ampere's law viz. $(\partial D/\partial t)$ is called the displacement current to distinguish it from J, the conduction current. By adding this term to Ampere's law, Maxwell assumed that the time reate of change of displacement produces a magnetic field just as a conduction current does.

Regarding Displacement Current it is Worthy to Note That

(i) Displacement current is a current only in the sense that it produces a magnetic field. It has none of the other properties of current. for example displacement current can have a finite value in perfect vacuum where there are no charges of any type.

(ii) The magnitude of the displacement current is equal to the time rate of change of electric displacement vector D.

(iii) Displacement current serves to make the total current continuous across discontinuties in conduction current. (See example 2 and problem 4).

(iv) The displacement current in a good conductor is negligible as compared to the condition current at any frequency lower than the optical frequencies ($\sim 10^{-15}$ Hertz). (See example 3).

(v) The addition of displacement current i.e. $(\partial D/\partial t)$ to Ampere's law i.e. curl H = J results in

$$\text{curl } H = J+\frac{\partial D}{\partial t}$$

i.e. displacement current relates the electric field vector E (as $D=\varepsilon E$) to the magnetic field vector H. This in turn implies that in case the dependent field is not posible to deal with electric and magnetic fields separately, but the two fields are interlinked and give rise to what are know as electromagnetic fields i.e. The addition of displacement current to Ampere's law result, in the unification of electric and magnetic phenomena.

It must be emphasized here that the ultimate justification for Maxwell's assumption of displacement current is in the experimental verification. Indeed the effects of the displacement current are difficult to observe directly except at very high frequencies. However indirect varification is afforded by predictions of many effects particularly in electromagnetic theory of light which are confirmed by experiments. We may therefore consider that Maxwell's form of Ampere's law has been subjected to experiement tests and has been found to be generally valid.

Note : Different Types of Current Densities—

(a) ***Conduction current density J*:** It arise due to the physical motion of true charge and is given by

$$J = \sigma E = nqv$$

(b) ***Polarisation current density* Jp:** It arises due to the polarisation of the material and is given by

$$Jp = (\partial P / dt)$$

(c) ***Magnetisation Currnet density* J_M:** It arises due to the magnetisation of the material and is given by

$$J_M = (\text{curl } M)$$

(d) ***Displacement current density* J_d :** It makes conduction current contiouous across as discoutinuity and given by

$$J_d = (\partial D / dt)$$

(e) ***Convection Current density* J_c:** It arises due to motion of a charged medium as a whole and is given by

$$J_c = \rho v$$

where ρ is the density of charge and v is the velocity of the medium. This current vanishes if the medium is stationary.

Example 14. (a) Apply Ampere's law to the closed curve C bounded by surface S in Fig. 6.24. (A) and (B), (b) Show that the circuit (A) and (B) yields the same result if we include the displacement current in Ampere's law.

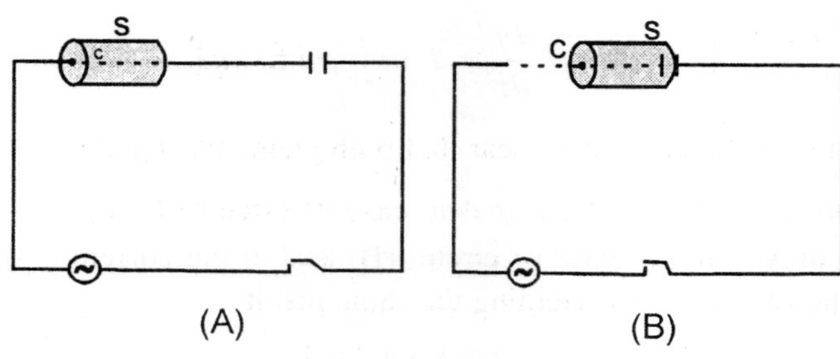

Fig. 6·24

Solution. (a) According to Ampere's circuital law

$$\oint_C H.dl = \int_S J.ds.$$

So for the circuit (A)

$$\oint_C H.dl = J.s = I \textbf{ Ans.}$$

And for the circuit (B)

$$\oint_C H.dl = \int_S 0.ds = 0. \textbf{ Ans.}$$

(b) If in Ampere's Law we include displacement current, the law becomes

$$\oint_C H.dl = \int_S \left(J + \frac{\partial D}{\partial t} \right) \bullet ds.$$

And for the circuit (A) as $\frac{\partial D}{\partial t} = 0$ it yields

$$\int_C H \bullet dl = \int J \bullet ds = I$$

While for the circuit (B) as J=0 it yields

$$\int_C H \bullet dl = \int_S \frac{\partial D}{\partial t} .ds$$

i.e.

$$\oint_C H \bullet dl = \frac{d}{dt} \int_S D \bullet ds$$

or

$$\int_C H \bullet dl = \frac{d}{dt} \int_\tau div\ D\, d\tau$$

$$\left(as \int_S A \bullet ds = \int div\ A d\tau \right)$$

or

$$\int_C H \bullet dl = \frac{d}{dt} \int_\tau \rho\, ds (as\ div\ D = \rho)$$

or

$$\oint_C H \bullet dl = \frac{dq}{dt} = I \left(as \int_\tau \rho d\tau = q \right) \quad (2)$$

From equation (1) and (2) it is clear that both yields the same result.

Note : From part (b) it is clear that in case of circuit (A) $\partial D / \partial t = 0$ the current is of conduction type only while in case of circuit (B) as J=0 the current is of displacement type only. Futher as the two cases are yielding the same result

$$J = J_d \text{ i.e. } I = I_d.$$

i.e. conduction current is equal to displacement current.

Example 15. If alternating field $E = E_0 \cos \omega t$ is applied to a conductor, show that the displacement current is negligible as compared to conduction current at any frequency lower than optical frequencies.

Solution. As the given fields is

$$E = E_0 \cos \omega t \tag{1}$$

And according to Ohm's law

$$J = \sigma E \tag{2}$$

So the conduction current will be given by

$$J = \sigma E_0 \cos\omega t. \tag{3}$$

However the displacement current will be

$$J_d = \frac{\partial D}{\partial t} = \frac{\partial}{\partial t}(\varepsilon E) \qquad [as\ D =_{\varepsilon} E]$$

i.e
$$J_d = \varepsilon_r \varepsilon_0 \frac{\partial}{\partial t}(E_0 \cos\omega t) \qquad [as\ \varepsilon = \varepsilon_r \varepsilon_0]$$

i.e.
$$J_d = -\omega\varepsilon_r \varepsilon_0 E_0 \sin\omega t$$

$$J_d = \omega\varepsilon_r \varepsilon_0 E_0 \cos\left(\omega t + \frac{\pi}{2}\right). \tag{4}$$

So from equation (3) and (4) it is clear that

$$\frac{J_d}{J} = \frac{\omega\varepsilon_r E_0 \cos\left(\omega t + \frac{\pi}{2}\right)}{\sigma E_0 \cos\omega t}$$

$$\left|\frac{J_d}{J}\right| = \frac{\omega\varepsilon_r \varepsilon_0}{\sigma} \tag{5}$$

But as for a good conductor

$$\varepsilon_r \approx 1^* \text{ and } \sigma \approx 10^7 \text{ ohm/meter}$$

$$\left|\frac{J_d}{J}\right| \approx \frac{2\pi \times f \times 9 \times 10^{-12}}{10^7} \sim f \times 10^{-17}$$

i.e. the displacement current in a good conductor is completely neligible compared to the conduction current at any frequency lower than optical frequencies (fop ~ 10^{15} Hertz).

Note : It is interesting to note that although conduction current (equation 3) is in phase with the electric field intensity (equation1), the displacement current (equation 4) leads the electric field by π/2 radians i.e. displacement current leads the conduction current by π/2 radians. In note (ii) of application (d) in § 2.3 we have shown that for electrostatic effects a conductor acts like a material of infinite dielectric constant. However in case of steady current as polarisation effects are completely overshadowed by dispersion of metals this result does not hold good. For purpose of estimation, in case of conduction through metal we usually take $\varepsilon_r = n^2 \rightarrow 1$.

Maxwell's Equations

(A) The equation :

There are four fundamental equation of electromagnetism and corresponds to a generalisation of certain experiement observations-regarding electricity and magnetism. The following for laws of electricity and magnetism constitutes the so called *'differential form'* of **Maxwell's** equation :

(i) Gauss' law for the electric field of charge yields

$$\text{div } D = \nabla.D = \rho$$

where D is electric displacement in coulombs/m^2 and ρ is the free charge density in coul/m^3.

(ii) Gauss' law for magnetic field yields

$$\text{div } B = \nabla.B = 0$$

where B is the magnetic induction in web/m^2.

(iii) Ampere's law in circuital form for the magnetic field accompanying a current when modified by Maxwell yields

$$\text{Curl } H = \nabla \times H = J + \frac{\partial D}{\partial t}$$

where H is the magnetic field intensity in amperes/m and J is the current density in amp/m^2.

(iv) Faraday's law in circuital form for the induced electromotive force produced by the rate of change of magnetic flux linked with the path yields

$$\text{Curl } E = \nabla \times E = -\frac{\partial B}{\partial t}$$

where E is the electric field intensity in Volts/*m*.

(B) Derivations

(i) Let us consider a surface S bounding a volume τ within a dielectric. Originally the volume τ contains no net charge but we allow the dielectric to be polarised say by placing it in an electric field. We also deliberately place some charge on the dielectric body. Thus we have two type of charges:

(a) real charge of density ρ (b) bound charge density ρ'.

Gauss' law then can be written as

$$\oint_S Eds = \frac{1}{\varepsilon_0} \int (\rho + \rho')d\tau$$

i.e $$\varepsilon_0 \oint_S E.ds = \int_\tau \rho\, d + \tau \int \rho'\, d\tau. \tag{i}$$

But as the bound charge density ρ'□ is defined as ρ' = –div P and

$$\oint_S E.ds = \int_\tau div\, E\, d\tau$$

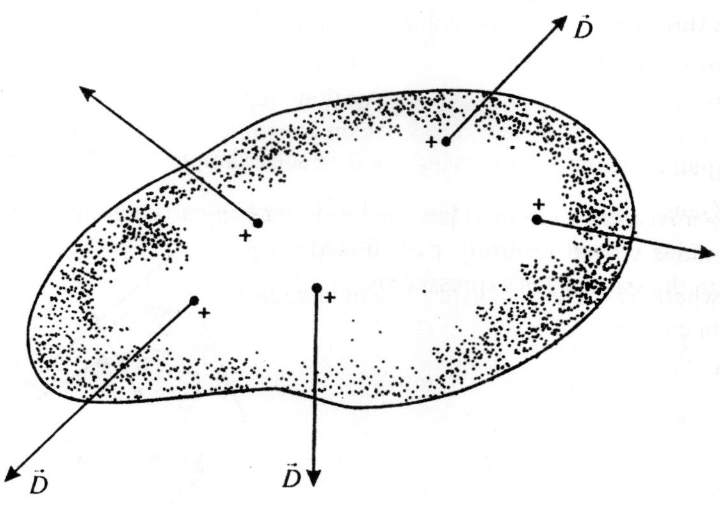

Fig. 6·25

So equation (1) becomes

$$\varepsilon_0 \int_\tau div\, E\, d\tau = \int_\tau \rho\, d\tau - \int_\tau div\, P\, d\tau$$

or $$\int_\tau div(\varepsilon_0 E + P) d\tau = \int_\tau \rho\, d\tau$$

or $$\int_\tau div\, D\, d\tau = \int_\tau \rho\, d\tau \quad (\text{as } D = \varepsilon_0 E + P)$$

or $$\int_\tau (div\, D - \rho)\, d\tau = 0.$$

Since this equation is true for all volumes, the integrand must vanish. Thus we have

div D = ∇. D = ρ.

(ii) Experiments to-date have shown that magnetic monopoles do not exist. This in turn implies that the magnetic lines of force are either closed group or go off to infinity. Hence the number of magnetic lines of force entering any arbitrary closed surface

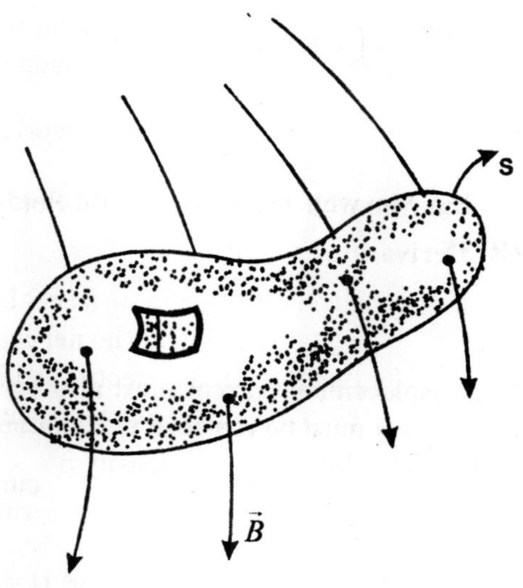

Fig. 6·26

is exactly the same leaving it. Therefore the flux of magnetic induction B across any closed surface is always zero.

Transforming this surface integral into volume integral by Gauss' theorem, we get

$$\int_{\tau} div\, B\, d\tau = 0.$$

But as the surface bounding the volume is quite arbitrary the above equation will be true only when the integrand vanishes i.e.

$$\text{div B} = \nabla.\ \text{B}=0. \qquad \text{(B)}$$

Note : For alternative methods of proving div B=0 see example 1 in chapter 3.

(iii) From Ampere's circuital law the work done in carrying unit magnetic poie once round a closed arbitrary path linked with the current I is expressed by

$$\oint_{C} H.dI = I$$

$$i.e.\ \oint_{C} H.dI = \int_{S} J \bullet ds$$

(as $I = \int J \bullet ds$)

where S is the surface bounded by the closed path C.

Now changing the line integral into surface integral by Stoke's theorem, we get

$$\int_{S} Curl\ H \bullet ds = \int_{\varepsilon} J \bullet ds$$

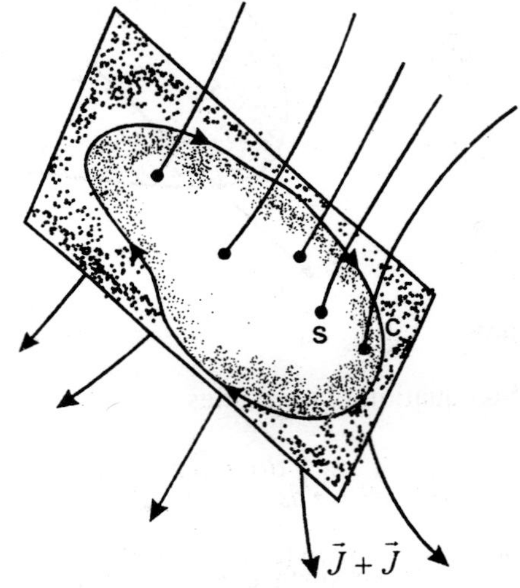

Fig. 6·27

i.e

$$\text{curl H=J.} \qquad (2)$$

But Maxwell found it to be incomplete for changing electric fields and assumed that a quantity

$$J_d = \frac{\partial D}{\partial t}$$

called displacement current must also be included into so that it may satisfy the continuity equation i.e J must be replaced in equation (2) by J+J_d so that the law becomes

$$\text{curl H} = \text{J} + \text{J}_d$$

i.e.

$$\text{curl H} = \text{J} + \frac{\partial D}{\partial t} \qquad \text{(C)}$$

(iv) According to Faraday's law of electromagnetic induction we know that the induced e.m.f. is proportional to the rate of change of flux i.e.

$$\in = -\frac{d\phi_B}{dt}$$

Now if E be the electric intensity at a point the work done in moving a unit charge through a small distance dI is E-dI. So the work done in moving the unit charge once round the circuit is $\oint_C E \bullet dI$. Now as e.m.f. is defined as the amount of work done in moving a unit charge once round the electric circuit.

$$\in = \oint_C E.dI \qquad (4)$$

So comparing equation (3) and (4), we get

$$\oint_C E \bullet dI = -\frac{d\phi_B}{dt}^*. \qquad (5)$$

But as $\phi_B = \int_S B \bullet ds$

So, $\oint_C E \bullet dI = -\frac{d}{dt}\int B \bullet ds.$

Transformating the line integral by Stoke's theorem into surface intrgral we get

$$\int_S curl\, E \bullet ds = -\frac{d}{dt}\int B \bullet ds.$$

Assuming that surface S is fixed in space and only B changes with time, above equation yields

$$\int_S \left(curl\, E + \frac{\partial B}{\partial t} \right) \bullet ds = 0$$

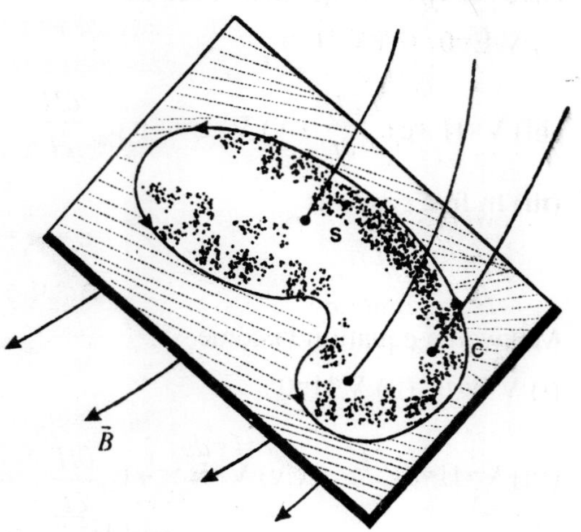

Fig. 6·28

*This equation shows that a changing magnetic field produced an electric field. However it has also been found that a changing electric field also produces a magnetic field. The magnetic field produced by changing electric field is given by

$$\oint_C B \bullet dI = \mu_0 \varepsilon_0 \frac{d\phi_E}{dt} = \frac{1}{c^2}\frac{d\phi_E}{dt} \qquad [\text{as } \mu_0 \varepsilon_0 = 1/c^2]$$

As the above integral is true for any arbitrary surface the integrand must vanish.

i.e

$$\text{curl E} = -\frac{\partial B}{\partial t}. \qquad (D)$$

(C) Particular Cases

(i) In a conducting medium of relative permitivity ε_r and permeability μ_r as

$$D = \varepsilon E = \varepsilon_r \varepsilon_0 E$$

And $B = \mu H = \mu_r \mu_0 H$.

Maxwell's equation reduce to

(i) $\nabla \cdot E = \rho / \varepsilon_r \varepsilon_0$ (ii) $\nabla \cdot H = 0$

(iii) $\nabla \times H = J + \varepsilon_r \varepsilon_0 \frac{\partial E}{\partial t}$ (iv) $\nabla \times E = -\mu_r \mu_0 \frac{\partial H}{\partial t}$.

(ii) In a non-conducting media of relative permitive ε_r and permeability μ_r as μ_r

$$\rho = \sigma = 0$$

So $$J = \sigma E = 0$$

and hence Maxwell's equations become

(i) $\nabla \cdot E = 0$ (ii) $\nabla \cdot H = 0$

(iii) $\nabla \times H = \varepsilon_r \varepsilon_0 \frac{\partial E}{\partial t}$ (iv) $\nabla \times E = -\mu_r \mu_0 \frac{\partial H}{\partial t}$.

(iii) In free space as

$$\varepsilon_r = \mu_r = 1$$

$$\rho = \sigma = 0.$$

Maxwell's equation become

(i) $\nabla \cdot E = 0$ (ii) $\nabla \cdot H = 0$

(iii) $\nabla \times H = \varepsilon_0 \frac{\partial E}{\partial t}$ (iv) $\nabla \times E = -\mu_0 \frac{\partial H}{\partial t}$

(D) Discussion

1. These equations are based on experimental observation. The equations: (A) and (C) correspond to electricity while (B) and (D) to magnetism.
2. These equations are general and apply to all electromagnetic phenomena in media which are at rest w.r.t. the coordinate system.
3. These equation are not independent of each other as from equation (D) we can derive (B) and from (C), (A) (See example-4). This is why equation (B) and (D) are called the first pair of Maxwell's equations while (A) and (C) are called the second pair.
4. The equation (A) represents Coulomb's law while (C) the law of conservation of charge i.e continuity equation (see example 5).
5. If we compare equation (A) with (B) and (C) with (D) we find that left hand sides are identical while right hand sides are not. This in turn implies that electric and magnetic

phenomena are assymmetric and this asymmetry arises due to the non-existance of monopoles.

Note: This asymmetry of electro-magnetism suggests that monopoles (a particle having either north or south magnetic charge) should exist as the concept of magnetic monopoles would bring to electricity and magnetsm a symmetry to which nature loves and is lacking in our present picture. Dirac has also proved on theoretical grounds that monopole should exist and presicted their properies. But so far the magnetic monopole has frustated all its investgators. The experiments have failed to find any sign of these. The theorists on the other hand have failed to find any sign of these. The theorists on the other hand have failed to find any good reason why monopoles should not exist.

Recently, American Institute of Physics and the University of California at Berkley jointly announced that monopole has been observed by a group of physists. If confirmed, the derection of monopoles will have a major on Physics and Technology.

(6) The correspondance of B and H with E and D through Maxwell equation (D) and (C) repectively implies that in case of time dependent fields the electric and magnetic fields are inseparably linked with each othe rgiving rise to what is known as electromagnetic field and it is not possible to deal separately with electric and magnetic fields in this situation.

(E) Physical Significance (or intehral Form)

By means of Gauss' and Stoke's Theorems we can write the Maxwell's field equations in integrl form and hence obtain their physical significance.

(i) Integrating Maxwell's first equation div D=ρ over an arbitrary volume τ we get

$$\int_\tau \nabla \bullet D\, d\tau = \int_\tau \rho\, d\tau$$

changing the vol. integral of L.H.S. into surface integral by Gauss' divergencetheorem and keeping in mind that $\int \rho\, dr = q$ we get

$$\oint D \bullet ds = q$$

So Maxwell's first equation signifies that the total *flux of electric displacement linked with a closed surface is equal to the total charge enclosed by the closed surface.*

(ii) Integrating Maxwell's second equation div B = 0 over an arbitrary vol. τ we get

$$\int_\tau \nabla \bullet B\, dr = 0.$$

Converting the vol. integral into surface integral with the help of Gauss' theorem we get

$$\oint B \bullet ds = 0 \qquad (B_1)$$

So Maxwell's II equation signifies *that the total flux of magnetic induction linked with a closed surface is zero.*

(iii) Integrating Maxwell's III equation curl H=J+$(\partial D / \partial t)$ over a surface S bounded by the loop C we get

$$\int curl\, H \bullet ds = \int\left(J + \frac{\partial D}{\partial t}\right)\bullet ds \qquad (C_1)$$

which signifies that *magnetomotive force around a closed path* $\left[\oint H \bullet dl\right]$ *is equal to the conduction current plus displacement current linked with that path.*

(iv) Integrating Maxwell's IV equation curl E = – ($\partial B / \partial t$) over a surface S bounded by the loop C we get

$$\int curl\, E \bullet ds = -\int \frac{\partial B}{\partial t}.ds$$

Converting the surface integral of L.H.S. into line integral with the help of Stoke's Theorem we get

$$\oint E \bullet dI = -\frac{\partial}{\partial t}\int B.ds \qquad (D_1)$$

which signifies that *the electromotive force i.e. line integral of electric intensity around a closed path is equal to the negative rate of change of magnetic flux linked with the path.*

Example 4. Starting with Maxwell's equations

$$curl\, E = -\frac{\partial B}{\partial t} \text{ and curl } H = J + \frac{\partial D}{\partial t}$$

respectively show that

div B =0 and div E = ρ

Solution. Taking the devergence of given Maxwell's equations we get

$$\text{div curl E} = -\text{ div}\frac{\partial}{\partial t}\text{ (B) and div curl H} = \text{div}\left(J + \frac{\partial D}{\partial t}\right)$$

Now as the div curl of any vector is zero and space and time operations are interchangable, above equations reduce to

or $$\frac{\partial}{\partial t}\text{(div B)} = 0 \text{ and } \frac{\partial}{\partial t}\text{(div D)}\ \frac{\partial \rho}{\partial t}$$

$$\left(as\ div\, J + \frac{\partial \rho}{\partial t} = 0\right)$$

Now if for each point of space div B and div D become zero at any time either in the past or in future, above equation on integration yields

div B = 0 and div D = ρ

Example 5. Starting from Mawell's equations prove (a) Coulomb's, iow, (b) Contnuity equation.

Solution. (a) From Maxwell's first equation we have

$$\text{div } D = \rho \tag{1}$$

Integrating this equation over a sphere of radius r we get

$$\int_\tau divD\, d_\tau = \int_\tau \rho\, dr$$

i.e.
$$\oint_S D \cdot ds = q$$

i.e.
$$\varepsilon_0 \oint_S D \cdot ds = q \quad (\text{as } D = \varepsilon_1 \varepsilon_2\ E = \varepsilon_0 E)$$

i.e.
$$\varepsilon_0 E(4\pi r^2) = q$$

i.e.
$$E = \frac{1}{4\pi\varepsilon_0} \frac{q}{r^2}$$

i.e.
$$E = \frac{1}{4\pi\varepsilon_0} \frac{q}{r^3} r \quad (\text{as E is radial}) \tag{2}$$

So the force experienced by a test charge q_0

$$F = q_0 E = \frac{1}{4\pi\varepsilon_0} \frac{q q_0}{r^2} r$$

This is the required result.

(b) From Maxwell's third equation we have

$$\text{Curl } H = J + \frac{\partial D}{\partial t}.$$

Now as the div curl of any vector is zero, above equation reduces to

$$\text{div}\left(J + \frac{\partial D}{\partial t}\right) = 0$$

i.e.
$$\text{div } J + \text{div}\left(\frac{\partial D}{\partial t}\right) = 0$$

i.e.
$$\text{div } J + \frac{\partial D}{\partial t}(\text{div } D) = 0 \tag{4}$$

or $$\text{div } J+\frac{\partial \rho}{\partial t}=0 \text{ (as from eqn. (1) } \nabla \bullet D = \rho) \quad (5)$$

This is the required result.

Energy in Electromagnetic Fields. (Poynting's Theorem)

Form Maxwell's equations it is possible to derive an important expression which we shall recognise at the energy prinecipple in an electromagnetic field.

For this consider Maxwell's equations (c) and (D) i.e. Ampere's and Faraday's laws in differential forms

$$\text{curl } H = J+\frac{\partial D}{\partial t} \quad (1)$$

and $$\text{curl } E = -\frac{\partial B}{\partial t}. \quad (2)$$

If we take the scalar product of equation (1) with E and of equation (2) with (-H) we get

$$E.\text{ curl } H = E\cdot J + E\cdot\frac{\partial D}{\partial t} \quad (3)$$

$$-H\cdot\text{curl } E = +H\bullet\frac{\partial B}{\partial t} \quad (4)$$

adding equations (3) and (4) we get

$$-H\cdot\text{curl } E + E\cdot\text{curl } H = J\cdot E + \left[E\bullet\frac{\partial D}{\partial t}+H\bullet\frac{\partial B}{\partial t}\right]$$

But by the vector idenity

$H\cdot\text{curl } E - E\cdot\text{curl } H = \text{div } (E\times H)$

The above equation reduced to

$$-\text{div } (E\times H) = J\cdot E + \left[E\bullet\frac{\partial D}{\partial t}+H\bullet\frac{\partial B}{\partial t}\right] \quad (5)$$

Now as $E\bullet\frac{\partial D}{\partial t} = \varepsilon_r\varepsilon_0\, E\bullet\frac{\partial E}{\partial t} = \frac{1}{2}\,\varepsilon_r\varepsilon_0\frac{\partial}{\partial t}(E.E) = \frac{1}{2}\,\frac{\partial}{\partial t}(E.D)$

and $H\bullet\frac{\partial B}{\partial t} = \mu_r\mu_0\, H\bullet\frac{\partial H}{\partial t} = \frac{1}{2}\,\mu\mu_0\frac{\partial}{\partial t}(H.H) = \frac{1}{2}\,\frac{\partial}{\partial t}(H.B)$

So equation (5) reduces to

$$J.E+\tfrac{1}{2}\frac{\partial}{\partial t}(E.D+H.B)+div(E\times H)=0. \tag{6}$$

Each term in the above equation can be given some physical meaning if it is multiplied by an element of volume dτ and integrated over a volume τ whose enclosing surface is S. Thus the result is

$$\int_\tau (J.E)d_\tau+\int_\tau \tfrac{1}{2}\frac{\partial}{\partial t}(E.D+H.B)dr+\oint_\tau div(E\times H)d\tau=0$$

But as $$\int_\tau div(E\times H)d\tau=\oint_S (E\times H)\bullet ds$$

$$\int_\tau (J.E)d_\tau+\int_\tau \tfrac{1}{2}\frac{\partial}{\partial t}(E.D+H.B)dr+\oint_\tau div(E\times H)d\tau=0 \tag{A}$$

To current distribution represented by the vector J can be considered as made up of various changes qt moving with velocity vt so that

$$\int J\bullet E\,dr=\int IdI\bullet E \quad [\text{as } J\,dr=I\,dI]$$

$$=\int dq\,v\bullet E \quad [\text{ as } I\,dI=(dq/dt)\,dI=dq\,v]$$

$$=\sum qi(vi\bullet Ei) \tag{7}$$

where E_i denotes the electric field at the position of charge q_i given by the *Lorentz expression*

$$F_i=q_i(E_i+V_i\times B_i).$$

So the work done per unit time on the charge qi by the field will be

$$\frac{\partial W_i}{\partial t}=F_i\bullet v_i \quad \left[\frac{dW}{dt}=\frac{F.dI}{dt}=F.v\right]$$

$$=q_i\,(E_i+v_i\times B_i)\cdot v_i \;(\text{ as } Fi=qi\,(Ei+vi\times Bi)$$

i.e. $$\frac{\partial W_i}{\partial t}=qi\;vi\bullet Ei \qquad [\text{as } vi\cdot(vi\times Bi)=(vi\times vi)\cdot Bi=0]$$

So the rate at which the work is doen by the field on the charges is

$$\frac{\partial W}{\partial t}=\sum\frac{\partial W_i}{\partial t}=\sum q_i v_i\bullet E_i. \tag{8}$$

Comparing equation (7) and (8) we find that

$$\int J\bullet E\,dr=\frac{dW}{dt} \tag{9}$$

i.e. the first term $\int (J \cdot E)\, dr$ *represents the rate at which work is doen by the field on the charges*

Note: It is worthly to note here that–

(i) In case of charged particles moving in free space with no external force acting, the work done by the field on the charges appears as kinetic energy of the particles as

$$\frac{dW}{dt} = \sum \frac{\partial W_i}{\partial t} = \sum F_i \cdot V_i$$

i.e
$$\frac{\partial W}{\partial t} \sum mi \frac{\partial V_i}{\partial t} \cdot vi \left(as Fi = mi \frac{\partial v_i}{\partial t} \right)$$

i.e
$$\frac{\partial W}{\partial t} = \frac{\partial}{\partial t}\left(\tfrac{1}{2} mi\ vi \cdot vi\right) = \sum \frac{\partial}{\partial t}\left(\tfrac{1}{2} mi\ vi^2\right)$$

i.e.
$$\frac{\partial W}{\partial t} = \frac{\partial}{\partial t} \sum \tfrac{1}{2} mi\ vi^2 = \frac{\partial T}{\partial t}.$$

(ii) In this matter, the work done by the field on the charges i.e. the kinetic encrgy is transferred to random motion, where it is described as heat energy or ohmic loss and is given by

$$\frac{\partial W}{\partial t} \int_\tau J \cdot E\, dr = \int_\tau \frac{J^2}{\sigma} dr \qquad \text{(as E=J/)}$$

i.e
$$\frac{\partial W}{\partial t} = \frac{J^2}{\sigma} \times S \times l \left(as \int_\tau d\tau = Sl \right)$$

i.e
$$\frac{\partial W}{\partial t} = \rho \frac{I^2}{S^2} Sl \left(as \sigma = 1/\rho \, and\, J = I/S \right).$$

i.e
$$\frac{\partial W}{\partial t} = I^2 \rho \frac{l}{S} = I^2 R \left(as\ R = \rho l / S \right)$$

(B) Interpretation of $\int_\tau \frac{1}{2} \frac{\partial}{\partial t} (E.D. + H.B)\, d\tau$

If we allow the volume to be arbitrary large the surface integral in eqn. (A) can be made to vanish by placing the surface S suffciently far away so that the field cannot propagate to this distance in any finite time i.e $\oint_S (E \times H) \cdot ds = 0$. So under these circumstances equation (A) reduces to

$$\frac{\partial}{\partial t}\int_{all\ space} \tfrac{1}{2}(E \bullet D + H \bullet B) d\tau + \frac{\partial W}{\partial t} = 0$$

i.e
$$\frac{\partial}{\partial t}\left[\int_{all\ space} \tfrac{1}{2}(E \bullet D + H \bullet B\, d\tau + W\right] = 0$$

Thus the quantity in the square bracket is conserved. Now consider a closed system in which the total energy is assumed to be constant. The system consists of the electromagnetic field and of all the charged particles present in the field. The term W represents the total kinetic energy of the particles. We are therefore led to associate the remaining energy term

$$\int_{all\ space} \tfrac{1}{2}(E \bullet D + H \bullet B)\, d\tau$$

with the energy of the electcro-magnetic field, i.e.

$$U = \int_{all\ space} \tfrac{1}{2}(E \bullet D + H \bullet B)\, d\tau. \tag{10}$$

the quantity U may be considered to be a kind of potential energy. One need not ascribe this potential energy to the charged particles and must consider this term as a field energy. A concept such as energy stored in the field-itself rather than residing with the particles is a basic concept of the theory of etectromagnetism.

Note : If we write equation (10) as

$$U = \int_{all\ space} u\, d\tau.$$

where u = $\tfrac{1}{2}(E \bullet D + H \bullet B)$ may be thought of as the energy density of the electromagnetic field.

Further as

$$u = \tfrac{1}{2}\, E \bullet D + \tfrac{1}{2}\, H \bullet B$$

first term on r.H.S. contains only electrical quantities while the second, one magnetic, we can have

$$u = u_e + u_m$$

with $u_e = \tfrac{1}{2}\, E \bullet D = \tfrac{1}{2}\, \varepsilon_r \varepsilon_0 E^2$ = energy density of electric field

ane $u_m = \tfrac{1}{2}\, H \bullet B = \tfrac{1}{2}\, \mu_r \mu_0 H^2$ = energy density of magnetic field

(C) Interpretation of $\oint_s (E \times H).ds.$

Instead of taking the volume integral in equations (A) over all space, let us now consider a finite volume. In this case the surface integral of (E×H) will not in general vanish and so this term must be retained. Let us construct the surface S in such a way that in the interval of time under consideration, none of the charged particles will cross this surface. Then for the conservation of energy

$$\frac{\partial U}{\partial t}+\frac{\partial W}{\partial t}=-\oint_s(E\times H)\bullet ds \tag{11}$$

The left hand side is the time rate of change of the energy of the field and of the particles contained within the volume τ. Thus the surface intergral $\oint_s(E\times H)\bullet ds$ must be considered as the energy flowing out of the volume bounded by the surface S per sec. But by hypothesis no particles are crossing the surface, so the vector (E×H) is to be interpreted as the amount of the field energy passing through unit area of the surface in unit time which is normal to the direction of energy flow. The vector (E×H) is called poynting vector* and is represented by S *i.e*

$$S = (E\times H) \tag{12}$$

Interpretation of the Energy Equation.

In the light of above, equation (6) in differential form can be written as

$$J\bullet E+\frac{\partial u}{\partial t}+\nabla\bullet S=0. \tag{13}$$

In the event that the medium has zero conductivity i.e. $J=\sigma E=0$, the above equation becomes exactly of the same form as the continuity equation which expresses the law of conservation of charge. We are led by this analogy that the physical meaning of equation 13, 11 or (A) 's to represent the law of conservation of energy for electromagnetic phenomena. According to equation (11) *the time rate of change of electromagnetic energy within a certain volume plus the rate at which the work is done by the field on the charges is equal to the energy flowing into the system through its bounding surface per unit time.*

Example 6. *Show that for a cylindrical resistor of length l, redius r and resistvity ρ, the rate of flow of energy P, at which energy flows into the resistor through its cylindrical surface (calculated by integrating the Poynting vector over this surface) is equal to the rate at which Joule heat is produced i.e.*

$$P = J^2R.$$

Solution. As in case of a conductor

$$J=\sigma E \text{ i.e. } E=\rho J. \text{ (as } \rho=1/\sigma) \tag{1}$$

The electric field is parallel to the direction of J. In addition to the above electric field there is also a magnetic field whose direction is given by 'right hand thumb' rule and magnitude by Ampere's law *i.e.*

$$\oint_C H\bullet dl=\int_s J\bullet ds$$

i.e.
$$H\,2\pi r=J\pi r^2$$

i.e.
$$H=\tfrac{1}{2}\,Jr \tag{2}$$

*For detalls see Article 4.5

The direction of *E and H are perpendicular to each other as shown in Fig. 6·29. so the Poynting vector S=E×H will be 'perpendicular to the plane of E and H and into the paper. Its magnitude will be*

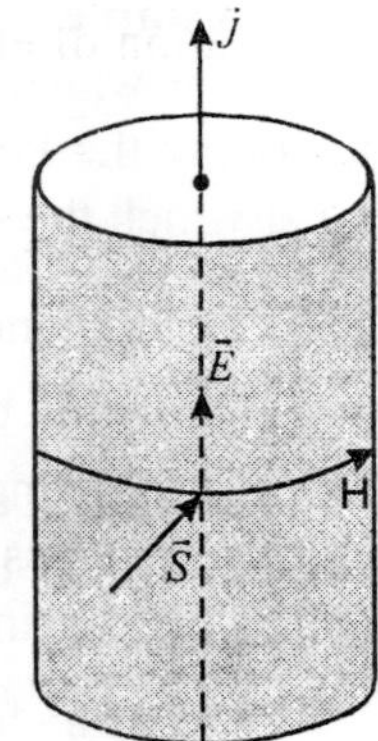

Fig. 6·29

$$S = EH = \rho J \times \frac{1}{2} Jr$$

i.e. $$S = \tfrac{1}{2} J^2 r \quad (3)$$

As a consequence of this rate of flow of field energy into the resistor through its cylindrical surface will be

$P = -\oint_s S \bullet ds = \oint_s S ds$ [as S and ds are anti-parallel]

i.e. $$\int \tfrac{1}{2}\rho J^2 r\, ds = \tfrac{1}{2}\,\rho J^2 r \times 2\pi r\, l \qquad \left[as \int ds = 2\pi r l\right]$$

i.e. $$P = \rho J^2\left(\pi r^2 l\right) = I^2 \rho \frac{l}{\pi a^2} \qquad \left[as\ J = \frac{I}{\pi r^2}\right]$$

i.e. $$P = I^2 R\left[as\ R = \rho \frac{I}{\pi r^2}\right]$$

Note: (i) This example clearly shows that according to field energy point of view, the energy that appears in a resistor as Joule heat does not enter it through the connecting wires but through the space around the wires and the resistor.

(ii) As in this example

$$U = \left(\tfrac{1}{2}\varepsilon_0 E^2 + \tfrac{1}{2}\,\mu_0 H^{2)}\right)\pi r^2 l$$

i.e. $$U = \left[\tfrac{1}{2}\varepsilon_0 (\rho J)^2 + \mu_0\left(\tfrac{1}{2} Jr\right)^2\right]\pi a^2 l$$

$$U = \left[\tfrac{1}{2}\varepsilon_0 \rho^2 + \tfrac{1}{2}\,\mu_0 r^2\right] J^2 \pi r^2 l = \text{constant}$$

or $$(\partial U / \partial t) = 0.$$

So this example is an illustration of

$$\frac{\partial W}{\partial t} = -\oint_s S.ds$$

of the general equation for conservation of field energy

$$\frac{\partial U}{\partial t} + \frac{\partial W}{\partial t} = -\oint_s S \bullet ds$$

Example 7. A parallel plate copacitor as shown in fig. 6.30 is being charged. Show that the rate at which energy flow into the capcitor from the surrounding space. (Calculated by integrating the Poynting vector over the cylindrical boundary of the volume) is equal to the rate at which the stored electrice energy increases i.e.

$$P = \pi r^2 h \varepsilon_0 E \frac{\partial E}{\partial t}.$$

Solution. Let E be the electric field at any instant when the capacitor is being charged. Naturally E will be perpendicular to the plates and will be directed from the positive to the negative plate.

Now as a changing electric field produces a magnetic field given by

$$\oint B \cdot dl = \mu_0 \varepsilon_0 \frac{d\phi_E}{dt}.$$

i.e.
$$\oint H \cdot dl = \frac{\partial}{\partial t} \int \varepsilon_0 E \cdot ds$$

$$\left[as\ B = \mu_0 H\ and\ \phi_E = \int E \cdot ds\right]$$

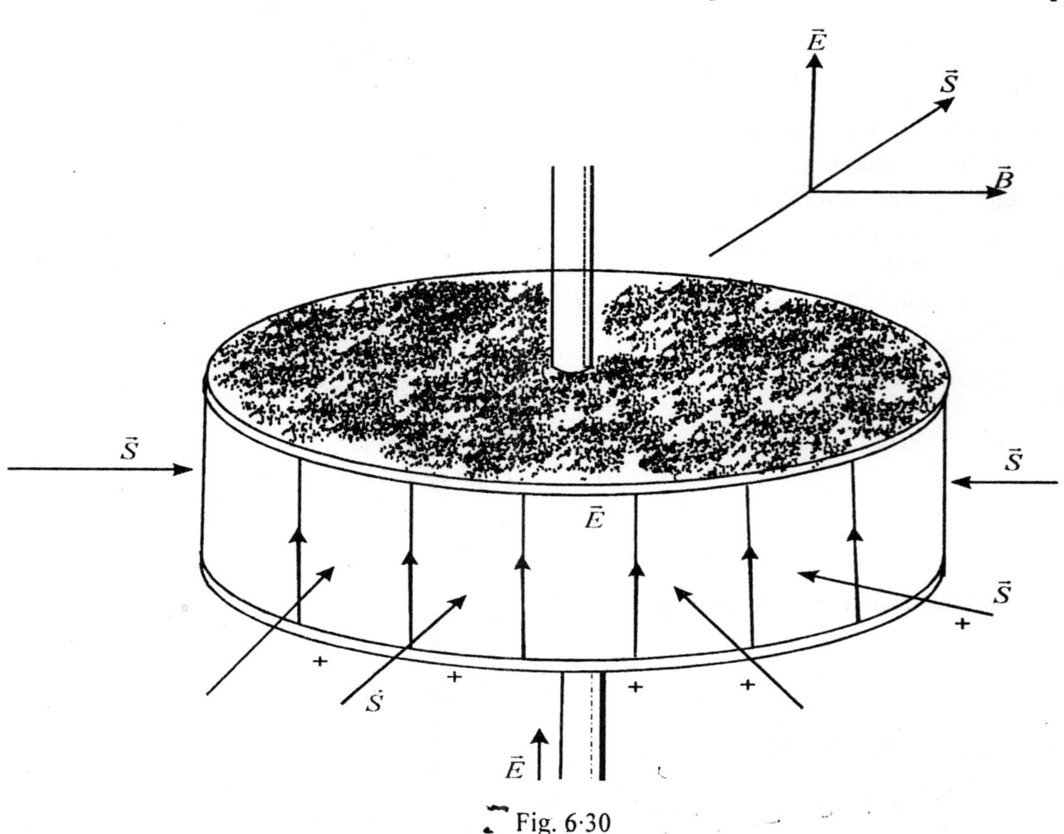

Fig. 6·30

or
$$H(2\pi r) = \frac{\partial}{\partial t}(\varepsilon_0 E)\pi r^2$$

or
$$H = \tfrac{1}{2}\,\varepsilon_0 r\frac{\partial E}{\partial t} \qquad (1)$$

The direction of H is determined by 'right hand thumb' rule and is shown in fig 6.30.

The directions of E and H are perpendicular to each other, so the Poynting vector S=E×H will be perpendicular to the plane of E and H and into the paper. Its magnitude will be

$$S = EH\sin 90 = \tfrac{1}{2}\,\varepsilon_0 rE\frac{\partial E}{\partial t} \qquad (2)$$

As a *consequence of this the rate of flow of field energy into the capacitor through its cylindrical surface will be*

$$P = -\int_s S \bullet ds = \oint_s Sds \text{ (as S and ds are anti-parallel)}$$

i.e
$$P = \int \tfrac{1}{2}\varepsilon_0 rE\frac{\partial E}{\partial t}ds = \tfrac{1}{2}\,\varepsilon_0 rE\frac{\partial E}{\partial t}2\pi rh$$

i.e
$$P = \pi r^2 h\varepsilon_0 E\frac{\partial E}{\partial t} \qquad (3)$$

This is the required result.

Note: (i) This example again tell us a peculiar thing that when we are charging a capacitor the energy is not coming down the wires but is coming through the curved surface of the capacitor. when the charges are far away there is a weak but enormously spread-out field that surrounds the capacitor. So the field energy which is way out moves toward the capacitor and eventually ends up between the plates.

(ii) In this example as

$U= \pi r^2 h \quad \left(\tfrac{1}{2}\varepsilon_0 E^2\right) \neq$ constant (as E is not consit).

i.e.
$$\frac{\partial U}{\partial t} = \pi r^2 h\frac{\partial}{\partial t}\left(\tfrac{1}{2}\varepsilon_0 E^2\right) = \pi r^2 h\varepsilon_0 E\frac{\partial E}{\partial t} \neq 0$$

and
$$J = \sigma E = 0 \text{ i.e. } \frac{\partial W}{\partial t} = \int (J \bullet E)d\tau = 0.$$

So this example is an illustration of field energy i.e.

$$\frac{\partial U}{\partial t} = -\oint_s S \bullet ds$$

of the general equation for the conservation of field energy i.e.

$$\frac{\partial W}{\partial t}+\frac{\partial U}{\partial t}=-\oint_s S \bullet ds$$

Poynting Vector.

In Fig. 6.28 we have seen that according to the law of conservation of energy in an electromagnetic field

$$\frac{\partial U}{\partial t}+\frac{\partial W}{\partial t}+\int (div\ S)\, d\tau = 0 \qquad (1)$$

with

$$S = E \times H. \qquad (2)$$

The vector S is known as Poynting vector. It is interpreted as *the amount of the field energy passing through unit area of the surface in a direction perpendicular to the plane containing E and H per unit time.* For example as in a plane electro-magnetic wave E and H are perpendicular to each other and also to the direction of wave propagation, S has a magnitude EH sin 90 = EH and points in the direction of wave propagation. The dimensions of Poynting vector are (energy/area ×time) so the units will be Joulel $m^2 \times$ sec or watt/m^2.

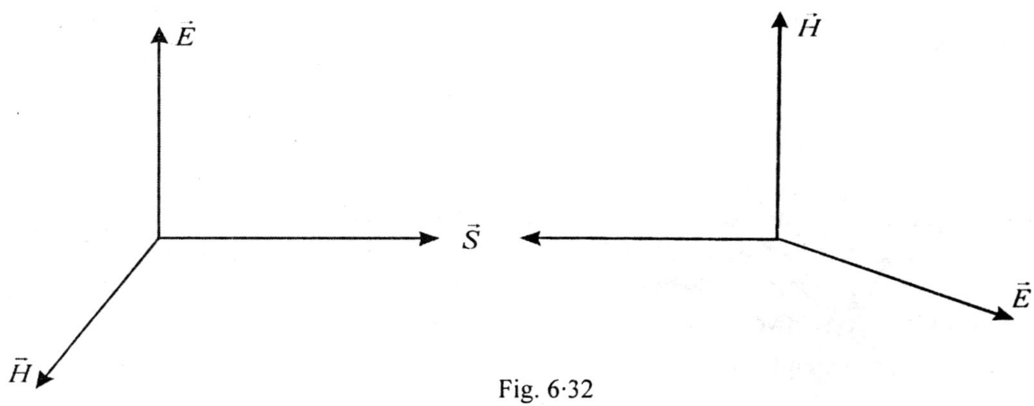

Fig. 6·32

Regarding Poynting vector it is worthy to note that—

(i) *Poynting vector at any orbitrary point in the field varies inversely as the square of the distance from the point source of radiation.* To undertstand it consider a source L of electromagnetic radiations which is emitting radiations at the rate of P watts and imagine two concentric spherical surface *A* and *B* of radii r_1 and r_2 respectively with source being at their common centre. If S_1 and S_2 are the poynting vector at any point on *A* and *B* respectively then

$$S_1 \times 4\pi r_1^2 = S_2 \times 4\pi r_2^2 = p$$

i.e.

$$S_1 = (P/4\pi r_1^2)$$

and

$$S_2 = (P/4\pi r_2^2).$$

So in general $S \propto (1/r^2)$

(ii) The definition of Poynting vector is not a mandatory. Since this vector has been introduced only by way of its divergence, the curl of any arbitrary vector can be added to it without altering the physical facts of the case i.e. it is arbitrary to the extent that curl or any vector field can be added to it i.e.

$$S = E \times H + G$$

where $\quad G = Curl\ M$

$\quad M$ = *any arbitrary field vector.*

However on various grounds, such an additional term has no physical consequence and the definition of S is retained as such. This definition is also found to be convenient particularly in electromagnetic theory.

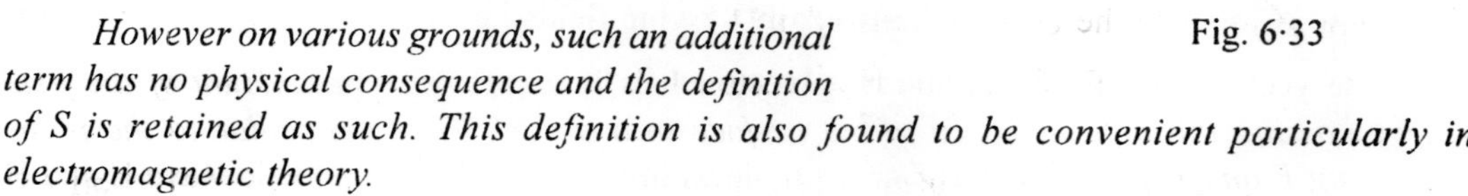

Fig. 6·33

(iii) If Poynting vector is zero then no electromagnetic energy can flow across a closed surface but if no net field energy is flowing across a closed surface the Poynting vector may or may not be zero. For example in case of the field due to a point charge in the presence of a magnet at rest as shown in fig. 6.34 (a) or a charged capacitor placed between the poles of a permanent magnet if E is not parallel to H, the Poynting vector is not zero as

$$|S| = |E \times H| = EH \sin\theta = 0.$$

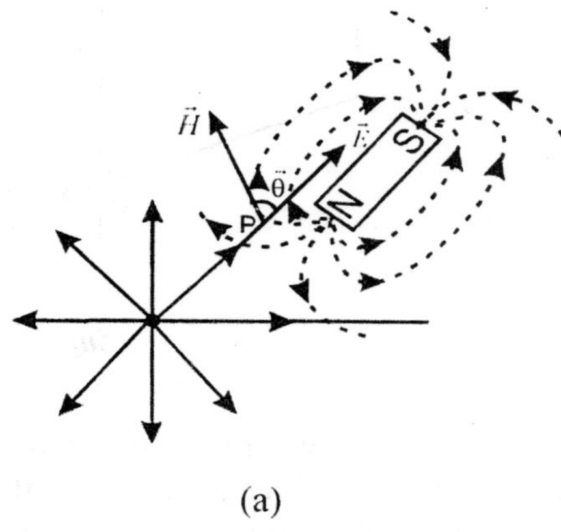

(a)

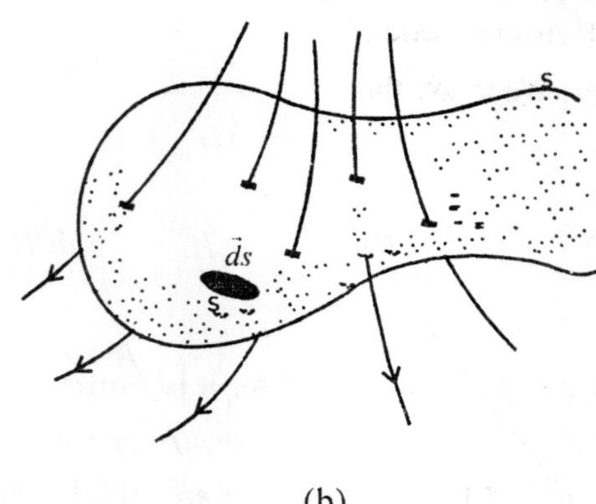

(b)

Fig. 6.34.

For steady fields $S \neq 0$ i.e. $S \cdot ds \neq 0$ but there is no flow of energy i.e. $\oint S \cdot ds = 0$ where the flow of energy across any closed surface is zero as

$$\oint_s S \cdot ds = \int_\tau \nabla \cdot S\, d\tau = \int_\tau \nabla \cdot (E \times H) d\tau$$

i.e. $$\oint_S S \cdot ds = \int_\tau (H \cdot curl\, E - E \cdot curl\, H) d\tau = 0$$

(as E and H are constant for steady fields).

This is because, S does not determine the rate of flow through small element ds at a point but implies that only the flux of S across a closed surface is significant.

(iv) *In case of time varying fields S=E×H givs the instantaneous value of the Poynting vector and the average value is defined as the average over one complete period i.e.*

$$< S >= \frac{1}{T}\int_0^{2\pi} (E \times H) dt.$$

For example in the case of sinusoidally varying fields i.e

$$E = E_0 \sin \omega t \text{ and } H = H_0 \sin \omega t$$

$$< S >= \frac{1}{T}\int_0^{T} (E_0 \sin \omega t) \times (H_0 \sin \omega t) dt$$

i.e $$< S >= \frac{1}{T}(E_0 \times H_0)\int_0^{2\pi/\omega} \sin^2 \omega t\, dt \, *$$

i.e. $$< S >= \tfrac{1}{2}(E_0 \times H_0) = \frac{E_0}{\sqrt{2}} \times \frac{H_0}{\sqrt{2}} = E_{av} \times H_{av}.$$

The importance of Poynting vector lies in the fact that with its help we can interpret various optical phenomena such as reflection, refraction, dispersion, scattering, differaction, etc., electromagnetically.

(v) We know that

$$B = \mu_0 H$$

i.e. $$\frac{\mu_0}{4\pi}\int I \frac{dI \times r}{r^3} = \mu_0 H, \qquad \left(as\ B = \frac{\mu_0}{4\pi}\int \frac{Id1 \times r}{r^3}\right)$$

or $$H = \frac{1}{4\pi}\int I \frac{dI \times r}{r^3}$$

$$\frac{*1}{T}\int_0^{2\pi/\omega} \sin^2 \omega t dt = \frac{\omega}{2\pi}\int_0^{2\pi/\omega} \frac{(1 - \cos 2\omega t)}{2} dt$$

$$= \frac{\omega}{4\pi}\left[t - \frac{\sin 2\omega t}{2\omega}\right]_0^{2\pi/\omega} = \frac{\omega}{4\pi} \times \frac{2\pi}{\omega} = \frac{1}{2}$$

or $$(H) = \frac{[A][L][L]}{[L^3]} = [AL^{-1}]$$

and $E=(F/q)$

i.e $$[E]=\frac{[MLT^{-2}]}{[AT]}=[MLT^{-3}A^{-1}]$$

So $$[S]=[E][H]=[MT^{-3}]$$

i.e. $$[S]=\frac{[ML^2T^{-2}]}{L^2T}=\frac{Energy}{area\times time}=\frac{Power}{area}$$

Example 8. (a) *Calculate the value of Poynting vector at the surface of the sun if the power radiated by the sun is 3.8×10^{26} watts and its radius is 7×10^{8}m.*

(b) *If the average distance between the sun and earth is 1.5×10^{11} meters. Show that the average solar energy incident on the earth is 2 cal/cm2, min (called solar constant). Calculate also the amplitudes of the electric and magnetic vectors at the surface of the earth assuming that the radiation is a plane wave.*

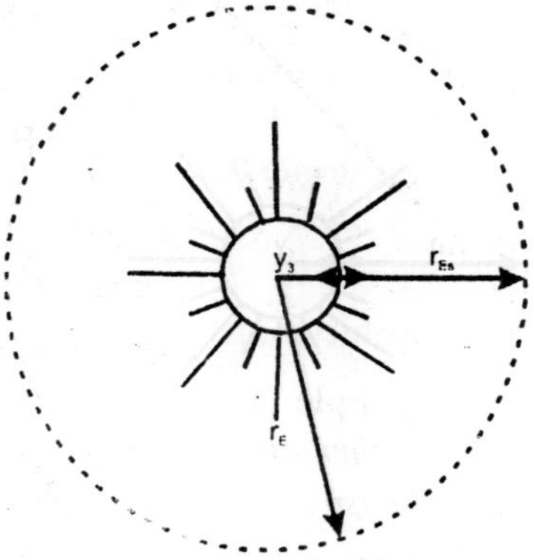

Fig 4.12

Solution. (a) If <Ss> is the average Poynting vector at the surface of the sun then by it is definition

$$\langle Ss\rangle\times4\pi r^2=P$$

i.e. $$\langle Ss\rangle=\frac{P}{4\pi r^2}\frac{3.8\times10^{26}}{4\pi\times(7\times10^{8})^2}$$

i.e $$\langle Ss\rangle=6.174\times10^{7}\text{ watt/meter}^2$$

(b) If $\langle S_E\rangle$ is the average value of Poynting vector at the surface of earth then

$$\langle S_E\rangle\,4\pi r_E^2=\langle Ss\rangle\,4 r_s^2=P$$

i.e $$\langle Se\rangle=\frac{rs^2}{r_E^2}\langle Ss\rangle=\frac{rs^2}{(rEs+rs)^2}\langle Ss\rangle\approx\frac{rs^2}{rEs^2}\langle Ss\rangle$$

i.e. $$\langle S_E\rangle\left[\frac{7\times10^{8}}{1.5\times10^{11}}\right]^2\times6.175\times10^{7}=1.5\times10^{3}\text{ watts/m}^2$$

i.e. $$\langle S_E\rangle=\frac{1.5\times10^{3}\times60}{4.2\times10^{4}}\text{cal./cm}^2\text{–min}=2.14\text{ cal./cm}^2\text{. min.}$$ **Ans.**

Now as in plane wave

$$H_{av}=\varepsilon_0 cE_{av}$$

So $$S_{av} = E_{av} H_{av} = \varepsilon_0 c E_a v^2 = \frac{1}{2}\varepsilon_0 c E_0^2$$ (as $E_{av} = E_0/\sqrt{2}$)

i.e. $$E_0 = \sqrt{\left(\frac{2S_{av}}{\varepsilon_0 c}\right)} = \sqrt{\left(\frac{2\times 1.5\times 10^3}{9\times 10^{-12}\times 3\times 10^8}\right)} \approx 1050 \text{ volts/mete}$$

and as $$B_0 = \mu_0 H_0 = \mu_0 \varepsilon_0 c E_0 = \frac{E_0}{c}$$

i.e. $$B_0{}^* = \frac{1050}{3\times 10^3} \text{ web/m}^2 = 3.5\times 10^{-6} \text{ web/m}^2.$$ **Ans.**

Example 9. *Assuming that in a plane electromagnetic wave E and H are perpendicular to each other and are related by the relation prove that the Poynting vector is c times the energy density of the field. What is the significance of this result?*

Solution. As the energy density of plane electromagnetic wave is given by

$$u_{av} \frac{1}{2}\varepsilon_0 E_{av}{}^2 + \frac{1}{2}\mu_0 H_{av}{}^2$$

or $$u av = \frac{1}{2}\varepsilon_0 E_{av}{}^2 + \frac{1}{2}\mu_0(\varepsilon_0 c E_{av})^2$$ [as $H_{av} = \varepsilon_0 c E_{av}$]

or $$u_{av} \frac{1}{2}\varepsilon_0 Eav^2 + \frac{1}{2}\varepsilon_0 Eav^2$$ [as $c = 1\sqrt{(w_0 \varepsilon_0)}$]

i.e. $$u_{av} = \varepsilon_0 E_{av}{}^2$$ ....(2)

Now as Poynting vector is given by

$$S_{av} = E_{av} \times H_{av}$$

i.e. $$S_{av} = E_{av} \times H_{av}$$ (as E and H mutully Perpendicular)

$$S_{av} = \varepsilon_0 c E_{av}{}^2$$ [as $H_{av} = \varepsilon_0 c E_{av}$] ...(2)

So comparing equatiobns (1) and (2) we get.

$$S_{av} = CU_{av}.$$

This is the desired result.

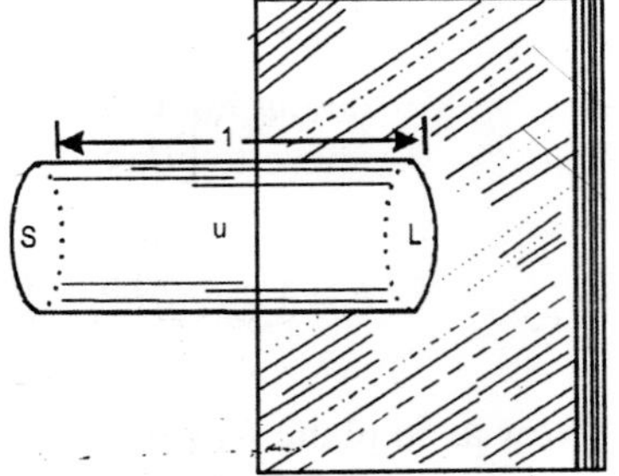

Fig 6.36

To understand what this relation means, consider the arrangement shown in the fig.6.36 in which a plane wave of limited extent falls on a black surface which absorbs all of it. Draw a cylinder of length l, area S and volume t and suppose it contains a quantity of energy U flowing towards the right with velocity v. At the end of time l/v this energy would all have flowed into the surface so the mean energy flowing per sec per unit area

$$S_{av} = \frac{U}{(l/v)S} = v\frac{U}{\tau} vu_{av.} \qquad ...(4)$$

*Here $H_0 = \frac{B_0}{\mu_0} = \frac{3.5\times10^{-6}}{4\pi\times10^{-7}} = \frac{35}{4\pi} = 2.8$ amp-turn/meter,

So comparing equations (3) and (4) we see that the electromagnetic wave carries its energy at the speed c, a result though not surprising is by no means obvious, since there are various kinds of waves which transport energy at a speed different from their wave velocity.

Mamentum in Electromagnetic Fields:

In Figure 6.37 it was shown that how it is possible and why it is useful to assign energy to the electromagnetic field. It will now be shown how it is possible to assign linear momentum to the fields.

Consider an arbitrary region of volume surrounded by a surface S and containing charges and currents. The field in the region of containing will then be governed completely and correctly by Maxwell's equations

$$\text{Div D} = \rho \qquad \text{(a) curl H}-\frac{\partial D}{\partial t} = J \qquad ...(b)$$

$$\text{Div B} = 0 \qquad \text{(c) curl E}+\frac{\partial B}{\partial t} = 0 \qquad ...(d)$$

and the total force on the matter by Lorentz force equation

$$F = \int_\tau (\rho E + J\times B)d\tau \qquad ...(e)$$

So substituting the values of ρ and J from Maxwell's equation (a) and (b) in Lorentz force equation (e), we get

$$F = \int_\tau \left\{E\, div\, D + \left(curl\, H - \frac{\partial D}{\partial t}\right)\times B\right\}d\tau \qquad ....(1)$$

Now taking the product of eqn. (c) with H and vector product of (–D) with equation (d) and adding these to equation (1) we get

$$F = \int_\tau \left\{E\, div\, D + \left(curl\, H - \frac{\partial D}{\partial t}\right)\times B - D\times\left(curl\, E + \frac{\partial B}{\partial t}\right) + H\, div\, B\right\}d\tau$$

i.e. $$F = \int_\tau \left\{(E\, div\, D - D\times curl\, E) + (H\, div\, B - B\times curl\, H) - \frac{\partial}{\partial t}(D\times B)\right\}d\tau$$

i.e. $$F = \int_\tau \left\{ \varepsilon_0 (E\, div\, E - E \times curl E) + \mu_0 (H\, div\, H - H \times curl\, H) - \varepsilon_0 \mu_0 \frac{\partial}{\partial t}(E \times H) \right\} d\tau$$

...(2)

To facilitate the avaluation of the integral on the right hand side of equation (2) consider the x-component only. Now as the x-component of

$$\text{E div E–E×curl E} = Ex\left(\frac{\partial E_x}{\partial x} + \frac{\partial E_y}{\partial y} + \frac{\partial E_z}{\partial z}\right) - \left\{ Ey\left(\frac{\partial E_y}{\partial x} - \frac{\partial E_x}{\partial y}\right) - Ez\left(\frac{\partial E_x}{\partial z} - \frac{\partial E_z}{\partial x}\right)\right\}$$

$$= \frac{\partial}{\partial x}(Ex^2) - E_x \frac{\partial E_x}{\partial x} - Ey\frac{\partial E_y}{\partial x} - E_z \frac{\partial E_z}{\partial x} + \frac{\partial}{\partial y}(E_x E_y) + \frac{\partial}{\partial z}(E_x E_z)$$

$$= \left[\frac{\partial}{\partial x}\left(E_x^2 - \tfrac{1}{2} E^2\right) + \frac{\partial}{\partial y}(E_x E_y) + \frac{\partial}{\partial z}(E_x E_z)\right] \quad ...(3)$$

Similar result holds for the term in H equation (2).

So x-component (4) can be written as

$$F_x = \int_\tau \left(\frac{\partial T_{xx}}{\partial x} + \frac{\partial T_{xy}}{\partial y} + \frac{\partial T_{xz}}{\partial z} - \mu_0 \varepsilon_0 \frac{\partial}{\partial t}(E \times H) \right) d\tau$$

with k=x, y and z

or $$F + \mu_0 \varepsilon_0 \int_\tau \frac{\partial}{\partial t}(E \times H) d\tau = \int \nabla . T_{jk} d\tau$$

or $$F + \int_\tau \frac{\partial}{\partial t} \frac{(E \times H)}{c^2} d\tau = \oint_s T_{jk}\, ds \left[as \int \nabla \bullet A\, d\tau = \oint_s A \bullet ds \right]$$

or $$F + \int_\tau \frac{\partial}{\partial t}\left(\frac{S}{c^2}\right) d\tau = \oint_s T_{jk}\, ds \; \{\text{as } S = (E\times H)\} \quad ....(A)$$

where T_{jk} is a by 3 matrix given by

$$T_{jk} = \begin{vmatrix} T_{xx} & T_{xy} & T_{xz} \\ T_{yx} & T_{yy} & T_{yz} \\ T_{zx} & T_{zy} & T_{zz} \end{vmatrix}$$

$T_{xx} = [\, \varepsilon_0 (E_x^2 - \tfrac{1}{2} E^2) + \mu_0 (H_x^2 - \tfrac{1}{2} H^2)]$

$T_{xy} = T_{yx} = \varepsilon_0 E_x E_y + \mu_0 H_x H_y$

$T_{yy} = [\varepsilon_0 (E_y^2 - \frac{1}{2} E^2) + \mu_0(H_y^2 - \frac{1}{2} H_2)]$

$T_{yz} = T_{zy} = \varepsilon_0 E_y E_x + \mu_0 H_y H_z$

$T_{zz} [\varepsilon_0 (E_y^{2} + \frac{1}{2} E^{2}) + \mu 0 (H_z^2 - \frac{1}{2} H^2)]$

$T_{zx} = T_{xz} = \varepsilon_0 E_z E_x + \mu_0 H^z H_x$

The abcve matrix is usually referred to as Electamagnetic stress Energy tensor to Maxwell Strees Tehson $\oint_s T_{jk} \bullet ds = 0$ (as there is no boundary surface) So equation (A) reduces to

$$F + \int_{allSpace} \frac{\partial}{\partial t}\left(\frac{S}{c^2}\right) dr = 0 \qquad ...(6)$$

Now in the light of conservation of linear momnetum for a systenm consisting of matter and field

$$P_{system} = P_{Matter} + P_{field}$$

or
$$\frac{\partial}{\partial t}(P_{matter}) + \frac{\partial}{\partial t}(P_{field}) = 0$$

or
$$F + \frac{\partial}{\partial t}(P_{field}) = 0 \left(as \frac{\partial}{\partial t} P_{matter} = F\right) \qquad ...(7)$$

Comparing equation (6) with (7) for conservation of linear momentum we conclude that

$$\text{P field} = \int_{allspace}\left(\frac{S}{c^2}\right) d\tau \qquad ...(B)$$

In the light of equation (B) equation (A) reduces to

$$F + \frac{\partial}{\partial t}(P_{field}) = \oint_s T_{jk} \bullet ds$$

or
$$\frac{\partial}{\partial t}(P_{matter} + P_{field}) = \oint_s T_{jk}.ds \left(as\ E = \frac{\partial}{dt} P_{matter}\right) \qquad ....(C)$$

Equation (C) represents the law of conservation of linear momentum for an electromagnetic field and according to it, *the time rate of change of the sum of particle momentum and field moentum is equal to the total force which would be exerted by the Maxwell stresses on the region considered.*

Note. It is also natural in the light of eqn. (A) to define the linear momentum density in the field as

$$g = \frac{S}{c^2} = \frac{u}{c} \text{ (as from example 9. S=uc)} \qquad ...(D)$$

The appearence of a field momentum becomes understandable when we consider that ir is possible for electromagnetic rediation to convey not only energy but also momentum from source of radiation to an absorber, and that this conveyed momentum is measurable as rediation pressure on the absorber.

For calculating the radiation pressure consider a surface element with normal in the positive x-direction. As the term $\oint T_{jk}.ds$ represents the force on the boundary

$$F_x = \oint_s T_{xk} n_x ds$$

i.e.
$$F_x = \oint_s \left(T_{xx} + T_{xy} + T_{xz}\right) n_x ds$$

i.e.
$$F_x = \oint_s T_{xx} ds$$

So rediation pressure which is force per unit area in direction opposite to that of F_x will be

$$P_{rad} = -T_{xx}$$

which on substituting the value of T_{xx} reduces to

$$P_{rad} = -[\varepsilon_0(E_x^2 - \tfrac{1}{2} E^2) + \mu_0 (H_x^2 - \tfrac{1}{2} H^2)] \quad \text{...(E)}$$

Now as for a light wave E and II are transverse to the direction of wave Pr=opa=gation (Which is here x-direction)

$$E_x = H_x = 0$$

$\therefore$
$$P_{rad} = [\tfrac{1}{2} \varepsilon_0 E^2 + \tfrac{1}{2} \mu_0 H^2] = u$$

i.e.
$$P_{rad} = u = \frac{S}{c} \text{ (as S=uc)} \quad \text{....(}E_1\text{)}$$

This equation holds good only if the surface absorbs all the radiation incident on it. However if it reflects all the incident radiation

$$P_{rad} = 2u = \frac{2S}{c} \quad \text{....(}E_2\text{)}$$

The factor of 2 being introduced because the wave is reflected with momentum equal to its initial momentum but in opposite direction.

For isotropically incident radiations (uniform from all directions as for example in Hohlraum radiation) we have for the average

$$E^2x = \frac{E^2}{3} \text{ and } H^2x = \frac{H^2}{3}$$

so from equation (E)

$$P_{\text{rad}} = -\left[\varepsilon_0\left(\frac{E^2}{3}-\frac{E^2}{2}\right)+\mu_0\left(\frac{H^2}{3}-\frac{H^2}{2}\right)\right]$$

i.e.
$$\frac{1}{6}\left[\varepsilon_0 E^2+\mu_0 H^2\right]=\frac{u}{3} \qquad \text{....}(E_3)$$

This equation holds good for all surface (either absorber or reflector) on account of perfect symmetry.

The radiation pressure of an eletromegnetic wave has been observed experimentally and has been found to agree with theory.

Radiation Pressure is very small indeed under the conditions of labortatory experiments ($\sim 10^{-5}$ Newton/m^2)* and hence is unimportant. However it is very important in stars. The high radiation density and so pressure in a star gives a large pressure tending to drive the matter outwards from the centre. This opposes gravitational force which tensds to make the star collapse.

One of the most striking phenomena resulting directly from the effects of radiationpressure is found in *comets*. A comet looks rather like a star with a long tail and moves in a long oval orbit around the sun. The tail of these objects are always pointing predominatly away from the sun. Because the material of such comets composed is extremely tenuous pressure from the sun is sufficient to elongate and direct the tail away from the sun.

Example 10. (a) *calculate the radiation pressure at the surface of earth and sun assuming that solar constant has a value of 2 co-cm²min at the surface of the earth, the radius of the sun is 1×10^8m and the average distance between earth and sun is 1.5×10^{11}m.*

(b) *If an interplanetary 'sail plane' had a sail of mass 10^{-4} gm/cm^2 and other weights neglible, what would be its maximum acceleration in centmeters per sec^2 due to solar radiation. (Solar constant is 2 call cm^2-min).*

Solution:

(a)
$$As\ S_E = \frac{2Cal}{cm^2\times\min} = \frac{2\times4.2}{0^{-4}\times60} = 1.5\times10^3\,\frac{Joules}{m^2-\sec}$$

$\therefore$
$$(\text{Prad})_E = \frac{S_E}{c} = \frac{1.5\times10^3}{3\times10^8} = 5\times10^{-3}\,N/m^2 \text{ Ans.}$$

Further as

$$S_s r_s^2 = S_E r_{ES}^2$$

i.e.
$$S_s = S_E\frac{r_{ES}^2}{r_s^2} = 1.5\times10^3\times\left(\frac{1.5\times10^{11}}{7\times10^8}\right)$$

$$\approx 6\times10^7 \text{ Watt/m}^2$$

$\therefore \quad (Prad)s = \frac{Ss}{c} = \frac{6 \times 10^7}{3 \times 10^8} 0.2n / m^2$ Ans.

(b) As in part (a)

$$(Prad)_E = 5 \times 10^{-6}\ N/m^2 = 5 \times 10\text{-}5 \times \frac{10^5\, dynes}{10^{-4}\, cm^2} = 5 \times 10^{-5} \frac{dynes}{cm^2}$$

Now as pressure $= \frac{Force}{area} = \frac{mass \times acc}{area}$

i.e. $\quad acc = \frac{pressure}{mass\ per\ unit\ area} = \frac{5 \times 10^{-6}}{10^{-4}} = 0.5cm / \sec^2$

Electromagnetic Potentials A and *f*.

The analysis of an electromagnetic field is often facilitated by the use of auxilary functions known as electromagnetic potentials. At every point of space the field vector satisfy the equations

$$\text{div } D = \rho \quad \text{....(A)}$$

$$\text{div } B = 0 \quad \text{....(B)}$$

$$\text{curl } H = J + \frac{\partial D}{\partial t} \quad \text{....(C)}$$

$$\text{curl } E = -\frac{\partial B}{\partial t} \quad \text{....(D)}$$

According to equation (B) the field of vector B is always solenoidal, consequently, B can be represented as the curl of another vector say A i.e.

$$B = \text{curl } A \quad \text{....(1)}$$

where A is a vector which is function of space (x, y, z) and time (t) both.

Now substituting the value of B in equation (D) from (1), we get

$$\text{curl } E = -\frac{\partial}{\partial t}(\text{curl } A)$$

or $$\text{curl}\left(E + \frac{\partial A}{\partial t}\right) = 0$$

i.e. the field of the vector $E + \frac{\partial A}{\partial t}$ is irrotational and must be equal to the gradient of some scale function i.e.

$$E + \frac{\partial A}{\partial t} = -\text{ grad } \phi$$

or $$\text{E} = -\text{grad } \phi - \frac{\partial A}{\partial t} \qquad ....(2)$$

Thus we have introduced a vector A and a sclar ϕ both being function of position and time. These are called electromagnetic potentials. The scalar ϕ is called the scalar potential and the vector A, vector potential. Regarding electromagnetic potentials it is worth nothing that

(i) These are mathematical functions which are not physically measurable.

(ii) These are not independent of each other.

(iii) These define the field vector E and B uniquely but themselves are nonunique (see Fig. 6.38).

(iv) These play an important role in relativiatic electrodynamics (See chapter 10).

Example 11. *Show that the potentials at the position defined by the vector r in uniform electric and magnetic field may be written as*

$$\phi = -\text{ E}\cdot\text{r and A} = \tfrac{1}{2}\,(\text{B}\times\text{r}).$$

Solution. (a) We know that

$$\text{E} = \text{i } E_x + \text{J } E_y + \text{k } E_z = (\text{E}\cdot\nabla)\text{ r}$$

But as E is uniform, above result can be written as

$$\text{E} = \nabla\,(\text{E}\cdot\text{r}). \qquad ....(1)$$

And also as in case of static electric field

$$\text{E} = -\text{grad}\phi = -\nabla\phi. \qquad ....(2)$$

So from equation (1) and (2), we have

$$\nabla(-\phi) = \nabla\,(\text{E}\cdot\text{r})$$

i.e. $$\phi = -\text{ E}\cdot\text{r} \qquad ...(3)$$

This is the desired result:

(b) We know that

curl (C × D) = C div D – (C · gard) D is C is a constant vector,

So for uniform B

$$\text{curl (B} \times \text{r)} = \text{B div r} - (\text{B}\cdot)\text{r}$$

i.e. $$\text{curl (B} \times \text{r)} = \text{B (3)} - \text{B} = 2\text{B} \qquad ....(4)$$

i.e. $$\text{B} = \tfrac{1}{2}\text{ curl (B} \times \text{r)}$$

And also according to definition of A

$$\text{B} = \text{curl A} \qquad ....(5)$$

So comparing equation (4) and (5)

$$\text{curl } A = \tfrac{1}{2} \text{ curl } (B \times r)$$

i.e. $$A = \tfrac{1}{2} (B \times r)$$

This is the desired result.

Example 12. *Express the Lorentz force formula in terms of Electromagnetic potentials.*

Solution. We know that force on a charged particle in an electromagnetic field is given by

$$F = q\,[E + v \times B] \qquad \text{...(1)}$$

And in terms of electromagnetic potentials (A and ϕ)the field vector E and B are given by

$$B = \text{curl } A \text{ and } E = -\text{ grad } \phi \frac{\partial A}{\partial t}. \qquad \text{...(2)}$$

So in the light of equation (2), (1) reduces to

$$F = q\left[\left(-\nabla\phi - \frac{\partial A}{\partial t}\right) + v \times (\nabla \times A)\right]$$

i.e. $$F = q\left[-\nabla\phi - \frac{\partial A}{\partial t} + \nabla(v \bullet A) - (v \bullet \nabla)A\right]$$

i.e. $$F = q\left[-\nabla\phi - \left\{\frac{\partial A}{\partial t} + (v \bullet \nabla)A\right\} + \nabla(v \bullet A)\right] \qquad \text{...(3)}$$

Now as the vector potential A is a function of space and time both

$$\frac{dA}{dt} = \frac{\partial A}{\partial t} + \frac{\partial A}{\partial x}.\frac{\partial x}{\partial t} + \frac{\partial A}{\partial y}.\frac{\partial y}{\partial t} + \frac{\partial A}{\partial z}.\frac{\partial z}{\partial t}$$

i.e. $$\frac{dA}{dt} = \frac{\partial A}{\partial t} + vx\frac{\partial A}{\partial x} + vy\frac{\partial A}{\partial y} + vz\frac{\partial A}{\partial t}$$

i.e. $$\frac{dA}{dt} = \frac{\partial A}{\partial t} + (v \bullet \nabla)A \qquad \text{....(4)}$$

So in the light of (4), (3) reduces to

$$F = q\left[-\nabla\phi - \frac{dA}{dt} + \nabla(v \bullet A)\right]$$

i.e. $$F = q\left[-\nabla(\phi - v \bullet A) - \frac{dA}{dt}\right] \qquad \text{...(5)}$$

This is the required result.

Example 13. *Show that*

(a) *The momentum of charged particle in an electromagnetic field is given by*

$$P = Mv+A,$$

(b) *The Lagrangian function of the charged particle is an electromagnetic field is given by*

$$L=\frac{1}{2} mv^2 - q (\phi - v\cdot A)$$

(c) *The Hamiltonion Function of the charged particle is an electromagnetic field is given by*

$$H = \frac{1}{2m}(p - qA)^2 + q\phi$$

Solution. (a) Force on a charged particle in an elctromagnetic field in terms opf electromagnetic potentials is given by

$$F = q\left[-\nabla(\phi - v \bullet A) - \frac{dA}{dt}\right] \qquad \text{[from example -12]}$$

i.e.
$$\frac{d}{dt}(mv) = q\left[-\nabla(\phi - v.A) - \frac{dA}{dt}\right]\left(as\frac{d}{dt}(mv) = F\right)$$

i.e.
$$\frac{d}{dt}(mv + qA) = -q\nabla(\phi - v.A) \qquad ...(1)$$

This equation has the general form of a set of Lagrangian eqations

$$\frac{d}{dt}\left(\frac{\partial L}{\partial x_i}\right) - \frac{\partial L}{\partial x_i} = 0. \qquad ....(2)$$

So comparing eqns. (1) and (2) we get

$$\frac{\partial L}{\partial x_i} = mv_i + qA \qquad ....(3)$$

$$\frac{\partial L}{\partial x_i} = \frac{\partial}{\partial x_i}\left[-q(\phi - v_i A_i)\right] \qquad ....(4)$$

But by definition of Lagrangian $\frac{\partial L}{\partial x_i}$ is the ith component ofmomentun so the ith component of momentum of the charged particle in an electromagnetic field will be given by

$$p_i = mv_i + qA_i$$

i.e.
$$p = mv + qA. \qquad ...(5)$$

This is the desired result.

(b) Integrating equation (3) and (4) we find

$$L = \frac{1}{2}\, mv^2 + q\,(v.A) + constt_1 \qquad ...(6)$$

$$L = constt_2 + q\,(v \cdot A) - q\phi \qquad ...(7)$$

where costant$_1$ is not a function of velocity while contant$_2$ is not a function of position.

A glance at equations (6) and (7) reveals that the proper Langangian for charged particle is

$$L = \frac{1}{2}\, mv^2 + q\,(v.A) - q$$

i.e.

$$L = \frac{1}{2}\, mv^2 - q\,(\phi - v.A). \qquad ...(8)$$

This is the desire result.

(c) As ha,iltonian function is definded as

$$H = \sum pi.xi - L\,. \qquad ...(9)$$

So subtituting the value of p and L from equation (5) and (8) in (9) we get

$$H = (mv + qA).\ v - [\tfrac{1}{2}\, mv^2 - q\,(\phi - v.A)]$$

i.e.

$$H = \frac{1}{2}\, mv^2 + q\phi$$

i.e.

$$H = \frac{(Mv)^2}{2m} + q\phi \qquad ...(10)$$

Now as from equation (5)

$$mv = p - qA$$

So equation (10) reduces to

$$H = \frac{1}{2m}(p - qA)^2 + q\phi. \qquad ...(11)$$

This is the desired result.

Maxwell Equations in Terms of Electromagnetic Potentials.

Now consider the Maxwell's equation (C) i.e.

$$\mu \text{ curl } H = \mu J + \mu \frac{\partial D}{\partial t}$$

or

$$\text{curl } B = \mu J + \varepsilon\mu \frac{\partial E}{\partial t}$$

By substituting of B and E in above we get

$$\text{curl (curl A)} = \mu J + \mu\varepsilon \frac{\partial}{\partial t}\left(-grad\phi - \frac{\partial A}{\partial t}\right)$$

i.e.
$$\text{grad div A} - \nabla^2 A = \mu J - \mu\varepsilon \frac{\partial}{\partial t}(grad\phi) - \mu\varepsilon \frac{\partial^2 A}{\partial t^2}$$

i.e.
$$\nabla^2 A - \mu\varepsilon \frac{\partial^2 A}{\partial t^2} - \text{grad}\left(divA + \mu\varepsilon \frac{\partial \phi}{\partial t}\right) = -\mu J. \quad ...(3)$$

Similarly if we consider equation (A) i.e.

$$\text{div D} = \rho$$

i.e.
$$\varepsilon \text{ div E} = \rho$$

i.e.
$$\text{div}\left(-grad\phi - \frac{\partial A}{\partial t}\right) = \frac{\rho}{\varepsilon}$$

i.e
$$\nabla^2 \phi + \frac{\partial}{\partial t}(divA) = -\frac{\rho}{\varepsilon}.$$

Adding and substracting $\mu\varepsilon \frac{\partial^2 \phi}{\partial t^2}$ it becomes

$$\nabla^2 \phi - \mu\varepsilon \frac{\partial^2 \phi}{\partial t^2} + \frac{\partial}{\partial t}\left(divA + \mu\varepsilon \frac{\partial \phi}{\partial t}\right) = -\frac{\rho}{\varepsilon}.$$

Equation (3) and (4) are field equations in terms of electromagnetic potentials, as equations (B) and (D) are satisfied in defining the scalar and vector potentials. So Maxwell equations are reduced from four to two by electromagnetic potentials, however these are couple.

Note: The reader may be happy to note that the concept of electromagnetic potential, reduces the number of Maxwell's equations from four to two, and even more or less of the same form. Actually this all is not a real simplicity but a trick which hides the complexity in the definition of the symbols A and ϕ.

To make the above statement more clear authors want to put something more interesting and fascinating that has been discovered recently. All the laws of physics can be contained in a single equation

$$U = 0$$

where U is a physical quantity called "*unwordlines*" of the situation and we have formula for it.

For example suppose we take the law of mechanics F = ma and rewrite it as F-- ma = 0 then $(F - ma)^2$ is called the ''mismatch'' of mechanics and the square of 'mismatch' i.e.

(F– $ma)^2$ is called the unwordliness of mechanics, in other words

$$U_1 = (F - ma)^2$$

In the same way we can calculate the unwordlines for every law of physics, then the total unword lines will be

$$U = \Sigma U_l$$

and the greatest law of nature will be

$$U = 0.$$

Obviously this law though appears to the beautifully simple, in notation is most complex.

Non-uniqueness of Electromagnetic Potentials and Concept of Gauge

In terms of electromagnetic potentials field vectors are given by

$$B = \text{Curl } A \qquad ...(1)$$

and

$$E = \text{grad } \phi - \frac{\partial A}{\partial t}. \qquad ...(2)$$

From equations (1) and (2) it is clear that for a given A and ϕ, each of the field vectors B and E has only one value i.e. A and ϕ determine B and E uniquely. However the converse is not true i.e. field vectors do not determine the potentials A and ϕ completely. This in turn implies that for a given A and ϕ there will be only one E and B while for a given E and B there can be an infinite number of A's and ϕ's. This is because the curl of the gradient of any scalaar vanishes identically and hence we may add to A the gradient of a scalar A without affecting B. That is a may be replaced by

$$A = A + \text{grad } \wedge. \qquad ...(3)$$

But if this is done, equation (2) becomes

$$E = -\text{ grad } \phi - \frac{\partial}{\partial t}(A' - \text{grad } \wedge)$$

i.e.

$$E = -\text{ grad } \left(\phi - \frac{\partial \wedge'}{\partial t}\right) - \frac{\partial A'}{\partial t}.$$

So if we make the transfornmation given by eqn. (3) we must also replace ϕ by

$$\phi' = \phi - \frac{\partial \wedge}{\partial}. \qquad ....(4)$$

The expressions the field vectors E and B remain unchanged under transformation (3) and (4) *i.e.*

$$B = \text{curl } A = \text{curl } (A' - \text{grad } \wedge) = \text{curl } A'$$

and

$$E = -\text{grad } \phi - \frac{\partial A}{\partial t} = -\left(\phi' + \frac{\partial \wedge}{\partial t}\right) - \frac{\partial}{\partial t}(A' - \text{ grad } \wedge) = \text{ grad } \phi' + \frac{\partial A'}{\partial t}$$

i.e. we get the same field vectopr whether we use the set (A, ϕ) or (A', ϕ'). *So electromagnetic potentials define the field vectors uniquely though they themselves are non-unique.*

The transformations given by equation (3) and (4) are called gauge transformations and the arbitrary scular gauge function. From the above it is also clear that even though we add the gradient of a scalar function, the field vectors remain unchanged. Now it is the field quantities and not the potentials that possess physical meaningfulness. We therefore say that the *fields vectors are invorient to gauge transformations* i.e. they are gauge invarient.

Because of the arbitrariness in the choice ofgauge i.e. non-unigueness of potentials, we are free to impose an anditional condition on A. We may state this in other terms: *a vector is not completely speacified by giving only its curl but if both the curl and the divergence of a vector are specified the vector is uniquely determined.* Clearly it is to our advantage to make a choice for div A in any convement manner that will provide a simplification for the particular problem under consideration. Generally div A is chosen in two ways (described in following articles) according as the field contains charge or not.

Lorentz Gauge

The Maxwell's field equations in terms of electromagnetic potentials are

$$\nabla^2 A - \mu\varepsilon \frac{\partial^2 A}{\partial t^2} - grad\left(divA + \mu\varepsilon \cdot \frac{\partial \phi}{\partial t}\right) = \mu J \quad \text{....(1)}$$

$$\nabla^2 \phi - \mu\varepsilon \frac{\partial^2 \phi}{\partial t^2} + \frac{\partial}{\partial t}\left(divA + \mu\varepsilon \frac{\partial \phi}{\partial t}\right) = \frac{\rho}{\varepsilon} \quad \text{....(2)}$$

A casual glance at equations (1) and (2) reveals that these equations will be much more simplified (i.e. will become identical and uncoupled) if we choose

$$\text{div A} + \mu\varepsilon \frac{\partial \phi}{\partial t} = 0. \quad \text{...(3)}$$

This requirement is called the Lorentz condition and when the vector and scalar potential satisfy it, the gauge is known as Lorentz gauge.

So with Lorentz condition field equation reduce to

$$\nabla^2 A - \mu\varepsilon \frac{\partial^2 A}{\partial t^2} = -\mu J. \quad \text{...(4)}$$

and

$$\nabla^2 \phi - \mu\varepsilon \frac{\partial^2 \phi}{\partial t^2} = -\rho / \varepsilon. \quad \text{...(5)}$$

But as

$$\mu\varepsilon = 1/v^{2*}.$$

So equations (4) and (5) can be written as

$$\Box^2 A = -\mu J \quad \text{...(6)}$$

$$\Box^2 = -\rho/\varepsilon \qquad ...(7)$$

with $$\Box^2 = \nabla^2 - \frac{1}{u^2}\frac{\partial^2}{\partial t^2}$$

Equations (6) and (7) are inhomogeneous wave equations and are known as D' Alemberian equations and can be so solved in general by a method similar to that we use to solve Poisson's equation. The potentials obtained by solving these equations are called retarted potentials and are discussed.

In order to determine the requirement that Lorentz condition places on $\wedge$, we substitute A' and ϕ'.

Now, $$\text{div}\,(A' - \text{grad}\wedge) + \mu\varepsilon\frac{\partial}{\partial t}\left(\phi' + \frac{\partial\wedge}{\partial t}\right) = 0$$

i.e. $$\text{div}\,A' + \mu\varepsilon\frac{\partial\phi'}{\partial t} = \nabla^2\wedge - \varepsilon\mu\frac{\partial^2\wedge}{\partial t^2}$$

So A' and will also satisfy equation (3) i.e. Lorentz condition provided that

$$\nabla^2\wedge - \varepsilon\mu\frac{\partial^2\wedge}{\partial t^2} = 0$$

i.e. $$\Box^2\wedge = 0. \qquad ...(8)$$

i.e., Lorentz condition is invariant under those gauge tranformations for which the gauge functions are solutions of the homogeneous wave equations.

(i) It makes the equations for A and ϕ independent of each other.

(ii) It leads to the wave equations which treat ϕ and A on equivalent footings.

(iii) It is a concept which is independent of the co-ordinate system chosen and so fits naturally into the considerations of special theory of relativity.

Coolomb gauge

$$\nabla^2 A - \mu\varepsilon\frac{\partial^2 A}{\partial t^2} - \text{grad}\left(divA + \mu\varepsilon\frac{\partial\psi}{\partial t}\right) = -\mu J$$

$$\nabla^2\phi - \mu\varepsilon\frac{\partial^2\phi}{\partial t^2} + \frac{\partial}{\partial - 1}\left(divA + \mu\varepsilon\frac{\partial\phi}{\partial t}\right) = -\rho/\varepsilon$$

$$\nabla^2\phi + \frac{\partial}{\partial t}(divA) = -\rho/\varepsilon$$

div A = 0

$$\left.\begin{aligned}\nabla^2\phi_{r,t} &= -\frac{\rho_{r,t}}{\varepsilon}\\ \phi &= \frac{1}{v\pi\varepsilon}\int_v \frac{\rho}{R}dv\end{aligned}\right] \quad (1)$$

$$\nabla^2 A - \mu\varepsilon\frac{\partial^2 A}{\partial + t^2} - \text{grad}\left(\mu\varepsilon\frac{\partial\phi}{\partial t}\right) = -\mu J$$

$$\nabla^2 A - \mu\varepsilon\frac{\partial^2 A}{\partial + t^2} = -\mu J + \mu\varepsilon\,\nabla.\left(\frac{\partial\phi}{\partial t}\right) \quad (2)$$

$$\nabla^2\left[\frac{1}{4\pi\varepsilon}\int\frac{\rho}{R}dv\right] = -\rho/\varepsilon$$

$$\nabla^2\left[\frac{1}{4\pi\varepsilon}\int\frac{J}{R}dv\right] = -J/\varepsilon$$

$$\nabla \times \nabla \times G = \nabla.\,(\nabla\cdot G) - \nabla^2 G$$

$$\Rightarrow \nabla^2 G = \nabla\,(\nabla.G) - \nabla \times \nabla \times G$$

Let $$G = \int\frac{J}{R}dv$$

$$\nabla^2\int\frac{J}{R}dv = \nabla\left\{\nabla.\int\frac{J}{R}dv\right\}$$

$$-\nabla\times\nabla\times\int\frac{J}{R}dv$$

$$-4\pi J = \nabla\left\{\nabla.\int\frac{J}{R}dv\right\}$$

$$-\nabla\times\nabla\times\int\frac{J}{R}dv$$

$$J = -\frac{1}{4\pi}\nabla\left\{\nabla.\int\frac{J}{R}dv\right\} + \frac{1}{4\pi}\nabla\times\nabla\times\int\frac{J}{R}dv$$

$J = J_e$	+	J_t
irroraional		soloncidal
as $\nabla\times J_e = 0$		$\nabla.\,J_t = 0$

$$eq_2, \nabla^2 A - \frac{1}{v^2}\frac{\partial^2 A}{\partial t^2} = -\mu(J_e + J_t) \ + \mu\varepsilon\nabla\left(\frac{\partial \phi}{\partial t}\right)$$

$$\Delta^2 A = -\mu J_e - \mu J_t + \mu\varepsilon\,\nabla\left[\frac{\partial}{\partial t}\left\{\frac{1}{4\pi\varepsilon}\int\frac{\rho}{R}dv\right\}\right]$$

$$= -\mu J_e - \mu J_t + \mu\,\nabla\left[\frac{\partial}{\partial t}\left\{\frac{1}{4\pi\varepsilon}\int\frac{\rho}{R}dv\right\}\right]$$

$$= -\mu J_e - \mu J_t + \mu\,\frac{1}{4\pi}\nabla\left[\int\frac{\nabla J_e}{R}dv\right]$$

$$= -\mu J_e - \mu J_t + \mu J_e \qquad \text{[as by continuity eq. } -\nabla J = \frac{\partial \rho}{\partial t}$$

$$= -\mu J_t$$

$$\Delta^2 A = -\mu J_t$$

So called transverse gauge

Free Space

$$\nabla . D = \rho$$

$$\nabla . B = 0$$

$$\left.\begin{array}{l} curl\ H = J + \dfrac{\partial \rho}{\partial t} \\ curl\ E = -\dfrac{\partial B}{\partial t} \end{array}\right] \begin{array}{l} J = \sigma E \\ D = \varepsilon E \\ B = \mu H \end{array}$$

For Free Space

$$\rho = \theta \qquad \mu_r = 1$$

$$\sigma = 0 \qquad \varepsilon_r = 1$$

$\Rightarrow$

$$\nabla . E = 0 \qquad \text{... (A)}$$

$$\nabla . H = 0 \qquad \text{... (B)}$$

$$\nabla \times H = \varepsilon_0 \frac{\partial E}{\partial t} \qquad \text{...(C)}$$

$$\nabla \times E = -\mu_0 \frac{\partial H}{\partial t} \qquad \text{...(D)}$$

$\nabla \times C$

$$\nabla \times (\nabla \times H) = \nabla \times \left[\varepsilon_0 \frac{\partial E}{\partial t}\right]$$

$$\nabla (\nabla . H) - \nabla^2 H = \varepsilon_0 \frac{\partial}{\partial t} (\nabla \times E)$$

$$\Rightarrow \quad -\nabla^2 H = -\mu_0 \varepsilon_0 \frac{\partial^2 H}{\partial t^2}$$

$$\nabla^2 H - \frac{1}{c^2} \frac{\partial^2 H}{\partial t^2} = 0$$

$\nabla \times D$

$$\nabla \times (\nabla \times E) = \nabla \times \left(\frac{-\partial B}{\partial t}\right)$$

$$\nabla(\nabla . E) - \nabla^2 E = -\frac{\partial}{\partial t}(\nabla \times B)$$

$$\nabla^2 E = \mu_0 \frac{\partial}{\partial t}(\nabla \times H)$$

$$\nabla^2 E = \mu_0 \varepsilon_0 \frac{\partial^2 E}{\partial t}$$

$$\nabla^2 E - \frac{1}{c^2} \frac{\partial^2 E}{\partial t^2} = 0$$

$\Rightarrow$ Analogy to

$$\nabla^2 \phi - \frac{1}{c^2} \frac{\partial^2 \phi}{\partial t^2} = 0$$

$$\phi = \phi_0 e^{-i(\omega t - \vec{k}.\vec{r})}$$

$$\Rightarrow \quad \begin{Bmatrix} E \\ H \end{Bmatrix} = \begin{Bmatrix} E_0 \\ H_0 \end{Bmatrix} e^{-i(\omega t - \vec{k}.\vec{r})}$$

For wave of above type operator

$\nabla \rightarrow ik$

curl $\rightarrow$ ikx

div → ik.

$$\frac{\partial}{\partial t} \to -i\omega$$

as $$\psi = \psi_0 e^{-i(\omega t - \vec{k} - \vec{r})}$$

$$\nabla \psi = \left(i\frac{d}{dx} + j\frac{d}{dy} + k\frac{d}{dz}\right)\left[\psi_0 e^{-i(\omega t - k_x x - k_y y - k_z z)}\right]$$

$$\nabla \psi = ik.\psi_0 e^{-i(\omega t \ldots\ldots)}$$

$$\frac{\partial}{\partial t}\psi = -i\omega(\psi)$$

herce

$$\vec{k}.\vec{E} = 0 \qquad K \perp E$$

$$E \perp H$$

$$\vec{k}.\vec{H} = 0 \qquad K \perp H$$

$$-\vec{k} \times \vec{H} = \varepsilon_0 \omega E$$

$$\vec{k} \times \vec{E} = \mu_0 \omega H$$

$$\vec{k} \times \vec{E} = \omega \mu_0 H$$

$$H = \frac{1}{\omega\mu_0}\left(\vec{k} \times \vec{E}\right) = \frac{K}{\omega\mu_0}\left(\vec{n} \times \vec{E}\right)$$

$$= \frac{1}{\mu_0 c}\left(\vec{n} \times \vec{E}\right) \quad K = \frac{\omega}{c}$$

$$\left[H \approx \varepsilon_0 c\left(\vec{n} \times \vec{E}\right)\right.$$

$$B = \frac{1}{c}\vec{n} \times \vec{E}$$

$$\left|\frac{E}{H}\right| = \frac{E_0}{H_0} = c\mu_0 = \frac{1}{\varepsilon_0 c} = Z_0$$

$Z_0 = c\mu_0$

$$= \sqrt{\frac{\mu_0}{\varepsilon_0}} = \sqrt{\frac{4\pi \times 10^{-7}}{1/4\pi \times 9 \times 10^9}} \approx 37752$$

Pointing Vector

$$S = E \times H$$

$$= E \times \frac{\vec{n} \times \vec{E}}{\mu_0 c}$$

$$= \frac{1}{\mu_0 c}\left[\vec{n}\left(\vec{E}.\vec{E}\right) - \vec{E}\left(\vec{E}.\vec{n}\right)\right]$$

$$= \frac{1}{\mu_0 c}.E^2\vec{n} \qquad as\ E\perp$$

$$S = \varepsilon_0 c\, E^2\vec{n}$$

$$S = \frac{1}{z_0}\, E^2\, \vec{n}$$

$$< S > = \frac{1}{2}\frac{1}{z_0}\, E^2\, \vec{n}$$

$$\frac{< u_e >}{< u_m >} = \frac{\frac{1}{2}\varepsilon_0 E^2}{\frac{1}{2}\mu_0 H^2} \qquad S = cu$$

$$= \frac{\varepsilon_0}{\mu_0}\frac{E^2}{H^2} = 1$$

In Isotopic Dielectric

$$\nabla . D = \rho$$

$$\nabla . B = 0 \qquad J = \sigma E$$

$$\nabla \times H = J + \frac{\partial D}{\partial t}, \qquad B = \mu H$$

$$\nabla \times E = -\frac{\partial B}{\partial t}, \; D = \varepsilon E$$

$$\sigma = 0 \qquad \rho = 0$$

$$\nabla . E = 0 \qquad -A$$

$$\nabla . H = 0 \qquad -B$$

$$\nabla \times H = \varepsilon \frac{\partial E}{\partial t} \qquad -C$$

$$\nabla \times E = -\mu \frac{\partial H}{\partial t} \qquad -D$$

$\nabla \times C$

$$\nabla \times (\nabla \times H) = \nabla \times \left(\varepsilon \frac{\partial E}{\partial t} \right)$$

$$\nabla(\nabla . H) - \nabla^2 H = \varepsilon \frac{\partial}{\partial t} (\nabla \times E)$$

$$-\nabla^2 H = -\varepsilon\mu \frac{\partial^2 H}{\partial t^2}$$

$$\nabla^2 H - \frac{1}{v^2} \frac{\partial^2 H}{\partial t^2} = 0$$

$\nabla \times D$

$$\nabla^2 E - \frac{1}{v^2} \frac{\partial^2 E}{\partial t^2} = 0$$

$$\begin{Bmatrix} E \\ H \end{Bmatrix} = \begin{Bmatrix} E_0 \\ H_0 \end{Bmatrix} e^{-i\left(\omega t - \vec{k}\vec{r}\right)}$$

speed $\quad U = \dfrac{1}{\sqrt{\varepsilon \mu}}$

$$U = \frac{C}{\sqrt{\varepsilon_r \mu_r}}$$

as $\quad n = \dfrac{C}{U} = \sqrt{\varepsilon_0 \mu_r}$

$$n = \sqrt{\varepsilon_r}$$

$$\nabla \to i k \quad \frac{\partial}{\partial t} \to -i\omega$$

$$K - E = 0 \qquad K \perp E$$
$$K - H = 0 \qquad K \perp H$$
$$> E \perp H$$

$$K \times E = \omega \varepsilon E$$

$$K \times E = \omega \mu H$$

$$H = \frac{K}{\omega \mu}(\vec{n} \times E)$$

$$\vec{H} = \frac{1}{\mu_v}\left(\vec{n} \times \vec{E}\right)$$

$$\vec{H} = \varepsilon_v \left(\vec{n} \times \vec{E}\right)$$

$$\left|\frac{\vec{E}}{\vec{H}}\right| = \mu_v = \sqrt{\frac{\mu}{\varepsilon}} = z\left[z = z_{0/n}\right]$$

$$\frac{Ue}{Um} = \frac{\frac{1}{2}\varepsilon E^2}{\frac{1}{2}\mu H^2} = \frac{\varepsilon}{\mu}\cdot\left(\frac{E_2}{H_2}\right)$$

$$= 1$$

$$U = Ue + Um$$

$$S = E \times H$$

$$= E \times \frac{\vec{n} \times \vec{E}}{Z} = \frac{1}{z}\left[\vec{n}\left(\vec{E}.\vec{E}\right) - \vec{E}\left(\vec{E}.\vec{n}\right)\right]$$

$$S = \frac{1}{2}E^2\,\vec{n}$$

$$\langle S \rangle = \frac{1}{2}S = \frac{1}{2} - \frac{1}{z}E^2\,\vec{n}$$

Check Yourself

1. One of the following is not a source of magnetostatic field:
   (a) A dc Current in a wire
   (b) A permanent magnet
   (c) An accelerated charge
   (d) An electric field linearly changing with time
   (e) A charged disk rotating at uniform speed

2. Consider points A, B, C, D, and E on a circle of radius 2. The items in the right list are the values of aϕ at different points on the circle. Match these items with the points in the list on the left.

(a) A (i) a_x

(b) B (ii) $-a_x$

(c) C (iii) a_y

(d) D (iv) $-a_y$

(e) D (v) $\frac{a_x + a_y}{\sqrt{2}}$ (vi) $\frac{-a_x - a_y}{\sqrt{2}}$

(vii) $\frac{-a_x + a_y}{\sqrt{2}}$ (viii) $\frac{a_x + a_y}{\sqrt{2}}$

3. The z-zxis carries filamentary current of 10A along az. Which of these is incorrect?

(a) $H = -a_x$ A/m at (0, 5, 0)

(b) $H = a_\phi$ A/m at (5, $\pi/4$, 0)

(c) $H = -0.8a_x - 0.6a_y$ at (−3, 4, 0)

(d) $H = -a_\phi$ at (5, $3\pi/2$, 0)

4. Plane y = 0 carries a uniform current of $30a_z$ mA/m. At (1, 10, −2), the magnetic field intensity is

(a) $-15a_x$ mA/m

(b) $15a_x$ mA/m

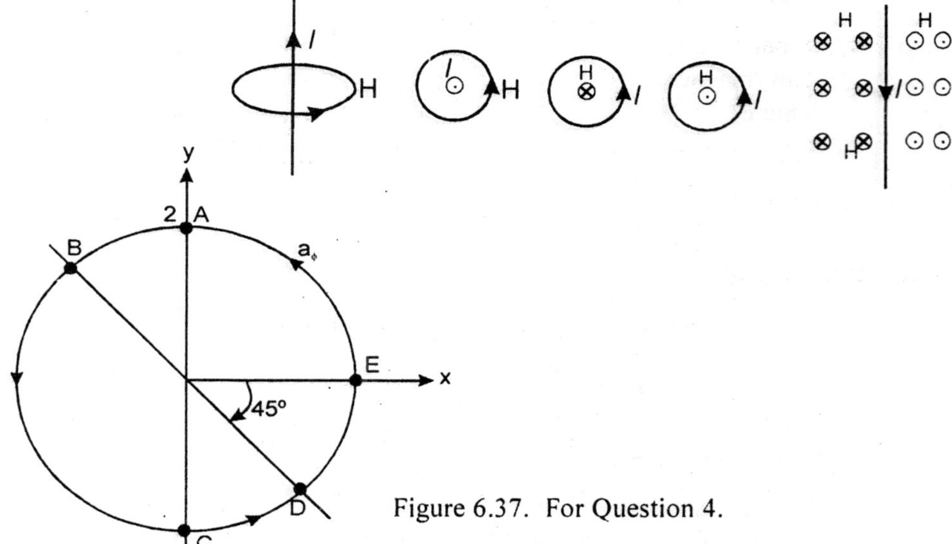

Figure 6.37. For Question 4.

(c) $477.5a_y$ μA/m

(d) $18.85a_y$ nA/m

(e) None of the above

5. Which of these statements is not characteristic of a static magnetic field?
   (a) It is solenoidal.
   (b) It is conservative.
   (c) It has no sinks or sources.
   (d) Magnetic flux lines are always closed.
   (e) The total number of flux lines entering a given region is equal is to the total number of flux lines leaving the region.

6. Two identical coaxial circular coils carry the same current *I* but in opposite directions. The magnitude of the magnetic field B at a point on the axis midway between the coils is
   (a) Zero
   (b) The same as that produced by one coil
   (c) Twice that produced by one coil
   (d) Half that produced by one coil.

7. One of these equations is not Maxwell's equation for a static electromagnetic field in a linear homogeneous medium.
   (a) $\nabla . B = 0$ (b) $\nabla \times D = 0$
   (c) $\oint B.dl = \mu_0 I$ (d) $\oint D.dS = Q$
   (e) $\nabla^2 A = \mu_0 J$

8. Two bar magnets with their north poles have strength Qm1 = 20 A.m and Qm2 = 10A.m (magnetic charges) are placed inside a volume. The magnetic flux leaving the volume is
   (a) 200 Wb (b) 30 Wb
   (c) 10 Wb (d) 0 Wb
   (e) -10Wb

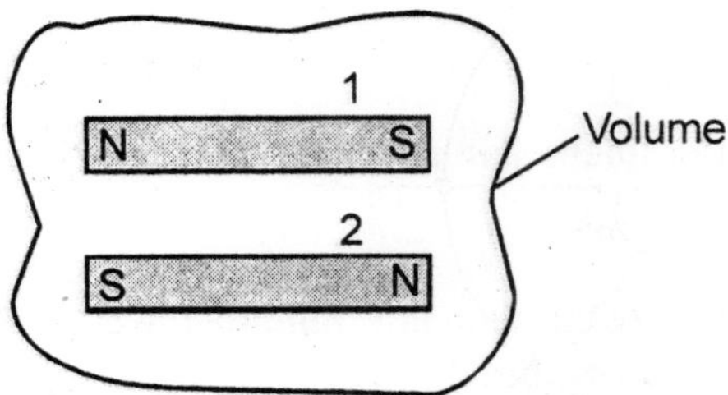

Figure 6.38. For Review Question 8.

Problem

1. (a) State Biot-savart's law

   (b) The y- and z-axes, respectively, carry filamentary currents 10A along a_y and 20A along $-a_z$. Find H at (-3, 4, 5).

Long Answer Questions

1. Explain how you can define magnetic field $\vec{B}$ without using the concept of magnetic pole.

2. Prove that the force acting on a test charge q_0 mving with a velocity $\vec{v}$ in a magnetic field of uniform flux density $\vec{B}$ is $q_0(\vec{v} \times \vec{B})$. How does it lead to the definition of $\vec{B}$.

   (K.U. 1992 S)

3. Describe the motion of a charged particle in a uniform magnetic field and electric field and hence derive the Lorentz force equation

   $$\vec{F} = q\left(\vec{E} + \vec{v} \times \vec{B}\right).$$ **(M.D.U. 1985; K.U. 1988, 89 A)**

4. Explain Biot-Savart law for the flux density at a point due to the current flowing in a current element. How will you find the direction of flux density.

   (K.U. 1986A, 88A, 92S)

5. Show that the magnetic intensity at a point due to a straight conductor carrying current *i* ampere is given by

   $$B = \frac{\mu_0 i}{2\pi a}$$

6. Derive Ampere circuital law from conductor.

   $$\oint_S \vec{B}.\vec{dl} = \mu_0 i$$ **(M.D.U. 1980; K.U. 1986 A, 92 S)**

7. Show that the divergence of a magnetic field vector $\vec{B}$ is zero every where.

   (K. U. 1993A)

8. Prove $\nabla \times \vec{B} = \mu_0 \vec{J}$ where the terms have their usual meaning.

   (M.D.U. 1986; K.U. 1986A, 93A, 95S)

9. What is meant by the solenoidal nature of the magnetic vector field $\vec{B}$.

   (K.U.1992 S)

10. State and deduce ampere circuital law, relating the current and the magnetic field that it produces in the neighbourhood.

    (K.U. 1988A)

11. Explain Ampere's circuital law and how will you use it to determine the flux density in the neighbourhood of a straight conductor carrying a current?

(K.U. 1983; M.D.U. 1983)

12. Calcutate the flux density at a point a point on the axis of a circular coil carrying steady current?

(M.D.U. 1986, 82; K.U. 1976, 79, 92, 93 A)

13. Find the torque acting on a current loop placed in a uniform magnetic field.

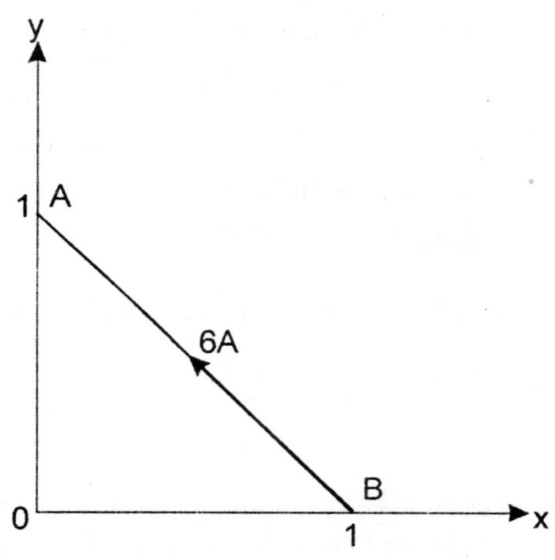

Figure 6.39. For Problem 3

2. A conducting filament carries current I from point A(0, 0, a) to point B(0,0,b). Show that at point P(x, y, 0).

$$H = \frac{I}{4\pi\sqrt{x^2 + y^3}}\left[\frac{b}{\sqrt{x^2 + y^2 + b}} - \frac{a}{\sqrt{x^2 + y^2 + a^2}}\right]ay$$

3. Consider AB in Figure 6.39 as part of an elecuit circuit. Find H at the origin due to AB.

4. Line x = 0, y = 0, ≤ z ≤ 10m carries current 2 A along az. Calculate H at points

   (a) (5, 0, 0) (b) (5, 5, 0)

   (c) (5, 15, 0) (d) (5, –15, 0)

5. (a) Find H at (0,0,5) due to side 2 of the triangular loop in figure 7.6 (a)

   (b) Find H at (0, 0, 5) due to the entire loop.

6. An infinitely long conductor is bent into an L shape as shown in Figure. If a direct current of 5 A flows in the current, find the magnetic field intensity at (a) (2, 2, 0), (b) (0, –2, 0), and (c) (0, 0, 2).

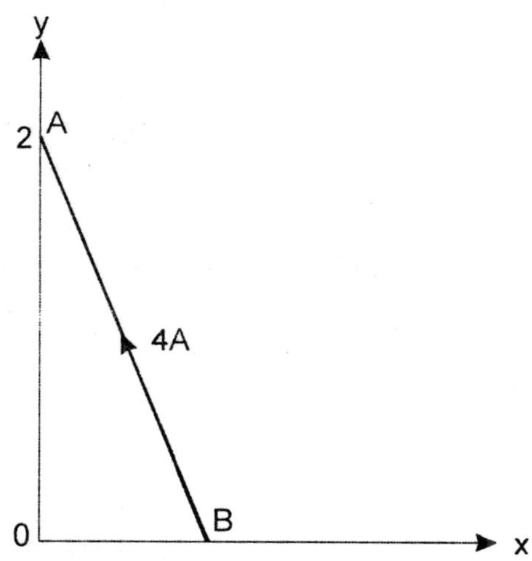

Figure 6.40. For Problem 3

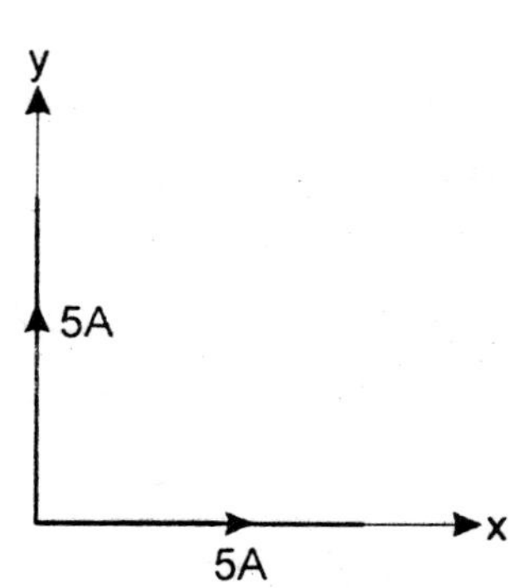

Figure 6.40. Current filament

7. A rectangular loop carrying 10 A of current is placed on z = 0 plane as shown in Figure 7.30. Evaluate H at

   (a) (2, 2, 0) (b) (4, 2, 0)

   (c) (4, 8, 0) (d) (0, 0, 2)

8. A square conducting loop of side $2a$ lies in the z = 0 plane and carries a currnet I in the counterclockwise direction. Show that at the center of the loop

$$H = \frac{\sqrt{2I}}{\pi a} a_z$$

9. Two indentical current loops have their centres at (0, 0,0) and (0, 0, 4) and their axes the same as the z-axis (so that the "Helmholtz coil" is formed). If each loop has radius 2m and carries current 5 A in $a\phi$, calculate H at

   (a) (0, 0, 0)
   (b) (0, 0, 2)

10. A 3-cm-long solenoid carries a current of 400 mA. If the solenoid is to produced a magnetic flux density of 5 m Wb/m^2, how many turns of wire are needed?

11. A solenoid of radius 4 mm and length 2cm has 150 turn/m and carries current 500 mA. Find: (a) |H| at the center, (b) |H| at the ends of the solenoid.

12. Plane x = 10 carries current 100 mA/m along a_z while line x = 1, y = –2 carries filamentary current 20π mA along a_z, Determine H at (4, 3, 2).

13. (a) State Ampere's circuit law.

    (b) A hollow conducting cylinder has inner radius a and outer radius b and carries current I along the positive z-direction. Find H everywhere.

14. (a) An infinitely long solid conductor of radius a is placed along the z-axis. If the conductor carries current I in the +z direction, show that

$$H = \frac{I\rho}{2\pi a^2} a_\phi$$

within the conductor. Find the corresponding current density.

(b) If I = 3 A and a = 2 cm in part (a), find H at (0, 1 cm, 0) and (0, 4 cm, 0).

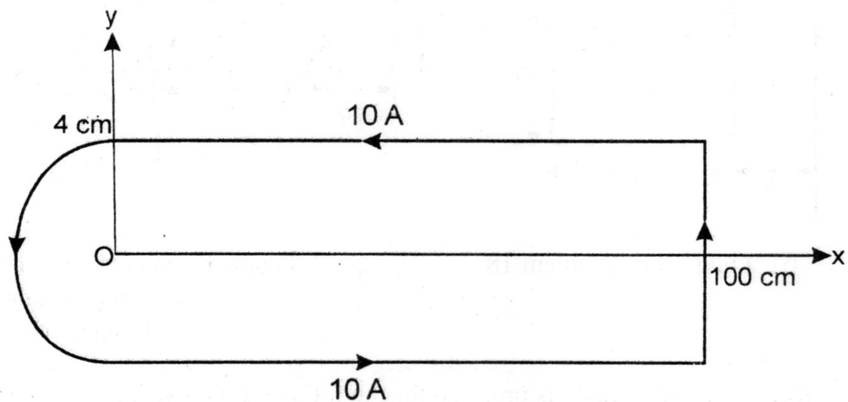

Figure Filamentary loop of Problem 15; not drawn to scale.

15. If H = yax – xay A/m on plane z = 0, (a) determine the current density and (b) verify Ampere's law taking the circulation of H around the edge of the rectangle z = 0, 0 < x < 3, –1 < y < 4.

16. In a certain conducting region,

$$H = yz(x^2 + y^2)a_x - y^2xza_y + 4x^2y^2a_z \text{ A/m}$$

(a) Determine J at (5, 2, –3)

(b) Find the current passing through x = –1, 0 < y, z < 2

(c) Show that ∇ . B = 0

17. An infinitely long filamentary wire carries a current of 2 A in the +z - direction. Calculate

(a) B at (–3, 4, 7)

(b) The flux through the square loop described by $2 \le \rho \le 6$, $0 \le z \le 4$, $\phi = 90°$

18. Consider the following arbitrary fields. Find out which of them can possibly represent electrostatic or magnetostatic field in free space.

(a) $A = y \cos axa_x + (y + e^{-x})a_z$

(b) $B = \dfrac{20}{\rho} a\rho$

(c) $C = r^2 \sin\theta \, a\phi$

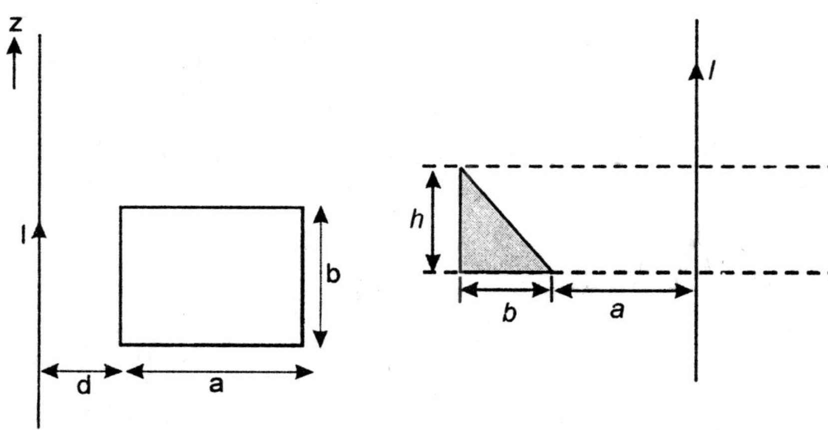

Figure, For Problem 18

Figure, Cross section of a brassring enclosing a long straight wire; for Problem 18.

19. Reconsider the previous problem for the following fields.

(a) $D = y^2 z a_x + 2(x + 1)yz a_y - (x + 1)z^2 a_z$

(B) $E = \dfrac{(z+1)}{\rho}\cos\phi\, a_\rho + \dfrac{\sin\phi}{\rho}$ (c) $F = \dfrac{1}{r^2}(2\cos\theta\, a_r + \sin\theta\, a_\theta)$

20. For a current distribution in free space.

$A = (2x^2y + yz)a_x + (xy^2 - xz^3)a_y - (6xyz - 2x^2y^2)a_z$ Wb/m

(a) Calculate B.

(b) Find the magnetic flux through a loop described by $x = 1, 0 < y, z < 2$.

(c) Show that $\nabla \cdot A = 0$ and $\nabla \cdot B = 0$.

21. The magnetic vector potential of a current distribution in free space is given by

$A = 15e^{-\rho}\sin\phi\, a_z$ Wb/m

Find H at (3, /4, –10). Calculate the flux through $\rho = 5, 0 \le \phi \le \pi/2, 0 \le z \le 10$.

22. A conductor of radius a carries a uniform current with $J = J_o a_z$. Show that the magnetic vector potential for $\rho \le a$ is

$$A = -\frac{1}{4}\mu_o J_o \rho^2 a_z$$